PRACTICE MASTERS

Merrill

Geometry

Applications and Connections

GLENCOE

Macmillan/McGraw–Hill

New York, New York
Columbus, Ohio
Mission Hills, California
Peoria, Illinois

TO THE TEACHER

These Practice Masters are also available as a consumable workbook, *Practice Workbook*. An answer key is not included in the *Practice Workbook*.

Send all inquiries to:
Glencoe Division, Macmillan/McGraw-Hill
936 Eastwind Drive
Westerville, OH 43081

ISBN: 0-02-824443-5
1 2 3 4 5 6 7 8 9 10 BANT 02 01 00 99 98 97 96 95 94

CONTENTS

Glencoe Division, Macmillan/McGraw-Hill

CONTENTS

1-1 Practice Worksheet

Connections from Algebra: The Coordinate Plane

Graph each point on the coordinate plane at the right.

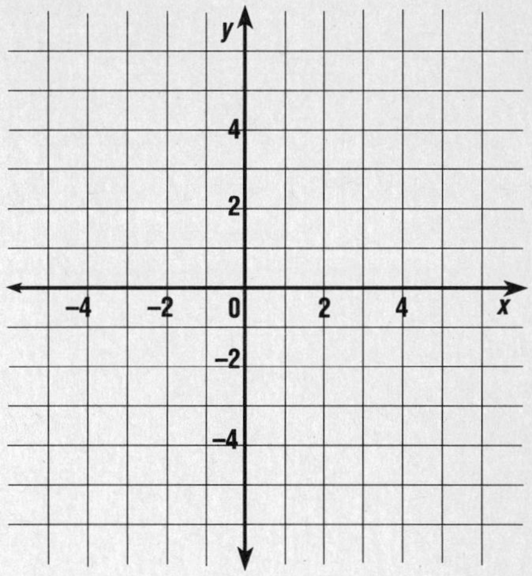

1. A (3, –5)
2. B (5, 0)
3. C (–2, –1)
4. D (0, –6)
5. E (4, 3)
6. F (–4, 3)

In the figure below, all of the segments shown are parallel to either the x- or y-axis. Determine the ordered pair that represents each point.

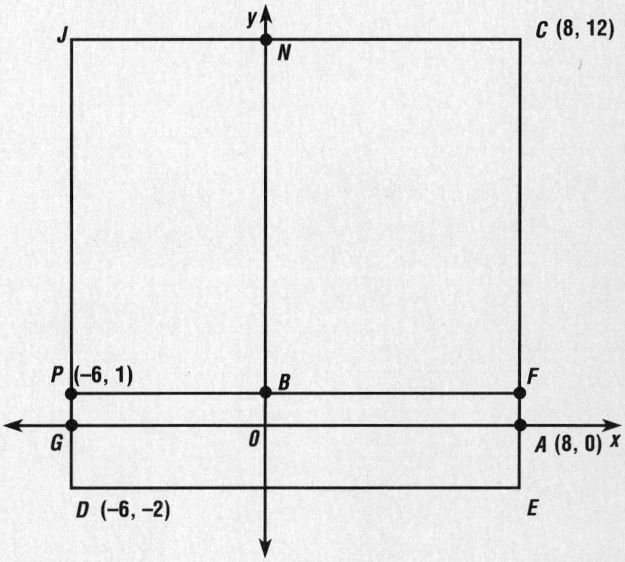

7. J
8. G
9. B
10. N
11. F
12. E

Points G (1, –1) and H (3, 3) lie on the line y = 2x – 3. Determine whether the following points are collinear to G and H.

13. I (0, –3)
14. J (2, 1)
15. K (–3, –8)
16. L (5, 7)
17. M (10, 16)
18. N (25, 48)
19. P (–10, –23)
20. Q (–35, –67)

1-1

NAME ________________________ DATE ____________

Practice Worksheet

Connections from Algebra: The Coordinate Plane

Graph each point on the coordinate plane at the right.

1. *A* (3, –5)   **2.** *B* (5, 0)

3. *C* (–2, –1)   **4.** *D* (0, –6)

5. *E* (4, 3)   **6.** *F* (–4, 3)

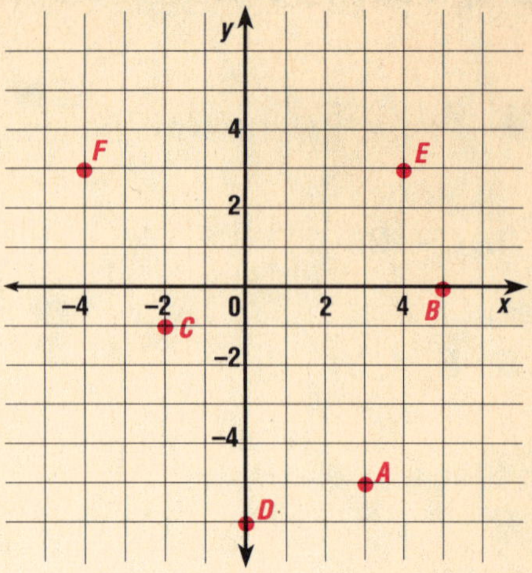

In the figure below, all of the segments shown are parallel to either the x- or y-axis. Determine the ordered pair that represents each point.

7. *J* **(–6, 12)**   **8.** *G* **(–6, 0)**

9. *B* **(0, 1)**   **10.** *N* **(0, 12)**

11. *F* **(8, 1)**   **12.** *E* **(8, –2)**

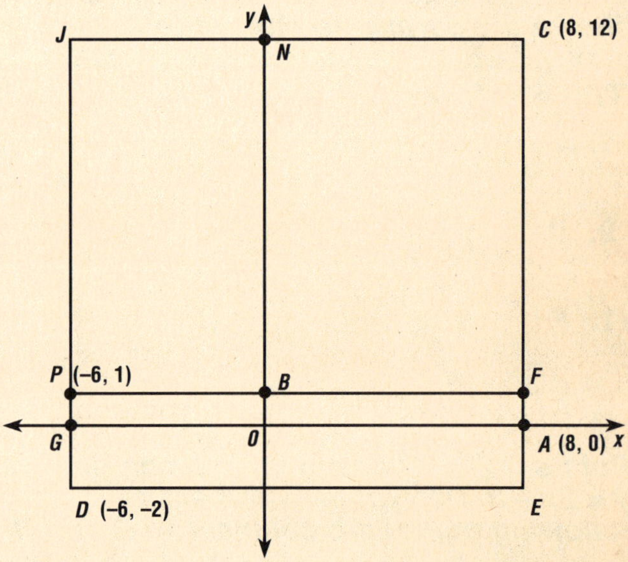

Points G (1, –1) and H (3, 3) lie on the line $y = 2x - 3$. Determine whether the following points are collinear to G and H.

13. *I* (0, –3) **yes**   **14.** *J* (2, 1) **yes**

15. *K* (–3, –8) **no**   **16.** *L* (5, 7) **yes**

17. *M* (10, 16) **no**   **18.** *N* (25, 48) **no**

19. *P* (–10, –23) **yes**   **20.** *Q* (–35, –67) **no**

1-2 NAME ______________________ DATE ____________

Practice Worksheet

Points, Lines, and Planes

Draw and label a figure for each relationship.

1. Lines ℓ, m and j intersect at P.

2. Plane $\mathcal{N}$ contains line ℓ.

3. Points A, B, C, and D are noncollinear.

4. Points A, B, C, and D are noncoplanar.

Refer to the figure at the right to answer each question.

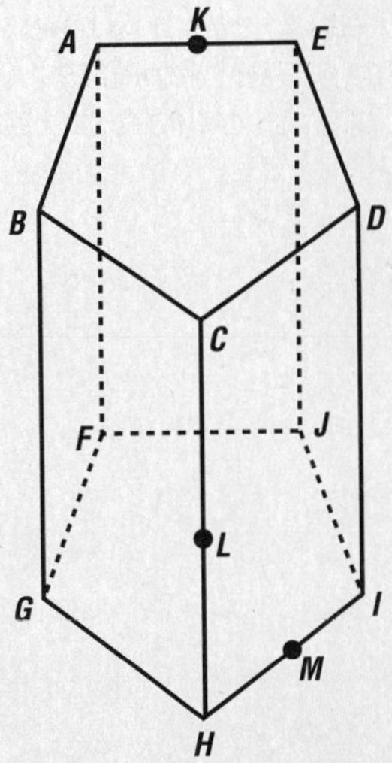

5. Are points H, M, I, and J coplanar?

6. Are points C, L, and H collinear?

7. How many planes are shown?

8. Name the intersection of planes ABG and CHG.

9. Name the intersection of plane ABC and HL.

10. Which segments are contained in all three of the planes GFH, CDI, and EDI?

11. If you drew line LM, in how many points would it intersect plane BGH?

NAME ______________________________ DATE ______________

1-2 Practice Worksheet

Points, Lines, and Planes

Draw and label a figure for each relationship.

Typical answers are given.

1. Lines ℓ, m and j intersect at P.

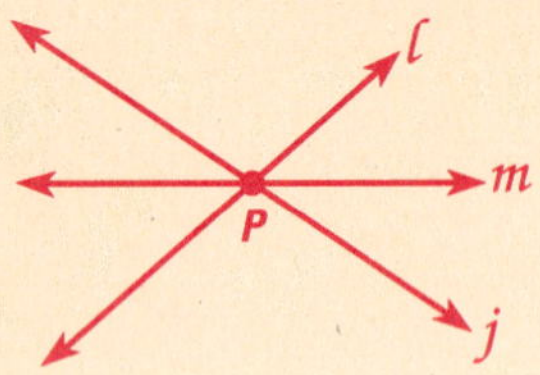

2. Plane $\mathcal{N}$ contains line ℓ.

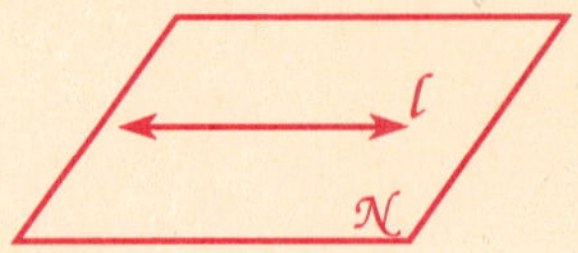

3. Points A, B, C, and D are noncollinear.

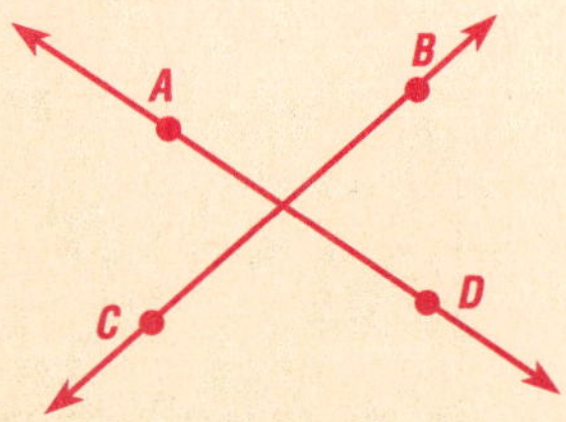

4. Points A, B, C, and D are noncoplanar.

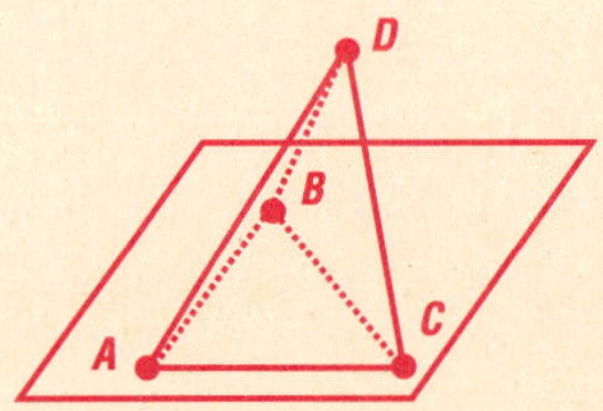

Refer to the figure at the right to answer each question.

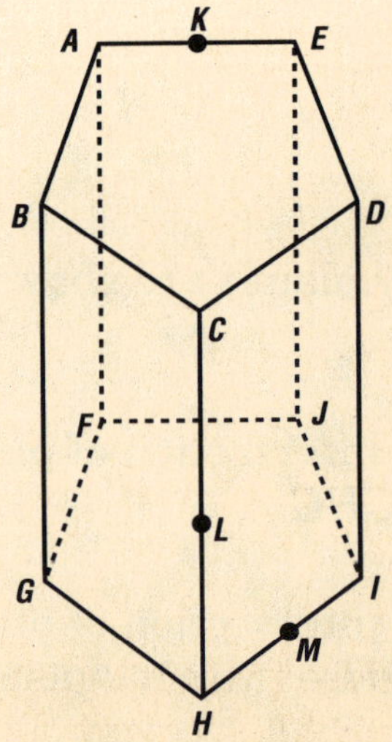

5. Are points H, M, I, and J coplanar? **yes**

6. Are points C, L, and H collinear? **yes**

7. How many planes are shown? **7**

8. Name the intersection of planes ABG and CHG. $\overleftrightarrow{BG}$

9. Name the intersection of plane ABC and HL. **C**

10. Which segments are contained in all three of the planes GFH, CDI, and EDI? **none**

11. If you drew line LM, in how many points would it intersect plane BGH? **1**

NAME ______________________ DATE ______________

1-3

Practice Worksheet

Problem–Solving Strategy: List the Possibilities

Solve. Use any strategy.

1. On a certain island, Borks always tell the truth, Beeks always lie, and Bicks sometimes tell lies and sometimes tell the truth. From the following statements by Fill, Foll, and Fall, determine who is the Bork, who is the Beek, and who is the Bick.

Fill: Foll is the Bork.
Foll: Fill is the Bick.
Fall: Fill is lying.

2. In the figure at the right, exactly two line segments can be moved to make two squares. Which two segments should be moved, and to where?

3. Tim can choose from a tan shirt, a blue shirt, and a green shirt. He can choose from black slacks or blue jeans. He can choose from a windbreaker, a sweatshirt, or a jacket. How many different outfits can he wear if he will not wear the sweatshirt with the black slacks?

Glencoe Division, Macmillan/McGraw-Hill

1-3

NAME ______________________ DATE ____________

Practice Worksheet

Problem–Solving Strategy: List the Possibilities

Solve. Use any strategy.

1. On a certain island, Borks always tell the truth, Beeks always lie, and Bicks sometimes tell lies and sometimes tell the truth. From the following statements by Fill, Foll, and Fall, determine who is the Bork, who is the Beek, and who is the Bick.

Fill: Foll is the Bork.
Foll: Fill is the Bick.
Fall: Fill is lying.

Fill is the Bick. Foll is the Bork. Fall is the Beek.

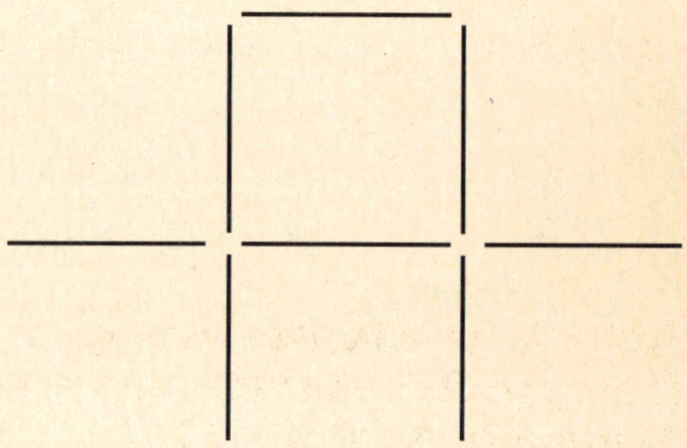

2. In the figure at the right, exactly two line segments can be moved to make two squares. Which two segments should be moved, and to where?

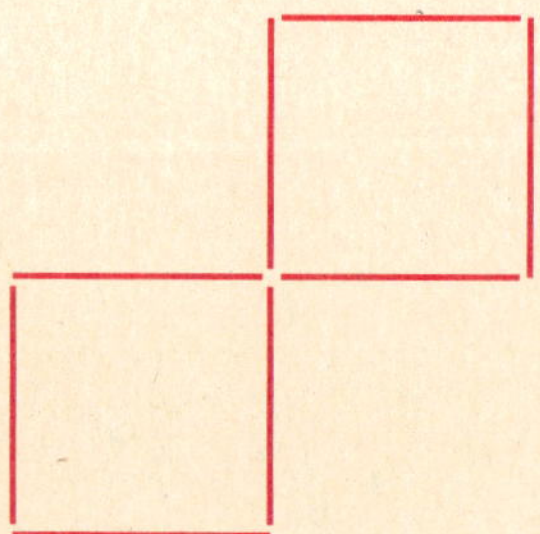

3. Tim can choose from a tan shirt, a blue shirt, and a green shirt. He can choose from black slacks or blue jeans. He can choose from a windbreaker, a sweatshirt, or a jacket. How many different outfits can he wear if he will not wear the sweatshirt with the black slacks? **15**

Glencoe Division, Macmillan/McGraw-Hill

NAME ________________________ DATE ____________

1-4

Practice Worksheet

Finding the Measure of Segments

Given that B is between A and C, find each missing measure.

1. $AB = 5.3$, $BC = \underline{?}$, $AC = 6.7$

2. $AB = 21$, $BC = 4.3$, $AC = \underline{?}$

3. $AB = \underline{?}$, $BC = 18.9$, $AC = 23$

4. $AB = 6\frac{3}{4}$, $BC = \underline{?}$, $AC = 10$

If B is between A and C, find the value of x and the measure of BC.

5. $AB = 3x$, $BC = 5x$, $AC = 8$

6. $AB = 3(x + 7)$, $BC = 2(x - 3)$, $AC = 50$

Refer to the coordinate plane at the right to find each measure. If the measure is not a whole number, round the result to the nearest hundredth.

7. AB

8. BD

9. AE

10. CE

11. AD

12. BE

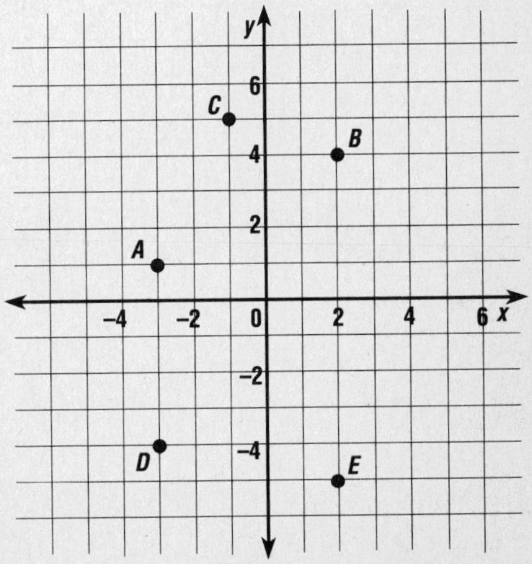

1-4

NAME ______________________ DATE ____________

Practice Worksheet

Finding the Measure of Segments

Given that B is between A and C, find each missing measure.

1. $AB = 5.3$, $BC = \underline{\ ?\ }$, $AC = 6.7$ **1.4**

2. $AB = 21$, $BC = 4.3$, $AC = \underline{\ ?\ }$ **25.3**

3. $AB = \underline{\ ?\ }$, $BC = 18.9$, $AC = 23$ **4.1**

4. $AB = 6\frac{3}{4}$, $BC = \underline{\ ?\ }$, $AC = 10$ **$3\frac{1}{4}$**

If B is between A and C, find the value of x and the measure of BC.

5. $AB = 3x$, $BC = 5x$, $AC = 8$ **1, 5**

6. $AB = 3(x + 7)$, $BC = 2(x - 3)$, $AC = 50$ **7, 8**

Refer to the coordinate plane at the right to find each measure. If the measure is not a whole number, round the result to the nearest hundredth.

7. AB **5.83**

8. BD **9.43**

9. AE **7.81**

10. CE **10.44**

11. AD **5**

12. BE **9**

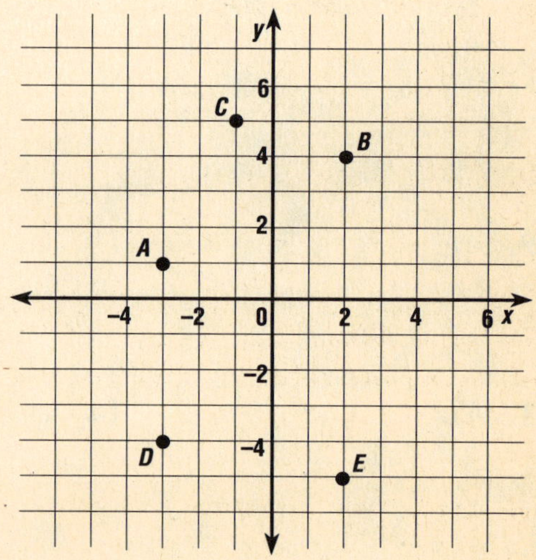

NAME ____________________ DATE ____________

1-5 Practice Worksheet

Segment Relationships

Refer to the figure below to determine whether each statement is true or false.

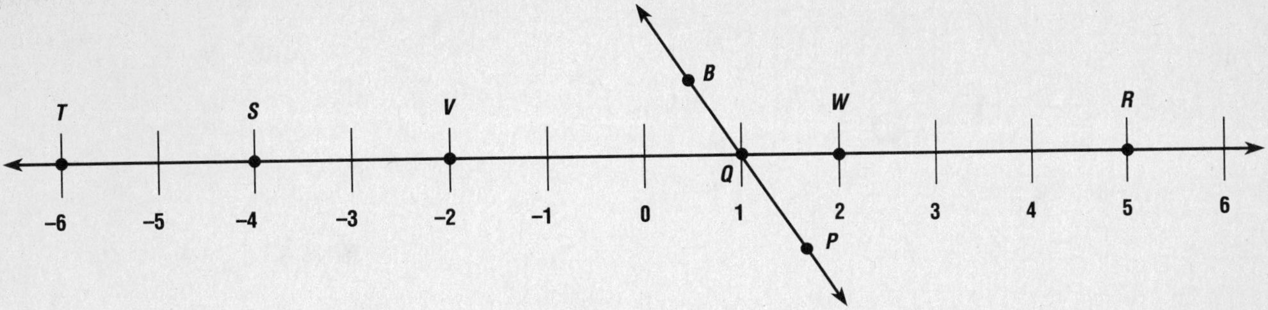

1. $\overline{PB}$ bisects $\overline{RS}$.

2. S is the midpoint of $\overline{TV}$.

3. $\overline{SV} \cong \overline{TS}$

4. V is the midpoint of $\overline{TW}$.

5. W bisect $\overline{VR}$.

6. $\overline{WR} \cong \overline{QV}$

7. $\overline{SW}$ is longer than $\overline{VR}$

8. $VW \leq TV$

Given the coordinate of one endpoint of $\overline{AB}$ and its midpoint, M, find the coordinates of the other endpoint.

9. $A(0, 9),\ M(2, 5)$

10. $B(-5, 1),\ M(1, -1)$

11. $A(-2, 3)\ M(0.5, 0.5)$

12. $A(4, 2),\ M(-2, 10)$

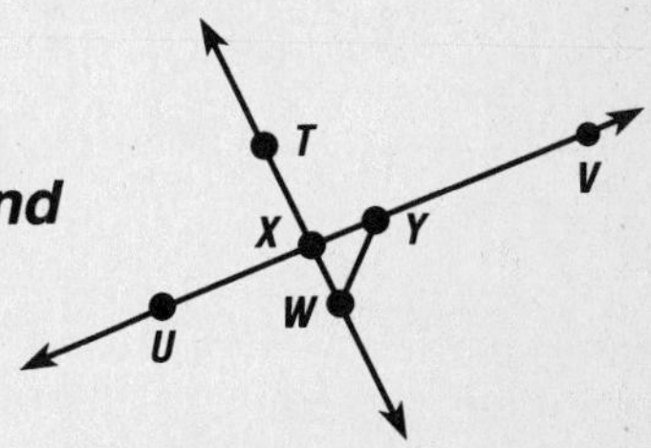

In the figure at the right, $\overline{WY}$ bisects $\overline{UV}$ at Y and $\overline{UY}$ bisects $\overline{TW}$ at X. Given the following conditions, find the value of x and the measure of the indicated segment.

13. $UY = 4x - 3,\ YV = x;\ UV$

14. $UV = x + 6,\ UY = x - 1;\ YV$

15. $TX = 2x + 1,\ XW = x + 7;\ TW$

16. $WX = x + 5,\ TW = 4x + 5;\ TX$

NAME ______________________ DATE ____________

1-5

Practice Worksheet

Segment Relationships

Refer to the figure below to determine whether each statement is true or false.

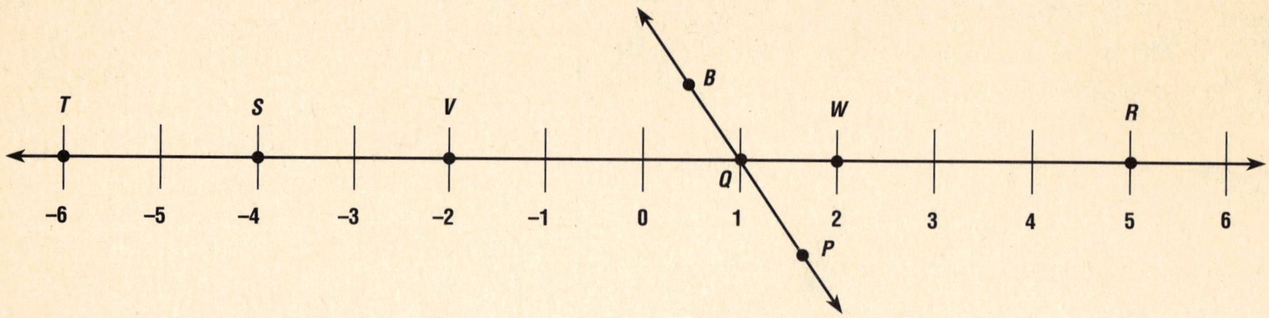

1. $\overline{PB}$ bisects $\overline{RS}$. **false**

2. S is the midpoint of $\overline{TV}$. **true**

3. $\overline{SV} \cong \overline{TS}$ **true**

4. V is the midpoint of $\overline{TW}$. **true**

5. W bisect $\overline{VR}$. **false**

6. $\overline{WR} \cong \overline{QV}$ **true**

7. $\overline{SW}$ is longer than $\overline{VR}$ **false**

8. $VW \leq TV$ **true**

Given the coordinate of one endpoint of $\overline{AB}$ and its midpoint, M, find the coordinates of the other endpoint.

9. $A(0, 9)$, $M(2, 5)$ **B (4, 1)**

10. $B(-5, 1)$, $M(1, -1)$ **A (7, –3)**

11. $A(-2, 3)$ $M(0.5, 0.5)$ **B (3, –2)**

12. $A(4, 2)$, $M(-2, 10)$ **B (–8, 18)**

In the figure at the right, $\overline{WY}$ bisects $\overline{UV}$ at Y and $\overline{UY}$ bisects $\overline{TW}$ at X. Given the following conditions, find the value of x and the measure of the indicated segment.

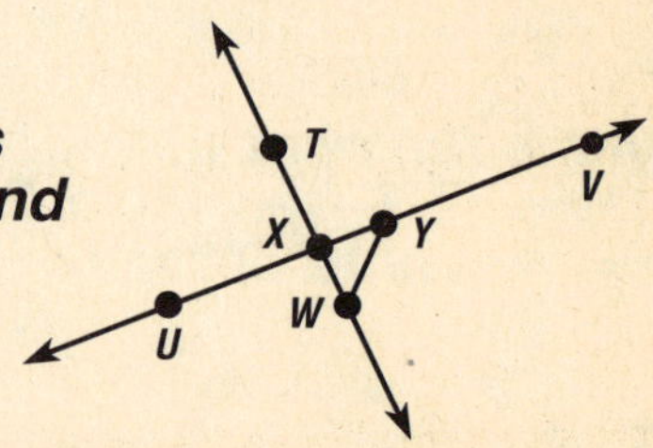

13. $UY = 4x - 3$, $YV = x$; UV **1, 2**

14. $UV = x + 6$, $UY = x - 1$; YV **8, 7**

15. $TX = 2x + 1$, $XW = x + 7$; TW **6, 26**

16. $WX = x + 5$, $TW = 4x + 5$; TX **2.5, 7.5**

NAME ______________________ DATE ____________

1-6 **Practice Worksheet**

Rays and Angles

Use the figure below to answer each question.

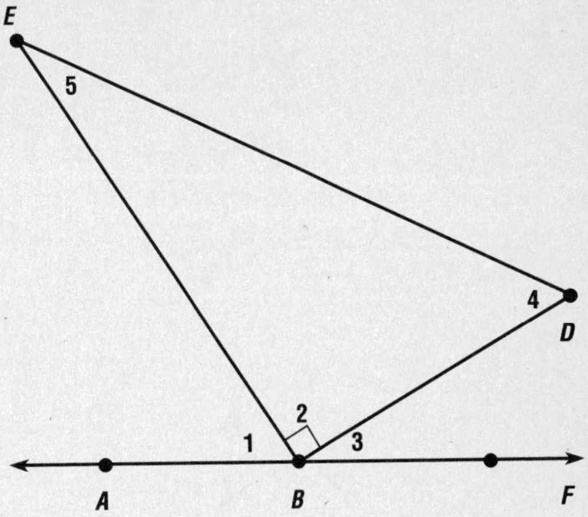

1. Give another name for $\angle 1$.

2. Name the vertex of $\angle EBD$.

3. What are the sides of 5?

4. Name the common side of $\angle 1$ and $\angle 2$.

5. Name a pair of opposite rays.

6. Name a point in the interior of $\angle EBF$.

7. Name three angles with $\overrightarrow{BE}$ as a side.

In the figure, $\overrightarrow{XP}$ and $\overrightarrow{XT}$ are opposite rays. Given the following conditions, find the value of x and the measure of the indicated angle.

8. $m\angle SXT = 4x + 1$, $m\angle QXS = 2x - 2$,
$m\angle QXT = 125$; $m\angle QXS$

9. $m\angle PXR = 3x$, $m\angle RXT = 5x + 20$, $m\angle RXT$

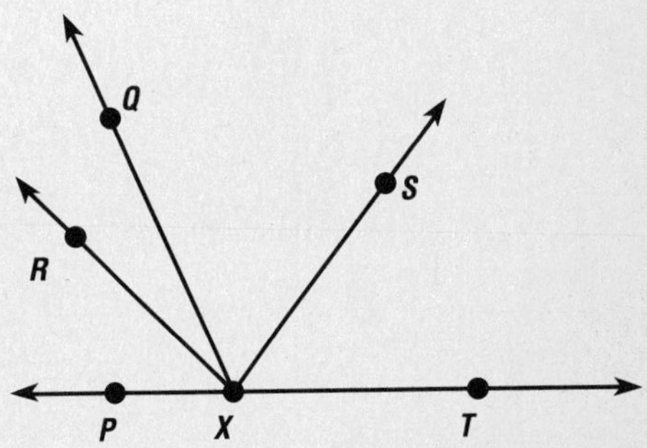

10. $m\angle RXQ = x + 15$, $m\angle RXS = 5x - 7$,
$m\angle QXS = 3x + 5$; $m\angle RXS$

11. $m\angle RXQ = 2x + 7$, $m\angle RXP = 3x - 11$,
$m\angle QXP = x + 40$; $m\angle QXT$

12. $m\angle TXS = x + 3$, $m\angle SXQ = 2x + 9$, $m\angle QXP = 4x - 7$;
$m\angle PXS$

NAME ______________________ DATE ____________

1-6 Practice Worksheet

Rays and Angles

Use the figure below to answer each question.

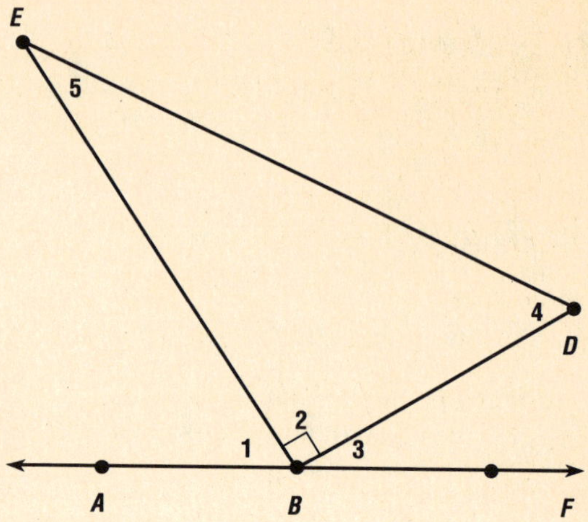

1. Give another name for ∠1.
   ∠ABE or ∠EBA
2. Name the vertex of ∠*EBD*. **B**
3. What are the sides of 5?
   $\overrightarrow{EB}$ and $\overrightarrow{ED}$
4. Name the common side of ∠1 and ∠2. **$\overrightarrow{BE}$**
5. Name a pair of opposite rays. **$\overrightarrow{BA}$ and $\overrightarrow{BF}$**
6. Name a point in the interior of ∠*EBF*. **D**
7. Name three angles with $\overrightarrow{BE}$ as a side.
   ∠FBE, ∠DBE, ∠ABE

In the figure, $\overrightarrow{XP}$ and $\overrightarrow{XT}$ are opposite rays. Given the following conditions, find the value of x and the measure of the indicated angle.

8. m∠*SXT* = 4*x* + 1, m∠*QXS* = 2*x* – 2,
   m∠*QXT* = 125; m∠*QXS* **21, 40**
9. m∠*PXR* = 3*x*, m∠*RXT* = 5*x* + 20, m∠*RXT* **20, 120**

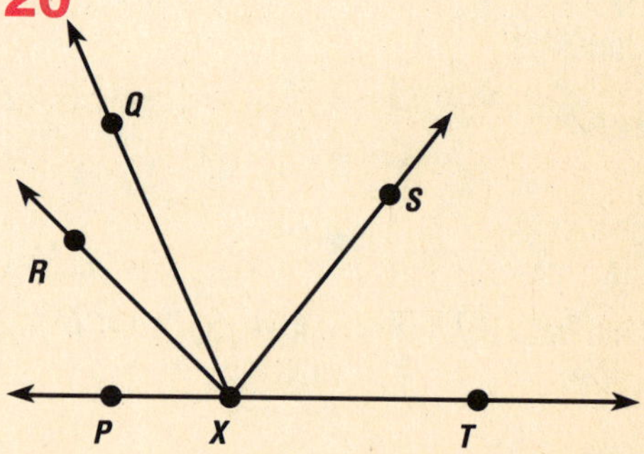

10. m∠*RXQ* = *x* +15, m∠*RXS* = 5*x* – 7,
    m∠*QXS* = 3*x* + 5; m∠*RXS* **27, 128**
11. m∠*RXQ* = 2*x* + 7, m∠*RXP* = 3*x* – 11,
    m∠*QXP* = *x* + 40; m∠*QXT* **11, 129**
12. m∠*TXS* = *x* + 3, m∠*SXQ* = 2*x* + 9, m∠*QXP* = 4*x* – 7;
    m∠*PXS* **25, 152**

NAME ______________________________ DATE ______________

1-7

Practice Worksheet

Classifying Angles

Refer to the figure to answer each question.

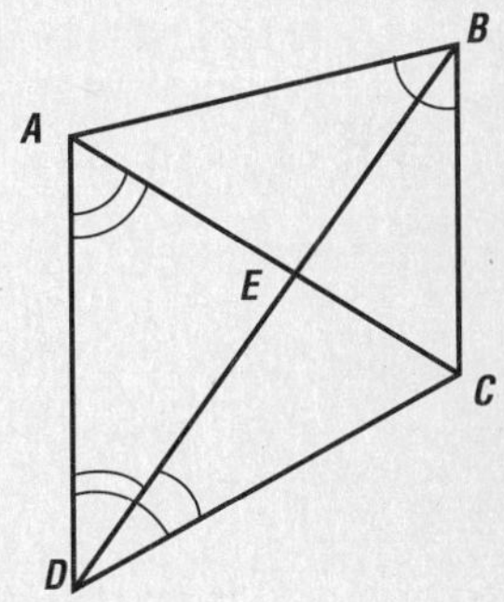

1. Name all pairs of angles that the figure identifies as congruent angles.

2. Name all segments that bisect angles and name the angles that they bisect.

3. If $m\angle ABD = 13x - 7$ and $m\angle CBD = 15x - 13$, find the value of x and $m\angle ABD$.

4. If $m\angle CAD = 10x + 12$ and $m\angle ADB = 20x - 8$, find the value of x and $m\angle DAC$.

In the figure, $\overrightarrow{TR}$ bisects $\angle PTS$ and $\overrightarrow{TQ}$ bisects $\angle PTR$. Given the following conditions, find the value of x and the measure of the indicated angle.

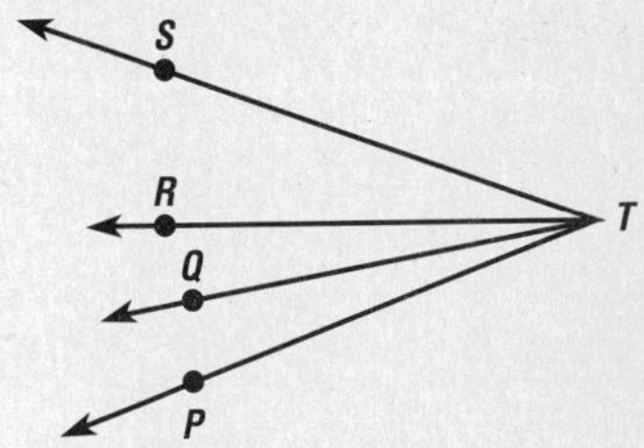

5. $m\angle PTR = 3x - 8$, $m\angle STR = 2x + 17$; $m\angle STP$

6. $m\angle PTS = x + 56$, $m\angle PTR = x - 23$; $m\angle RTS$

7. $m\angle PTQ = 30 - 3x$, $m\angle QTR = 5x - 18$; $m\angle PTS$

8. $m\angle STR = 73 - x$, $m\angle PTR = 94 - 2x$; $m\angle PTQ$

9. $m\angle PTS = 6x - 12$, $m\angle QTR = x + 5$; $m\angle PTR$

10. Suppose $\angle DEF$ is a right angle and H is in the interior of $\angle DEF$. If $m\angle DEH$ is five times $m\angle HEF$, find $m\angle HEF$.

NAME ______________________ DATE ____________

1-7

Practice Worksheet

Classifying Angles

Refer to the figure to answer each question.

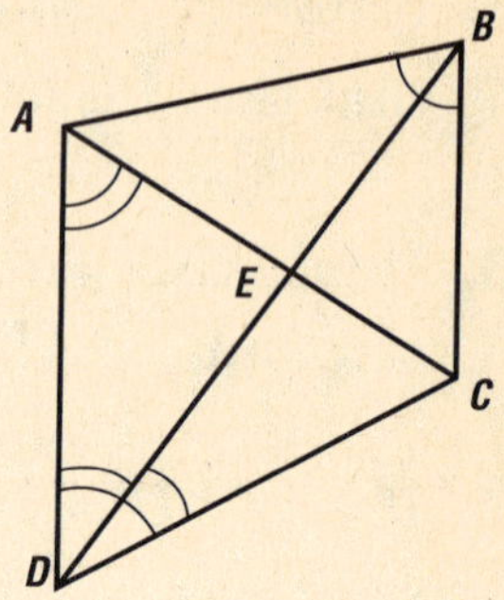

1. Name all pairs of angles that the figure identifies as congruent angles. **∠DAC and ∠ADB, ∠ABD and ∠DBC, ∠DAC and ∠BDC, ∠ADE and ∠CDE**

2. Name all segments that bisect angles and name the angles that they bisect. **$\overline{BD}$, ∠ABC; $\overline{BD}$, ∠ADC**

3. If m∠*ABD* = 13*x* – 7 and m∠*CBD* = 15*x* – 13, find the value of *x* and m∠*ABD*. **3, 32**

4. If m∠*CAD* = 10*x* + 12 and m∠*ADB* = 20*x* – 8, find the value of *x* and m∠*DAC*. **2, 32**

In the figure, $\overrightarrow{TR}$ bisects ∠PTS and $\overrightarrow{TQ}$ bisects ∠PTR. Given the following conditions, find the value of x and the measure of the indicated angle.

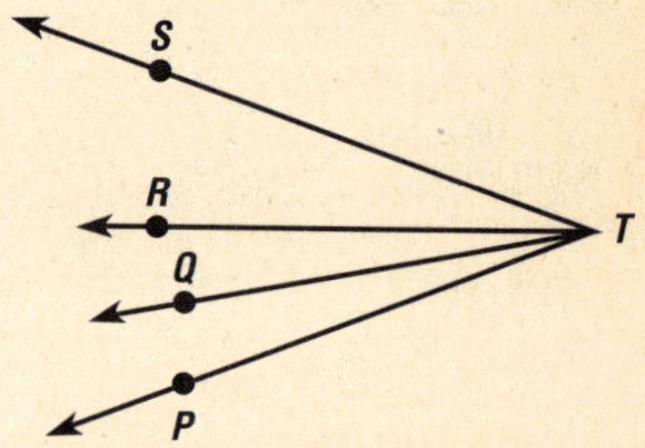

5. m∠*PTR* = 3*x* – 8, m∠*STR* = 2*x* + 17; m∠*STP* **25, 134**

6. m∠*PTS* = *x* + 56, m∠*PTR* = *x* – 23; m∠*RTS* **102, 79**

7. m∠*PTQ* = 30 – 3*x*, m∠*QTR* = 5*x* – 18; m∠*PTS* **6, 48**

8. m∠*STR* = 73 – *x*, m∠*PTR* = 94 – 2*x*; m∠*PTQ* **21, 26**

9. m∠*PTS* = 6*x* – 12, m∠*QTR* = *x* + 5; m∠*PTR* **16, 42**

10. Suppose ∠*DEF* is a right angle and *H* is in the interior of ∠*DEF*. If m∠*DEH* is five times m∠*HEF*, find m∠*HEF*. **15**

Glencoe Division, Macmillan/McGraw-Hill

NAME ________________________________ DATE ______________

1-8

Practice Worksheet

Pairs of Angles

Identify each pair of angles as adjacent, vertical, complementary, supplementary, and/or as a linear pair.

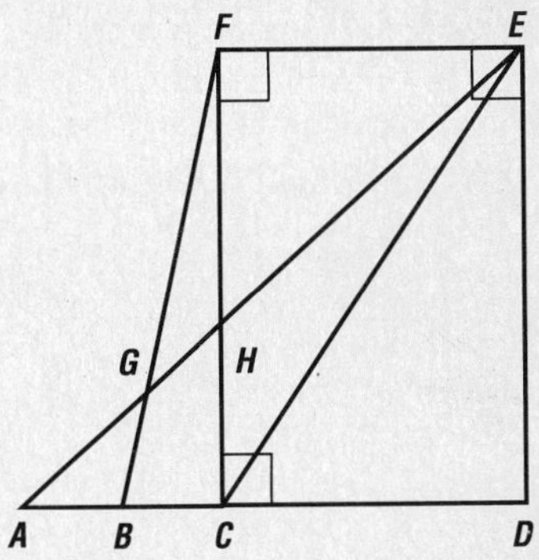

1. $\angle FGE$ and $\angle EGB$

2. $\angle ECH$ and $\angle HCA$

3. $\angle FHG$ and $\angle EHC$

4. $\angle EHF$ and $\angle FHG$

5. $\angle HCE$ and $\angle ECD$

6. $\angle FCD$ and $\angle FCB$

Find the value of x and m∠ABC.

7.

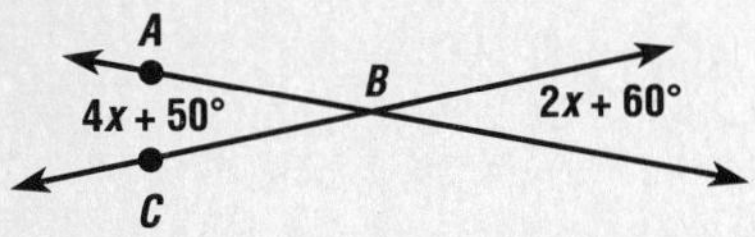

8.

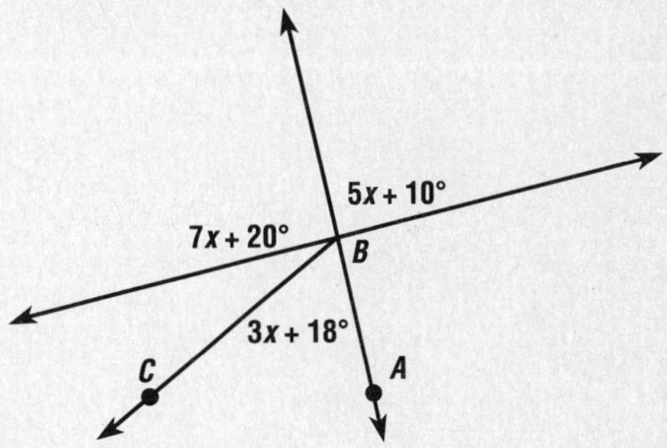

9.

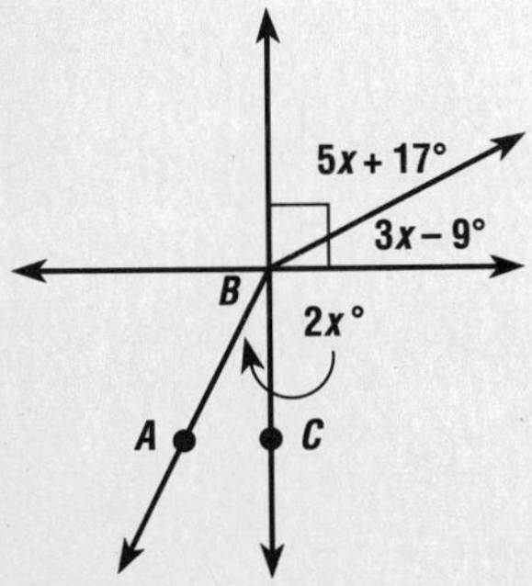

10.

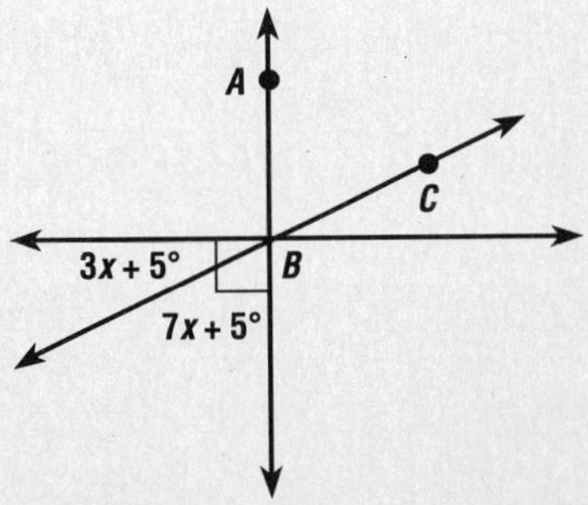

Glencoe Division, Macmillan/McGraw-Hill

NAME ______________________ DATE ____________

1-8

Practice Worksheet

Pairs of Angles

Identify each pair of angles as adjacent, vertical, complementary, supplementary, and/or as a linear pair.

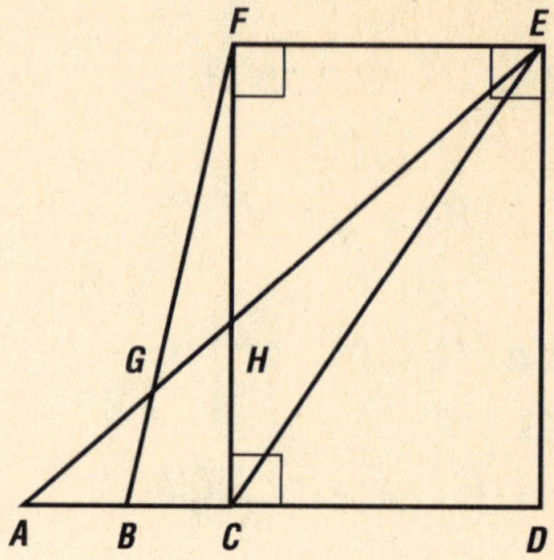

1. $\angle FGE$ and $\angle EGB$
   adjacent, supplementary, linear pair
2. $\angle ECH$ and $\angle HCA$
   adjacent
3. $\angle FHG$ and $\angle EHC$
   vertical
4. $\angle EHF$ and $\angle FHG$
   adjacent, supplementary, linear pair
5. $\angle HCE$ and $\angle ECD$
   adjacent, complementary
6. $\angle FCD$ and $\angle FCB$
   adjacent, supplementary, linear pair

Find the value of x and m∠ABC.

7.

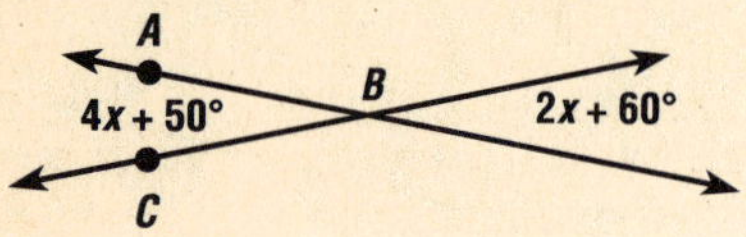

5, 70

8.

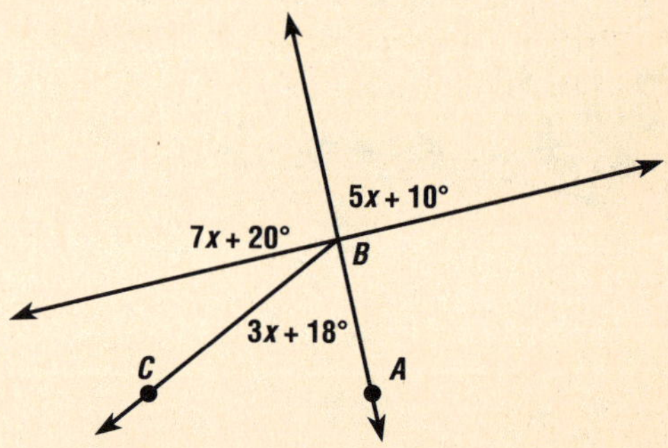

12.5, 55.5

9.

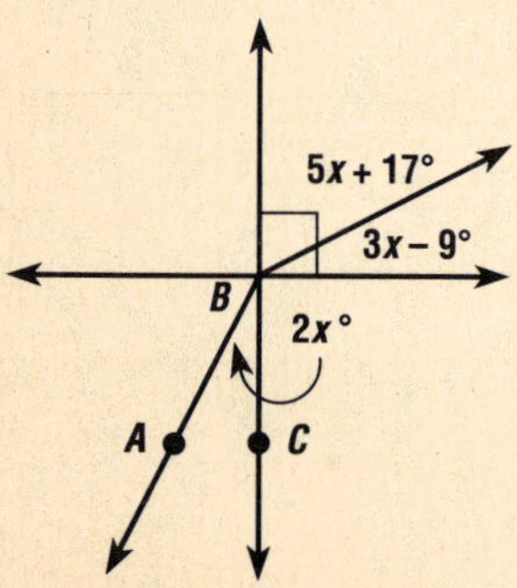

10.25, 20.5

10.

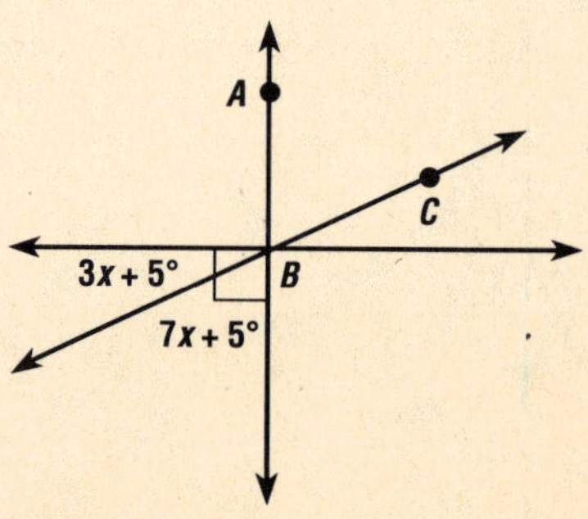

8, 61

NAME ______________________________ DATE ______________

1-9 Practice Worksheet

Right Angles and Perpendicular Lines

For each figure, find the value of x. Then determine if $\overline{AB} \perp \overline{CD}$.

1.

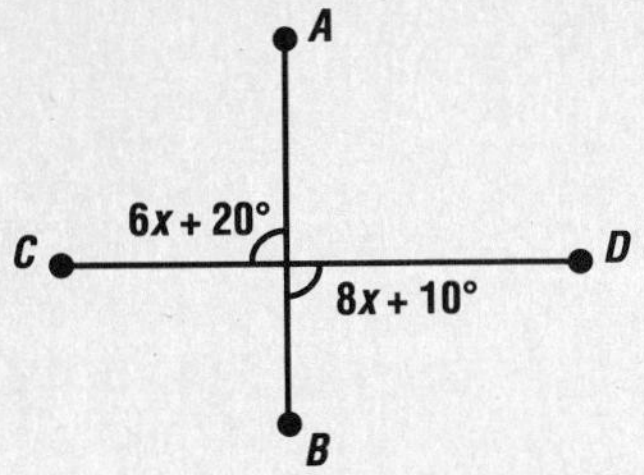

2.

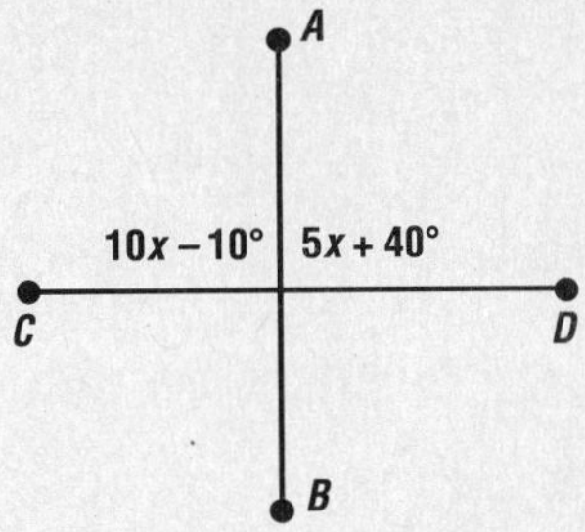

3.

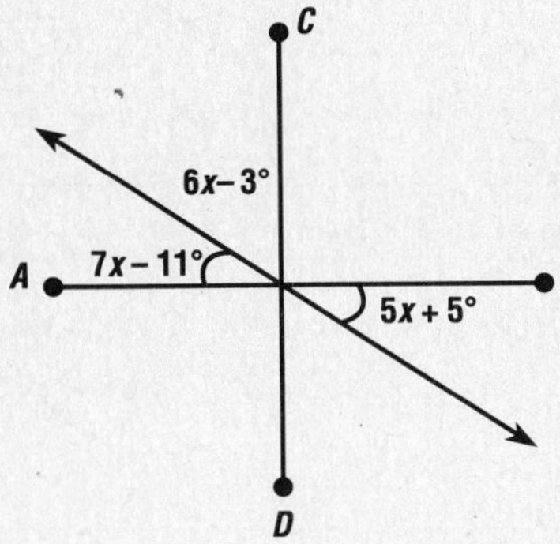

4.

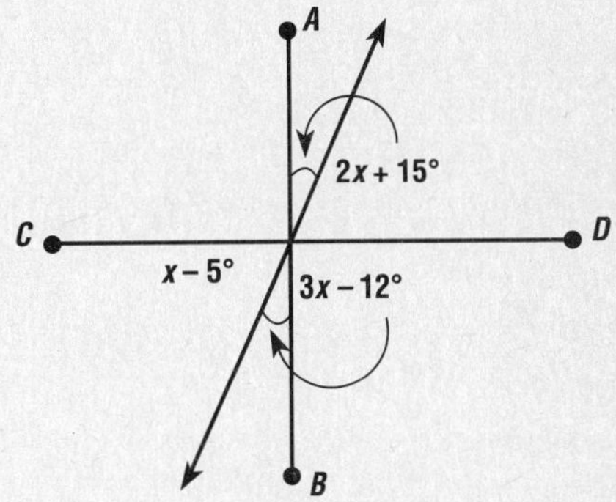

Determine whether each relationship can be assumed from the figure. Write yes or no.

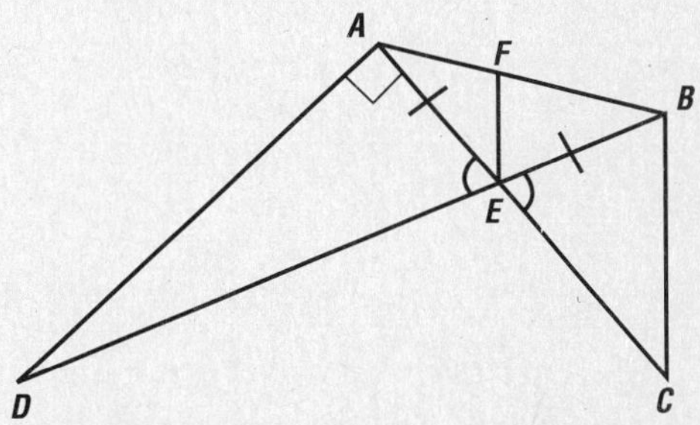

5. $\angle BFE$ is a right angle.

6. $\overline{AD} \perp \overline{AC}$

7. $\overline{ED} \cong \overline{EC}$

8. F bisects $\overline{AB}$.

9. $\angle AEF$ and $\angle BEF$ are congruent.

10. $AC = AE + EC$

11. $\overline{BC} \perp \overline{BD}$

12. $\angle AED \cong \angle BEC$

1-9

NAME ______________________ DATE ____________

Practice Worksheet

Right Angles and Perpendicular Lines

For each figure, find the value of x. Then determine if $\overline{AB} \perp \overline{CD}$.

1.

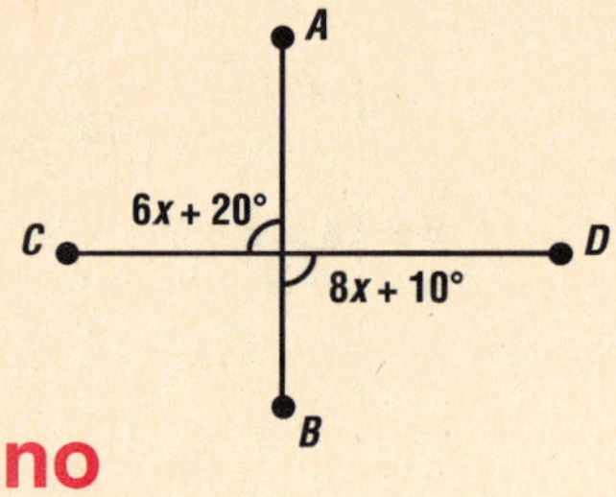

5, no

2.

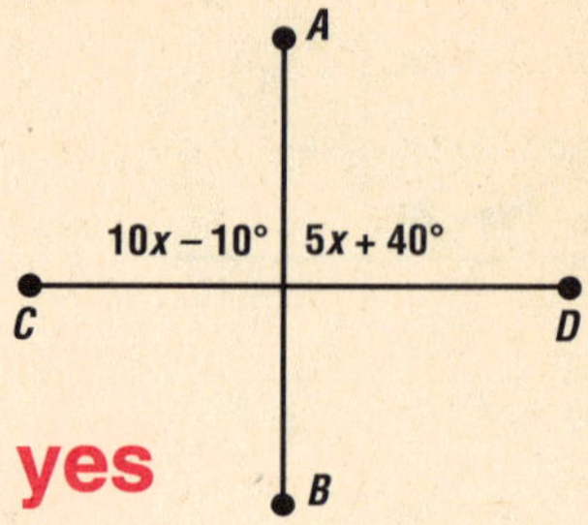

10, yes

3.

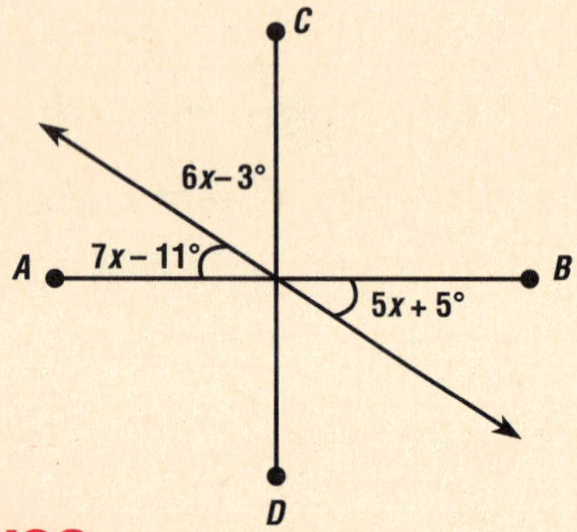

8, yes

4.

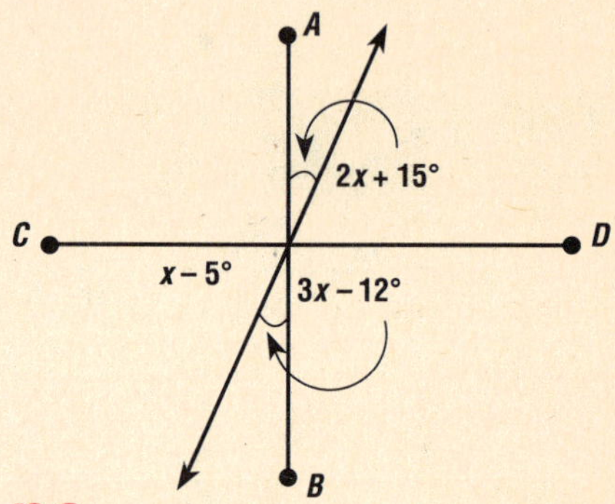

27, no

Determine whether each relationship can be assumed from the figure. Write yes or no.

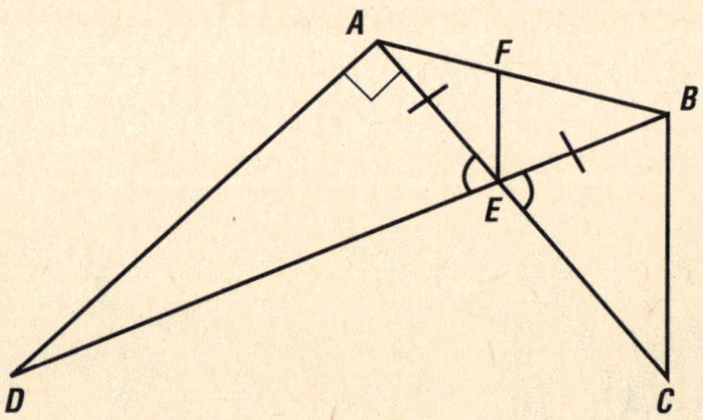

5. $\angle BFE$ is a right angle. no

6. $\overline{AD} \perp \overline{AC}$ yes

7. $\overline{ED} \cong \overline{EC}$ no

8. F bisects $\overline{AB}$. no

9. $\angle AEF$ and $\angle BEF$ are congruent. no

10. $AC = AE + EC$ yes

11. $\overline{BC} \perp \overline{BD}$ no

12. $\angle AED \cong \angle BEC$ yes

NAME ______________________ DATE ____________

2-1

Practice Worksheet

Inductive Reasoning and Conjecturing

Determine if the conjecture is true or false based on the given information. Explain your answer and give a counterexample for any false conjecture.

1. Given: noncollinear points A, B, C, and D
Conjecture: A, B, C, and D are coplanar.

2. Given: A, B, C, and D are collinear points.
Conjecture: $AB + BC + CD = AD$

3. Given: $\angle A$ and $\angle B$ are right angles.
Conjecture: $\angle A \cong \angle B$

4. Given: Point C between H and V.
Conjecture: $\angle HCV$ is a straight angle.

Write a conjecture based on the given information. If appropriate, draw a figure to illustrate your conjecture.

5. $\overline{AB}$, $\overline{CD}$, and $\overline{EF}$ intersect at X, and X is not an endpoint of any of the three segments.

6. $\angle MNO$ and $\angle PNO$ are adjacent angles.

7. $A\ (3, 1)$, $B\ (3, -5)$, $C\ (3, 7)$

8. A, B, C, and D are coplanar points.

NAME ______________________ DATE ____________

2-1 Practice Worksheet

Inductive Reasoning and Conjecturing

Determine if the conjecture is true or false based on the given information. Explain your answer and give a counterexample for any false conjecture.

1. Given: noncollinear points A, B, C, and D
Conjecture: A, B, C, and D are coplanar. **false;**

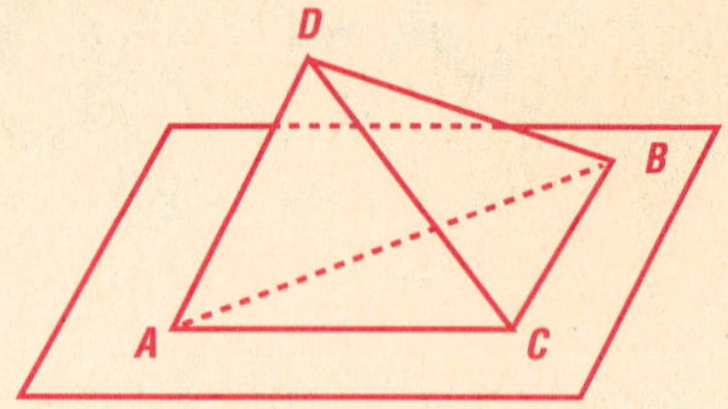

2. Given: A, B, C, and D are collinear points.
Conjecture: $AB + BC + CD = AD$
false;

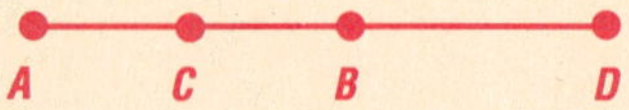

$AB + BC + CD > AD$

3. Given: $\angle A$ and $\angle B$ are right angles.
Conjecture: $\angle A \cong \angle B$
true; $m\angle A = 90$, $m\angle B = 90$

4. Given: Point C between H and V.
Conjecture: $\angle HCV$ is a straight angle. **true;**

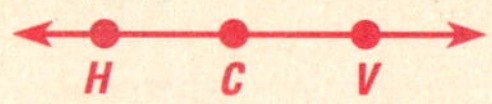

$\overrightarrow{CH}$ and $\overrightarrow{CV}$ are opposite rays.

Write a conjecture based on the given information. If appropriate, draw a figure to illustrate your conjecture.

Answers for Exercises 5-8 may vary. Sample answers are given.

5. $\overline{AB}$, $\overline{CD}$, and $\overline{EF}$ intersect at X, and X is not an endpoint of any of the three segments.
A, B, C, D, E, F, and X are noncollinear. (true)

6. $\angle MNO$ and $\angle PNO$ are adjacent angles.
Point O is not between M and P. (true)

7. A (3, 1), B (3, –5), C (3, 7)
Points A, B, and C lie on the same vertical line. (true)

8. A, B, C, and D are coplanar points.
$\angle ABC$ and $\angle ABD$ are adjacent angles. (false)

NAME ______________________ DATE ____________

2-2
Practice Worksheet

If-Then Statements, Converses, and Postulates

Identify the hypothesis and conclusions of each conditional statement.

1. If $3x - 1 = 7$, then $x = 2$.

2. If Carl scores 85%, then he passes.

Write each conditional statement in if-then form.

3. All students like vacations.

4. The game will be played provided it doesn't rain.

Write the converse of each conditional. Determine if the converse is true or false. If it is false, give a counterexample.

5. If it rains, then it is cloudy.

6. If x is an even number, then x is divisible by 2.

In the figure, P, Q, R, and S are in plane $\mathcal{N}$. Use the postulates you have learned to determine whether each statement is true or false.

7. R, S, and T are collinear.

8. There is only one plane that contains all the points R, S, and Q.

9. $\angle PQT$ lies in plane $\mathcal{N}$.

10. $\angle SPR$ lies in plane $\mathcal{N}$.

11. If X and Y are two points on line m, then $\overleftrightarrow{XY}$ intersects plane $\mathcal{N}$ at P.

12. Point K is on plane $\mathcal{N}$.

13. $\mathcal{N}$ contains $\overline{RS}$.

14. T lies in plane $\mathcal{N}$.

15. R, P, and T are coplanar.

16. ℓ and m intersect.

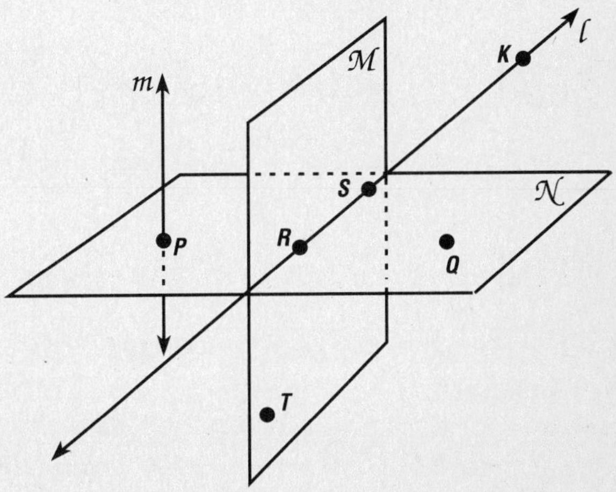

2-2

NAME ______________________ DATE __________

Practice Worksheet

If-Then Statements, Converses, and Postulates

Identify the hypothesis and conclusions of each conditional statement.

1. If $3x - 1 = 7$, then $x = 2$.
Hypothesis: $3x - 1 = 7$
Conclusion: $x = 2$

2. If Carl scores 85%, then he passes.
Hypothesis: Carl scores 85%.
Conclusion: he passes.

Write each conditional statement in if-then form.

3. All students like vacations. **If a person is a student, then that person likes vacations.**

4. The game will be played provided it doesn't rain. **If it doesn't rain, then the game will be played.**

Write the converse of each conditional. Determine if the converse is true or false. If it is false, give a counterexample.

5. If it rains, then it is cloudy.
If it is cloudy, then it rains.; false; It can be cloudy without raining.

6. If x is an even number, then x is divisible by 2.
If x is divisible by 2, then x is an even number.; true

In the figure, P, Q, R, and S are in plane $\mathcal{N}$. Use the postulates you have learned to determine whether each statement is true or false.

7. R, S, and T are collinear. **false**

8. There is only one plane that contains all the points R, S, and Q. **true**

9. $\angle PQT$ lies in plane $\mathcal{N}$. **false**

10. $\angle SPR$ lies in plane $\mathcal{N}$. **true**

11. If X and Y are two points on line m, then $\overleftrightarrow{XY}$ intersects plane $\mathcal{N}$ at P. **true**

12. Point K is on plane $\mathcal{N}$. **true**

13. $\mathcal{N}$ contains $\overline{RS}$. **true**

14. T lies in plane $\mathcal{N}$. **false**

15. R, P, and T are coplanar. **true**

16. ℓ and m intersect. **false**

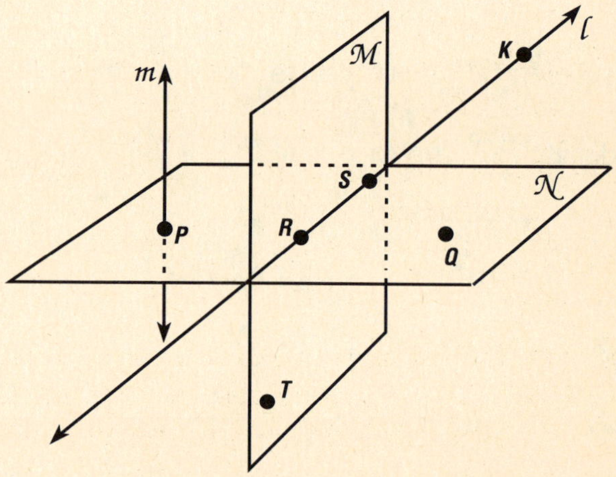

Glencoe Division, Macmillan/McGraw-Hill

NAME ______________________ DATE ____________

2-3

Practice Worksheet

Deductive Reasoning

Determine if a conclusion can be reached from the two given statements using the law of detachment or the law of syllogism. If a conclusion is possible, state it and the law that is used. If a conclusion does not follow, state "no conclusion".

1. If Jim is a Texan, then he is an American.
Jim is a Texan.

2. If spot is a dog, then he has four legs.
Spot has four legs.

3. If Rachel lives in Tampa, then Rachel lives in Florida.
If Rachel lives in Florida, then Rachel lives in the U.S.A.

4. If October 12 is a Monday, then October 13 is a Tuesday.
October 12 is a Monday.

5. If Henry studies his algebra, then he passes the test.
If Henry passes the test, then he will get a good grade.

Determine if statement (3) follows from statements (1) and (2) by the law of detachment or the law of syllogism. If it does, state which law was used.

6. (1) If the measure of an angle is greater than 90, then it is obtuse.
(2) $m\angle T$ is greater than 90.
(3) $\angle T$ is obtuse.

7. (1) If Pedro is taking history, then he will study about World War II.
(2) Pedro will study about World War II.
(3) Pedro is taking history.

8. (1) If Julie works after school, then she works in a department store.
(2) Julie works after school.
(3) Julie works in a department store.

9. (1) If William is reading, then he is reading a magazine.
(2) If William is reading a magazine, then he is reading a magazine about computers.
(3) If William is reading, then he is reading a magazine about computers.

10. (1) A vocalist can read music.
(2) Ann Marie can read music.
(3) Ann Marie is a vocalist.

Glencoe Division, Macmillan/McGraw-Hill

2-3

NAME ________________ DATE ________

Practice Worksheet

Deductive Reasoning

Determine if a conclusion can be reached from the two given statements using the law of detachment or the law of syllogism. If a conclusion is possible, state it and the law that is used. If a conclusion does not follow, state "no conclusion".

1. If Jim is a Texan, then he is an American.
Jim is a Texan. **Jim is an American.; detachment**

2. If spot is a dog, then he has four legs.
Spot has four legs. **no conclusion**

3. If Rachel lives in Tampa, then Rachel lives in Florida.
If Rachel lives in Florida, then Rachel lives in the U.S.A. **If Rachel lives in Tampa, then Rachel lives in the U.S.A.; syllogism**

4. If October 12 is a Monday, then October 13 is a Tuesday.
October 12 is a Monday. **October 13 is a Tuesday.; detachment**

5. If Henry studies his algebra, then he passes the test.
If Henry passes the test, then he will get a good grade. **If Henry studies his algebra, then he will get a good grade.; syllogism**

Determine if statement (3) follows from statements (1) and (2) by the law of detachment or the law of syllogism. If it does, state which law was used.

6. (1) If the measure of an angle is greater than 90, then it is obtuse.
(2) $m\angle T$ is greater than 90.
(3) $\angle T$ is obtuse. **yes; detachment**

7. (1) If Pedro is taking history, then he will study about World War II.
(2) Pedro will study about World War II.
(3) Pedro is taking history. **no**

8. (1) If Julie works after school, then she works in a department store.
(2) Julie works after school.
(3) Julie works in a department store. **yes; detachment**

9. (1) If William is reading, then he is reading a magazine.
(2) If William is reading a magazine, then he is reading a magazine about computers.
(3) If William is reading, then he is reading a magazine about computers. **yes; syllogism**

10. (1) A vocalist can read music.
(2) Ann Marie can read music.
(3) Ann Marie is a vocalist. **no**

NAME ______________________ DATE ____________

2-4

Practice Worksheet

Connecting Algebra to Proof

Name the property of equality that justifies each statement.

1. If $m\angle A = m\angle B$, then $m\angle B = m\angle A$.

2. If $x + 3 = 17$, then $x = 14$.

3. $xy = xy$

4. If $7x = 42$, then $x = 6$.

5. If $XY - YZ = XM$, then $XM + YZ = XY$.

6. $2(x + 4) = 2x + 8$

7. If $m\angle A + m\angle B = 90$, and $m\angle A = 30$, then $30 + m\angle B = 90$.

8. If $x = y + 3$ and $y + 3 = 10$, then $x = 10$.

Complete each proof by naming the property that justifies each statement.

9. Prove that if $2(x - 3) = 8$, then $x = 7$.
Given: $2(x - 3) = 8$
Prove: $x = 7$

Statements	Reasons
a. $2(x - 3) = 8$	**a.** ____________
b. $2x - 6 = 8$	**b.** ____________
c. $2x = 14$	**c.** ____________
d. $x = 7$	**d.** ____________

10. Prove that if $3x - 4 = \frac{1}{2}x + 6$, then $x = 4$.
Given: $3x - 4 = \frac{1}{2}x + 6$
Prove: $x = 4$

Statements	Reasons
a. $3x - 4 = \frac{1}{2}x + 6$	**a.** ____________
b. $\frac{5}{2}x - 4 = 6$	**b.** ____________
c. $\frac{5}{2}x = 10$	**c.** ____________
d. $x = 4$	**d.** ____________

Glencoe Division, Macmillan/McGraw-Hill

2-4

NAME ______________________ DATE ____________

Practice Worksheet

Connecting Algebra to Proof

Name the property of equality that justifies each statement.

1. If $m\angle A = m\angle B$, then $m\angle B = m\angle A$.
symmetric

2. If $x + 3 = 17$, then $x = 14$.
subtraction

3. $xy = xy$
reflexive

4. If $7x = 42$, then $x = 6$.
division

5. If $XY - YZ = XM$, then $XM + YZ = XY$.
addition

6. $2(x + 4) = 2x + 8$
distributive

7. If $m\angle A + m\angle B = 90$, and $m\angle A = 30$, then $30 + m\angle B = 90$.
substitution

8. If $x = y + 3$ and $y + 3 = 10$, then $x = 10$.
transitive

Complete each proof by naming the property that justifies each statement.

9. Prove that if $2(x - 3) = 8$, then $x = 7$.
Given: $2(x - 3) = 8$
Prove: $x = 7$

Statements	Reasons
a. $2(x - 3) = 8$	**a.** Given
b. $2x - 6 = 8$	**b.** Distributive property of equality
c. $2x = 14$	**c.** Addition property of equality
d. $x = 7$	**d.** Division property of equality

10. Prove that if $3x - 4 = \frac{1}{2}x + 6$, then $x = 4$.
Given: $3x - 4 = \frac{1}{2}x + 6$
Prove: $x = 4$

Statements	Reasons
a. $3x - 4 = \frac{1}{2}x + 6$	**a.** Given
b. $\frac{5}{2}x - 4 = 6$	**b.** Subtraction property of equality
c. $\frac{5}{2}x = 10$	**c.** Addition property of equality
d. $x = 4$	**d.** Multiplication property of equality

NAME ______________________ DATE ______________

2-5

Practice Worksheet

Problem-Solving Strategy: Process of Elimination

Solve. Use any strategy.

1. Barky, Spot, and Tiger are dogs. One is a black labrador retriever, one is a multi-colored collie, and one is a spotted dalmation. One is owned by a doctor, one is owned by a lawyer, and one is owned by an insurance salesperson. The lawyer's dog does not have spots. The doctor's dog has more than one color. The insurance salesperson's dog is solid in color. Who owns which dog?

2. In a three-digit number, the sum of the hundreds digit and the tens digit is equal to the ones digit. How many such three-digit numbers are there?

3. Shannon placed the numbers 1, 2, and 3 on a number cube. She placed each number on the cube twice. Four people were seated at a cardtable, as shown.

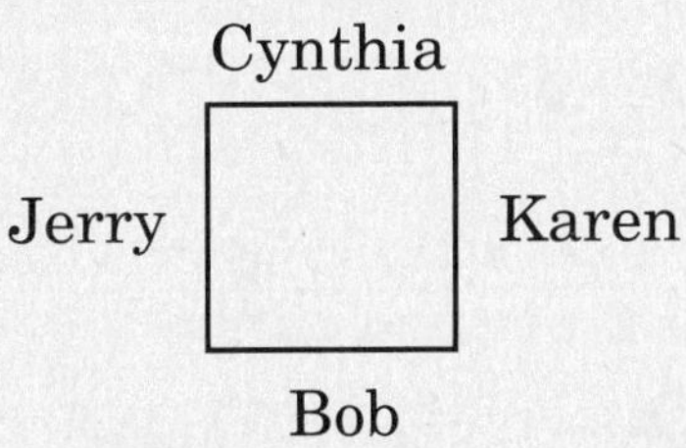

The number cube was placed on the table at an angle so that each person could see the top of the number cube, and two of its sides. The four people made the following comments:

Cynthia: I see a 1, and the other two numbers are even.
Karen: I can see two 2's.
Bob: I see three different numbers.
Jerry: I see three different numbers, also.

What three numbers can each of the people see? What number is face down?

Glencoe Division, Macmillan/McGraw-Hill

NAME ______________________ DATE ______________

2-5

Practice Worksheet

Problem-Solving Strategy: Process of Elimination

Solve. Use any strategy.

1. Barky, Spot, and Tiger are dogs. One is a black labrador retriever, one is a multi-colored collie, and one is a spotted dalmation. One is owned by a doctor, one is owned by a lawyer, and one is owned by an insurance salesperson. The lawyer's dog does not have spots. The doctor's dog has more than one color. The insurance salesperson's dog is solid in color. Who owns which dog?
   doctor–dalmation; lawyer–collie; insurance salesperson–labrador retriever

2. In a three-digit number, the sum of the hundreds digit and the tens digit is equal to the ones digit. How many such three-digit numbers are there? **45**

3. Shannon placed the numbers 1, 2, and 3 on a number cube. She placed each number on the cube twice. Four people were seated at a cardtable, as shown.

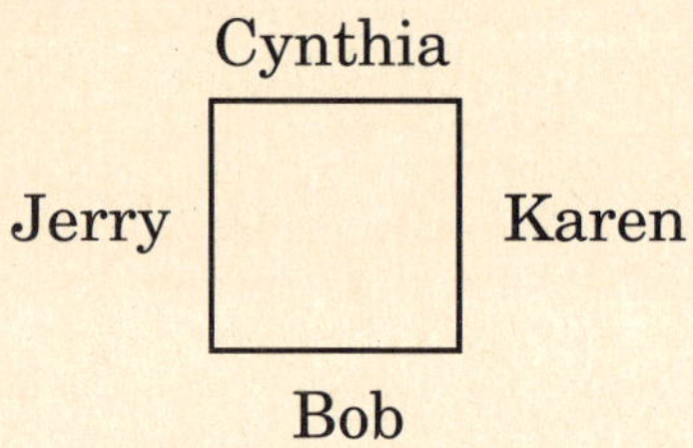

   The number cube was placed on the table at an angle so that each person could see the top of the number cube, and two of its sides. The four people made the following comments:

   Cynthia: I see a 1, and the other two numbers are even.
   Karen: I can see two 2's.
   Bob: I see three different numbers.
   Jerry: I see three different numbers, also.

   What three numbers can each of the people see? What number is face down?
   Cynthia: 1, 2, 2; Karen: 1, 2, 2;
   Bob: 1, 2, 3; Jerry: 1, 2, 3; 3 is face down.

NAME ______________________________ DATE ______________

2-6 Practice Worksheet

Two-Column Proofs with Segments

Complete each proof.

1. Given: $AD = 2AB + BC$
Prove: $\overline{AB} \cong \overline{CD}$

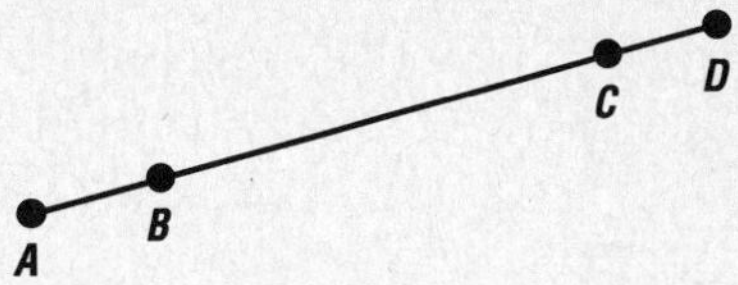

Statements	Reasons
a. $AD = 2AB + BC$	**a.** ______________
b. $AD = AB + BC + CD$	**b.** ______________
c. $2AB + BC = AB + BC + CD$	**c.** ______________
d. $AB = CD$	**d.** ______________
e. $\overline{AB} \cong \overline{CD}$	**e.** ______________

2. Given: B is between A and D.
C is between A and D.
Prove: $AB + BD = AC + CD$

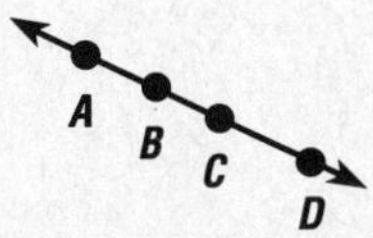

Statements	Reasons
a. B is between A and D. C is between A and D.	**a.** ______________
b. $AB + BD = AD$	**b.** ______________
c. $AC + CD = AD$	**c.** ______________
d. $AD = AC + CD$	**d.** ______________
e. $AB + BD = AC + CD$	**e.** ______________

Write a two-column proof.

3. Given: B is the midpoint of $\overline{AC}$.
Prove: $AB + CD = BD$

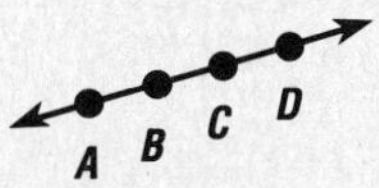

Statements	Reasons

Glencoe Division, Macmillan/McGraw-Hill

NAME ______________________ DATE ____________

Practice Worksheet

Two-Column Proofs with Segments

Complete each proof.

1. Given: $AD = 2AB + BC$
   Prove: $\overline{AB} \cong \overline{CD}$

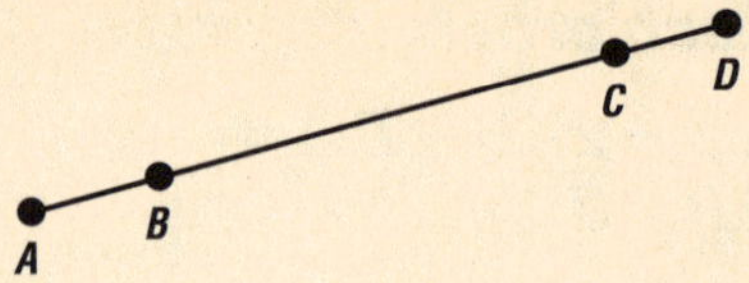

Statements	Reasons
a. $AD = 2AB + BC$	**a.** Given
b. $AD = AB + BC + CD$	**b.** Segment addition postulate
c. $2AB + BC = AB + BC + CD$	**c.** Substitution
d. $AB = CD$	**d.** Subtraction property of equality
e. $\overline{AB} \cong \overline{CD}$	**e.** Definition of congruent segments

2. Given: B is between A and D.
   C is between A and D.
   Prove: $AB + BD = AC + CD$

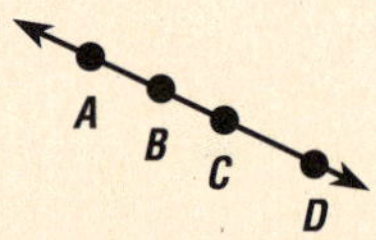

Statements	Reasons
a. B is between A and D. C is between A and D.	**a.** Given
b. $AB + BD = AD$	**b.** Segment addition
c. $AC + CD = AD$	**c.** Segment addition
d. $AD = AC + CD$	**d.** Symmetric property of equality
e. $AB + BD = AC + CD$	**e.** Transitive property of equality

Write a two-column proof.

3. Given: B is the midpoint of $\overline{AC}$.
   Prove: $AB + CD = BD$

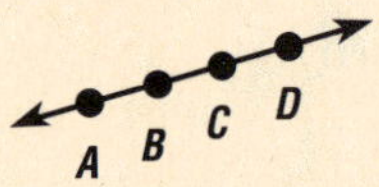

Statements	Reasons
a. B is the midpoint of $\overline{AC}$.	a. Given
b. $AB = BC$	b. Definition of midpoint
c. $BC + CD = BD$	c. Segment addition
d. $AB + CD = BD$	d. Substitution property of equality

NAME ______________________ DATE ____________

2-7 Practice Worksheet

Two-Column Proof with Angles

Complete each statement if m∠BGC = 43 and m∠DGE = 56.

1. ∠*FGA* ≅ ___?___

2. ∠*BGF* and ___?___ are supplementary.

3. m∠*CGD* = ___?___

4. m∠*AGF* = ___?___

5. ∠*EGC* and ___?___ are supplementary.

6. m∠*AGB* = ___?___

7. m∠*AGC* = ___?___

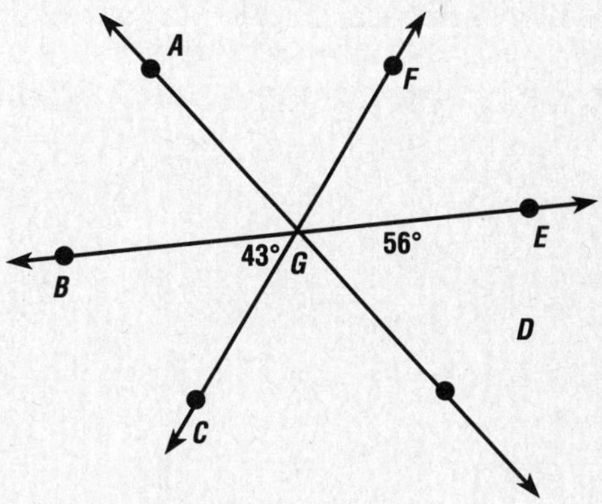

Write a two-column proof.

8. Given: ∠*AEC* ≅ ∠*DEB*
Prove: ∠*AEB* ≅ ∠*DEC*

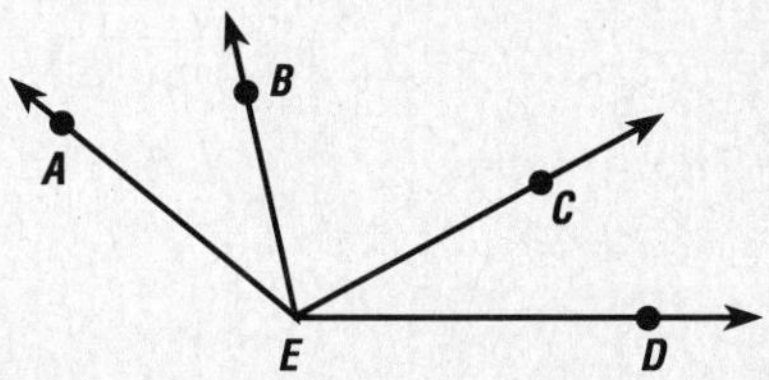

Statements	Reasons

NAME ______________________ DATE ____________

2-7

Practice Worksheet

Two-Column Proof with Angles

Complete each statement if m∠BGC = 43 and m∠DGE = 56.

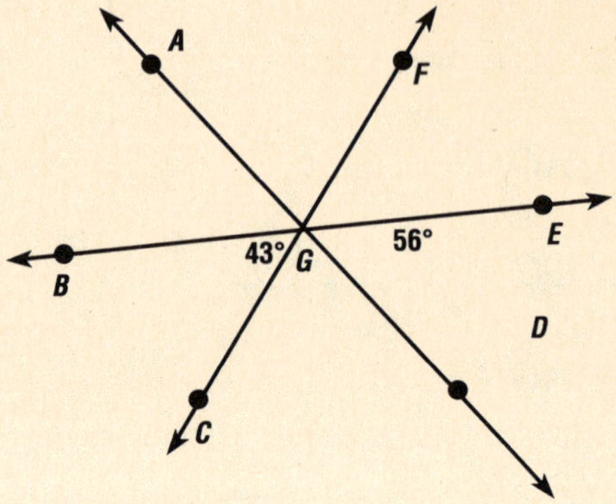

1. ∠FGA ≅ ___?___ ∠DGC

2. ∠BGF and ___?___ are supplementary.
   ∠FGE or ∠BGC

3. m∠CGD = ___?___ 81

4. m∠AGF = ___?___ 81

5. ∠EGC and ___?___ are supplementary.
   ∠BGC or ∠FGE

6. m∠AGB = ___?___ 56

7. m∠AGC = ___?___ 99

Write a two-column proof.

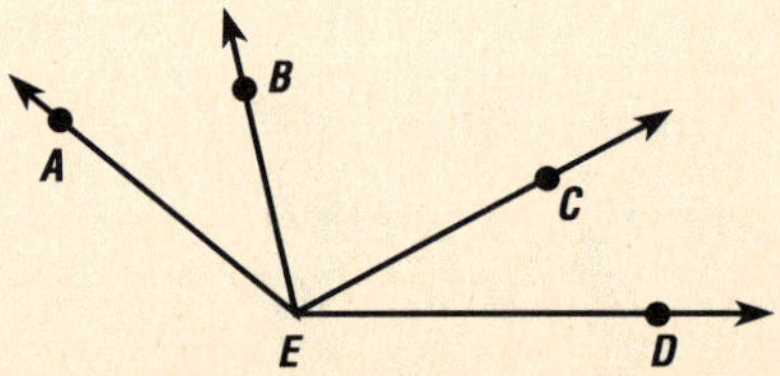

8. Given: ∠AEC ≅ ∠DEB
   Prove: ∠AEB ≅ ∠DEC

Statements	Reasons
a. ∠AEC ≅ ∠DEB	a. Given
b. m∠AEC = m∠DEB	b. Definition of congruent angles
c. m∠AEC = m∠AEB + m∠BEC, m∠DEB = m∠DEC + m∠BEC	c. Angle addition postulate
d. m∠AEB + m∠BEC = m∠DEC + m∠BEC	d. Transitive property of equality
e. m∠AEB = m∠DEC	e. Subtraction property of equality
f. ∠AEB ≅ ∠DEC	f. Definition of congruent angles

3-1

NAME ____________________ DATE ____________

Practice Worksheet

Problem-Solving Strategy: Draw a Diagram

Solve. Use any strategy.

1. Draw and label the diagram that you would use to prove the theorem *angles complementary to the same angle are congruent.* Then state the given information and the statement to be proved.

2. At a town's bicentennial celebration, men dressed up as settlers and tipped their hats whenever they met another man. At a town meeting, ten men were present. How many times were pairs of hats tipped as two men met for the first time?

3. In how many ways can you make change for a one-dollar bill if none of the coins is a penny?

4. Consider a circle with 1000 points marked inside it. Is there a straight line across the circle with exactly 500 of the points to one side and the other 500 points to the other side? Explain.

3-1

NAME ______________________ DATE ____________

Practice Worksheet

Problem-Solving Strategy: Draw a Diagram

Solve. Use any strategy.

1. Draw and label the diagram that you would use to prove the theorem *angles complementary to the same angle are congruent.* Then state the given information and the statement to be proved.

Typical answer:

Given: $\angle DBA$ and $\angle ABC$ are complementary.
$\angle EFG$ and $\angle ABC$ are complementary.
Prove: $\angle DBA \cong \angle EFG$

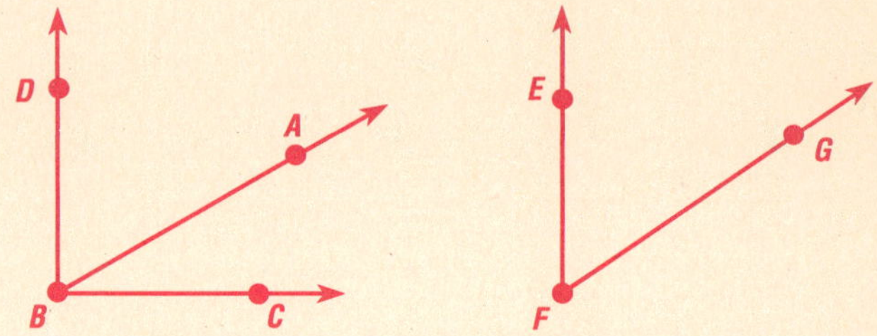

2. At a town's bicentennial celebration, men dressed up as settlers and tipped their hats whenever they met another man. At a town meeting, ten men were present. How many times were pairs of hats tipped as two men met for the first time? **45**

3. In how many ways can you make change for a one-dollar bill if none of the coins is a penny? **40**

4. Consider a circle with 1000 points marked inside it. Is there a straight line across the circle with exactly 500 of the points to one side and the other 500 points to the other side? Explain.
yes; Consider all the lines determined by pairs of the 1000 points. Pick point X outside the circle and draw a line ℓ through X that does not intersect the circle. Rotate ℓ clockwise about X. Eventually it will cross the circle and intersect the 1000 points one at a time. Any position of ℓ between that for the 500th and 501st point intersected by ℓ will give a line of the kind that is wanted.

NAME ________________ DATE ________

3-2

Practice Worksheet

Parallels and Transversals

Identify the special angle pair name for the angles.

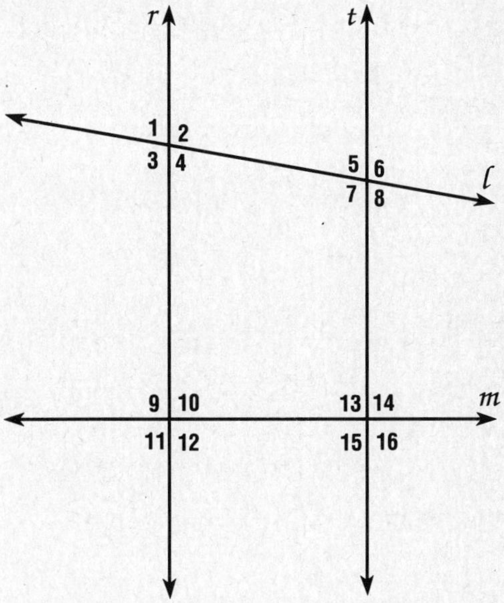

1. ∠1 and ∠4

2. ∠2 and ∠10

3. ∠4 and ∠9

4. ∠6 and ∠3

5. ∠14 and ∠10

6. ∠7 and ∠13

The three-dimensional figure shown at the right is called a right pentagonal prism.

7. Identify all segments joining points marked in plane JIH that appear to be skew to $\overline{EB}$.

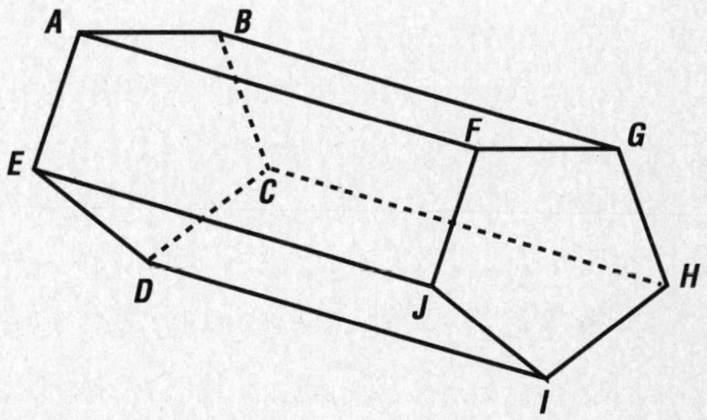

8. Which segments seem parallel to $\overline{BG}$?

9. Which segments seem parallel to $\overline{GH}$?

10. Identify all planes that appear parallel to plane *FGH*.

3-2

NAME ________________ DATE ________

Practice Worksheet

Parallels and Transversals

Identify the special angle pair name for the angles.

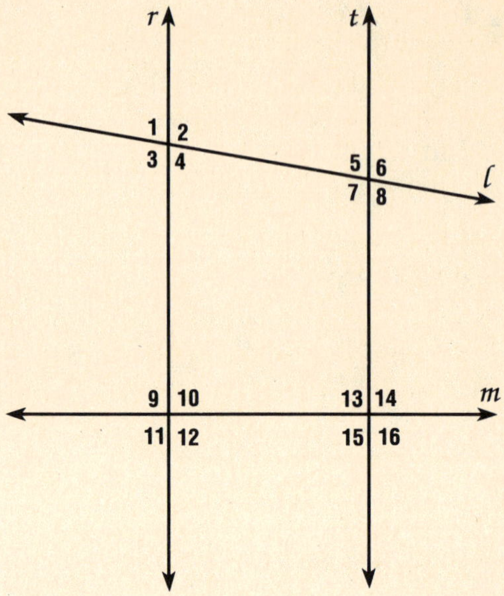

1. ∠1 and ∠4 vertical angles

2. ∠2 and ∠10 corresponding angles

3. ∠4 and ∠9 alternate interior angles

4. ∠6 and ∠3 alternate exterior angles

5. ∠14 and ∠10 corresponding angles

6. ∠7 and ∠13 consecutive interior angles

The three-dimensional figure shown at the right is called a right pentagonal prism.

7. Identify all segments joining points marked in plane JIH that appear to be skew to $\overline{EB}$.
$\overline{FG}$, $\overline{FJ}$, $\overline{FH}$, $\overline{FI}$, $\overline{GH}$, $\overline{GI}$, $\overline{HJ}$, $\overline{IJ}$

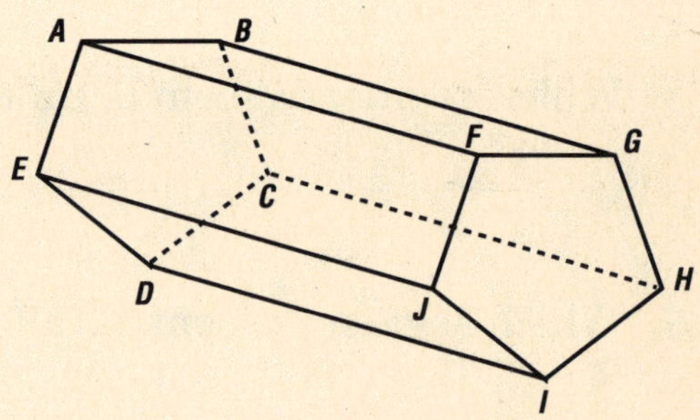

8. Which segments seem parallel to $\overline{BG}$?
$\overline{AF}$, $\overline{EJ}$, $\overline{DI}$, $\overline{CH}$

9. Which segments seem parallel to $\overline{GH}$? $\overline{BC}$

10. Identify all planes that appear parallel to plane *FGH*. plane *ABC*

3-3

NAME ________________________ DATE ____________

Practice Worksheet

Using Parallel Lines

In the figure, $\ell \parallel m$.

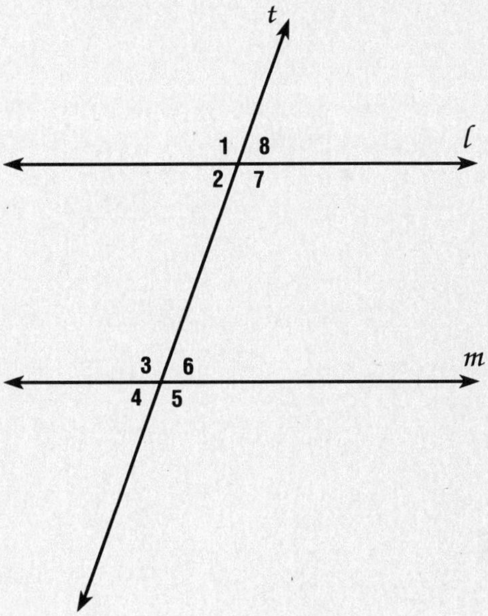

1. If $m\angle 7 = 100$, find $m\angle 3$.

2. If $m\angle 7 = 95$, find $m\angle 6$.

3. If $m\angle 1 = 120$, find $m\angle 5$.

4. If $m\angle 4 = 20$, find $m\angle 7$.

5. If $m\angle 3 = 140$, find $m\angle 8$.

6. If $m\angle 4 = 30$, find $m\angle 1$.

7. If $m\angle 4 = 40$, find $m\angle 2$.

8. If $m\angle 7 = 125$, find $m\angle 4$.

9. If $\ell \perp t$, find $m\angle 3$.

10. If $m\angle 1 + m\angle 3 = 230$, find $m\angle 6$.

In the figure, $s \parallel t$. ***Find the measure of each angle.***

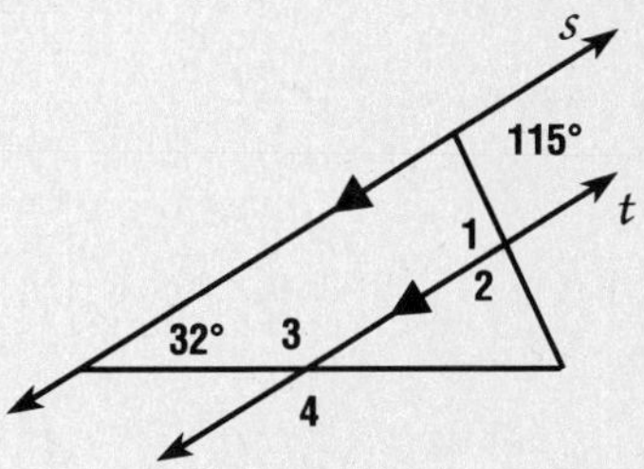

11. $m\angle 1$

12. $m\angle 2$

13. $m\angle 3$

14. $m\angle 4$

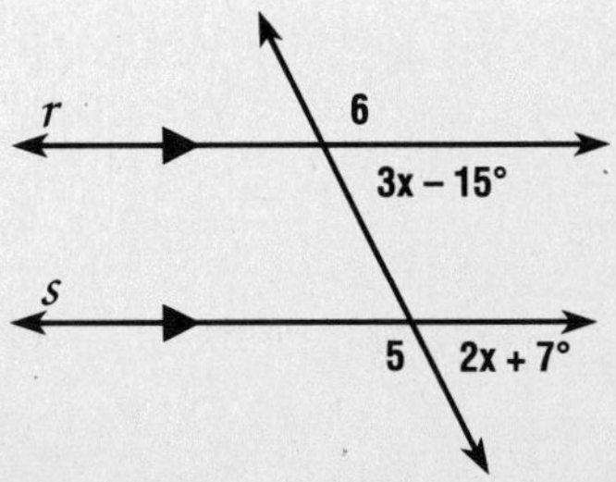

15. In the figure, $r \parallel s$. Find the value of x.

Glencoe Division, Macmillan/McGraw-Hill

3-3

NAME ____________________ DATE ____________

Practice Worksheet

Using Parallel Lines

In the figure, $\ell \parallel m$.

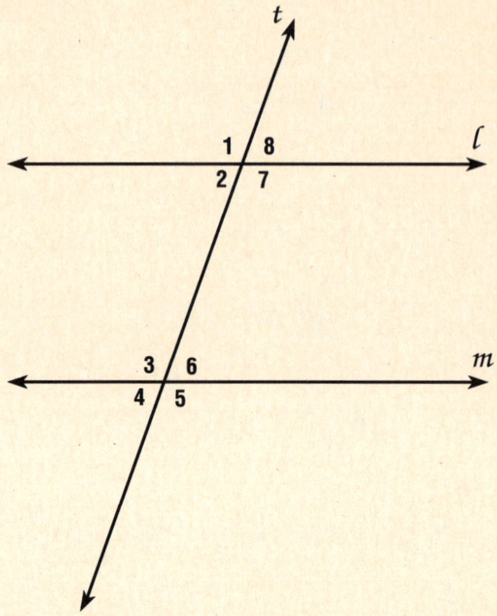

1. If m∠7 = 100, find m∠3. **100**

2. If m∠7 = 95, find m∠6. **85**

3. If m∠1 = 120, find m∠5. **120**

4. If m∠4 = 20, find m∠7. **160**

5. If m∠3 = 140, find m∠8. **40**

6. If m∠4 = 30, find m∠1. **150**

7. If m∠4 = 40, find m∠2. **40**

8. If m∠7 = 125, find m∠4. **55**

9. If $\ell \perp t$, find m∠3. **90**

10. If m∠1 + m∠3 = 230, find m∠6. **65**

In the figure, $s \parallel t$. Find the measure of each angle.

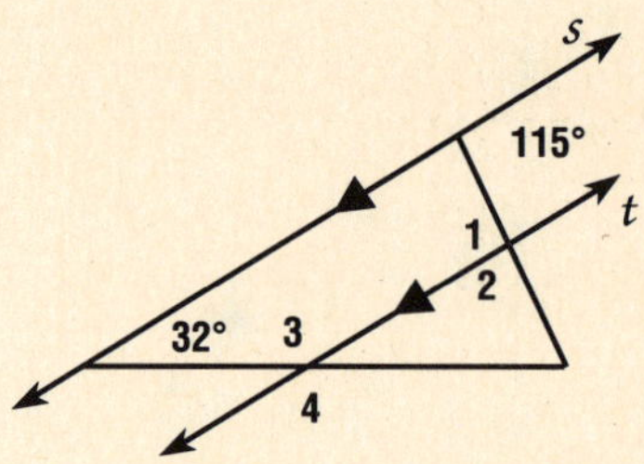

11. m∠1 **115**

12. m∠2 **65**

13. m∠3 **148**

14. m∠4 **148**

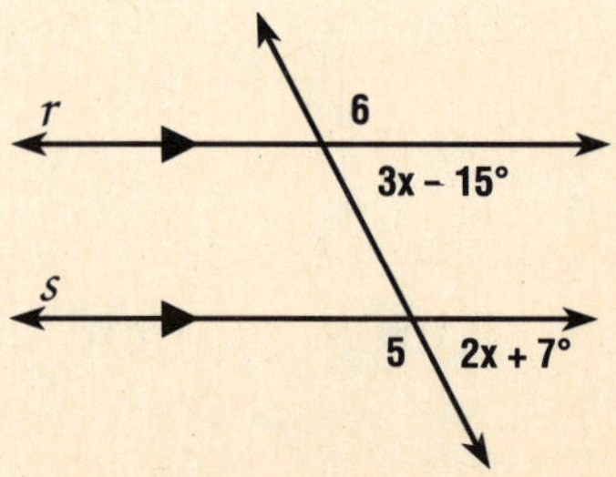

15. In the figure, $r \parallel s$. Find the value of x. **22**

Glencoe Division, Macmillan/McGraw-Hill

NAME ______________________ DATE ____________

3-4 Practice Worksheet

Proving Lines Parallel

For exercises 1 to 6, find the value of x so that $\ell \parallel m$.

1.

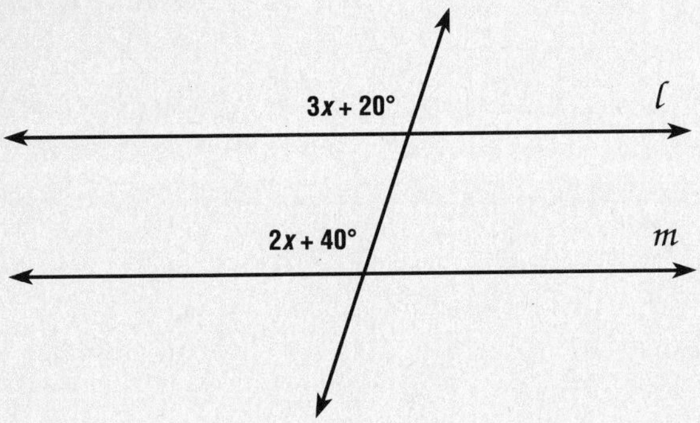

2.

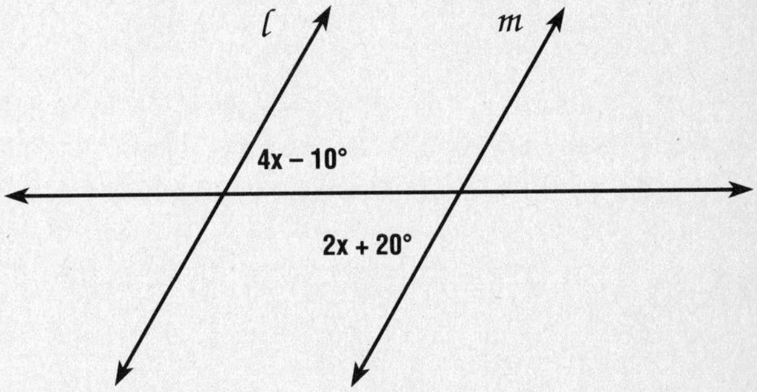

3.

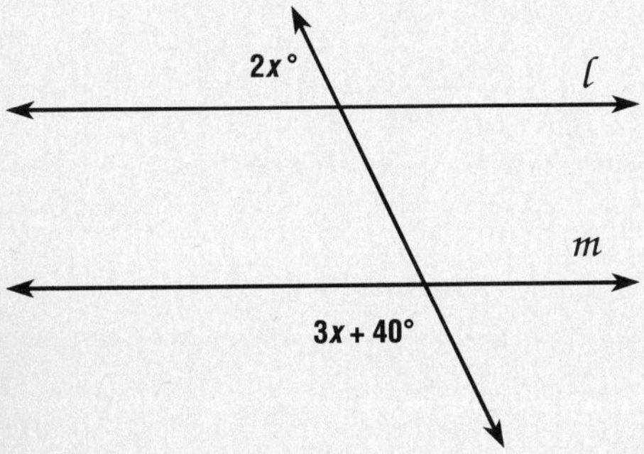

4.

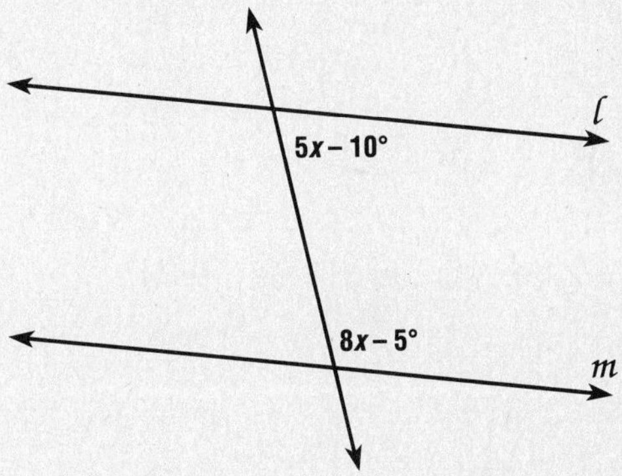

5.

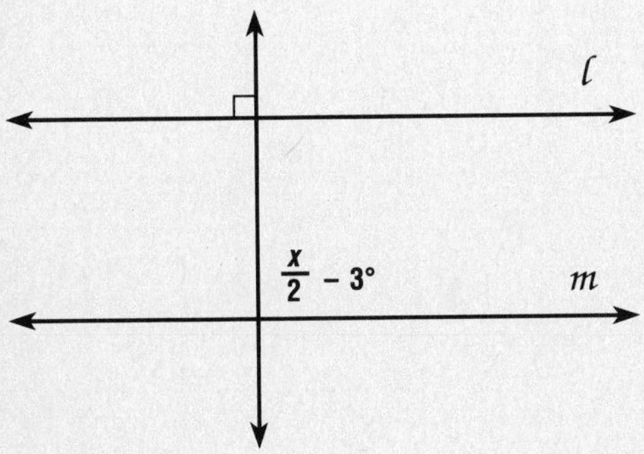

6.

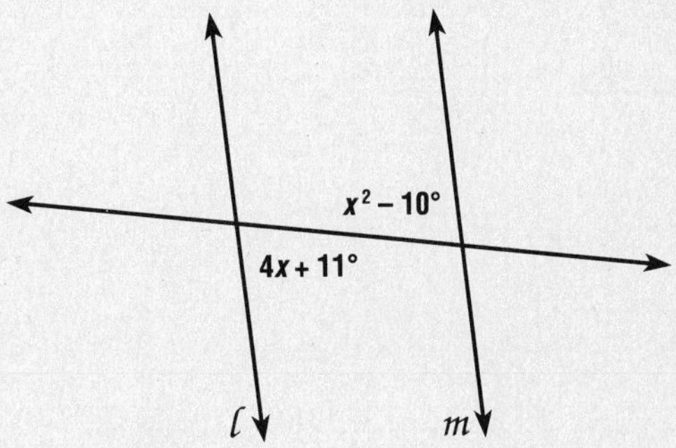

7. If $\ell \nparallel m$, can $x = 50$? Justify your answer.

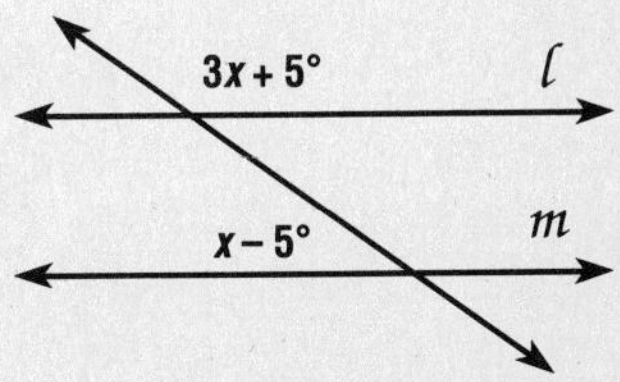

8. Find m∠1 for the figure at the right.

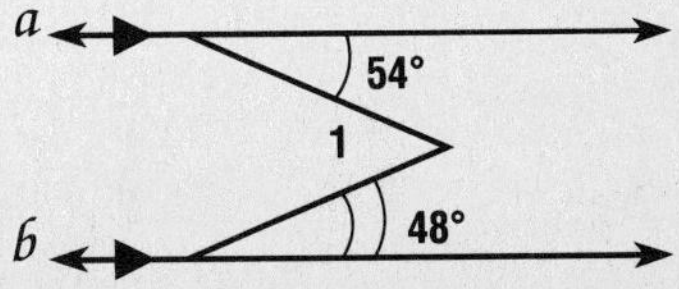

NAME ______________________ DATE ____________

3-4 Practice Worksheet

Proving Lines Parallel

For exercises 1 to 6, find the value of x so that $\ell \parallel m$.

1.

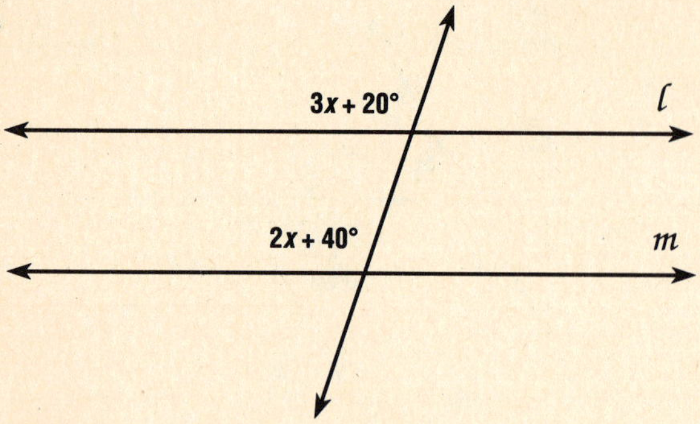

20

2.

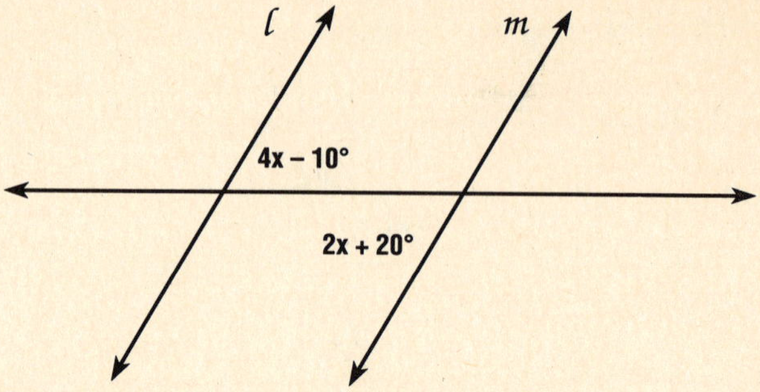

15

3.

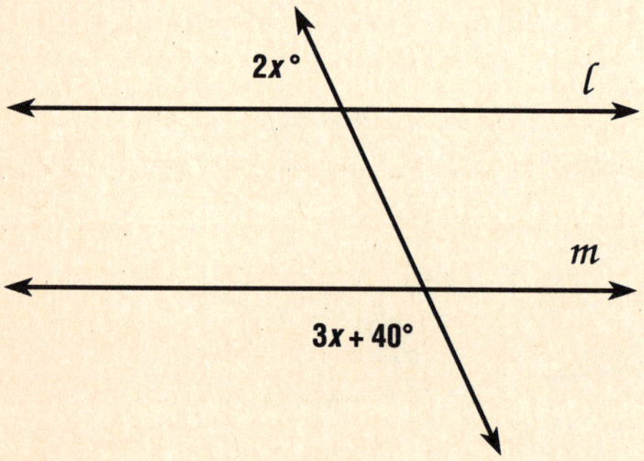

28

4.

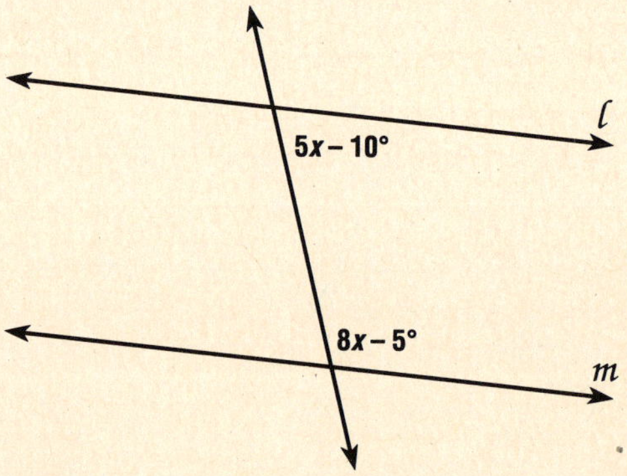

15

5.

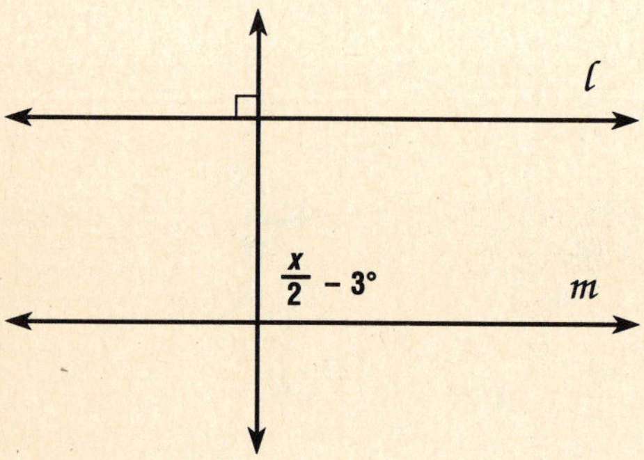

186

6.

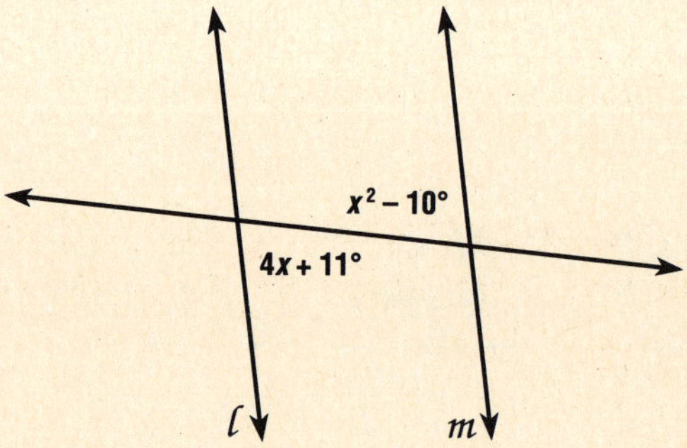

7

7. If $\ell \nparallel m$, can $x = 50$? Justify your answer.
yes; The lines are parallel only if $x = 45$.

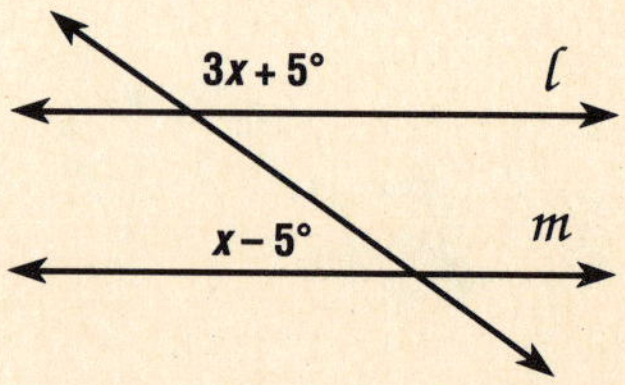

8. Find m∠1 for the figure at the right. 102

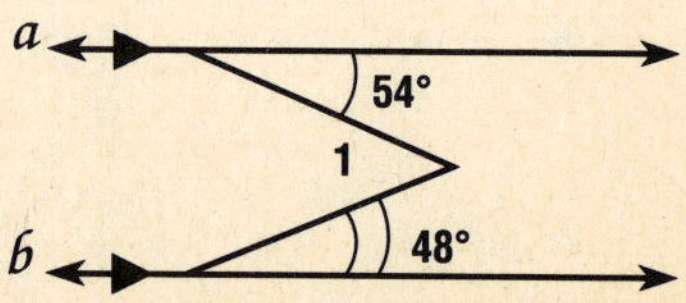

3-5

NAME ______________________________ DATE ______________

Practice Worksheet

Slopes of Lines

Determine the slope of each line named below.

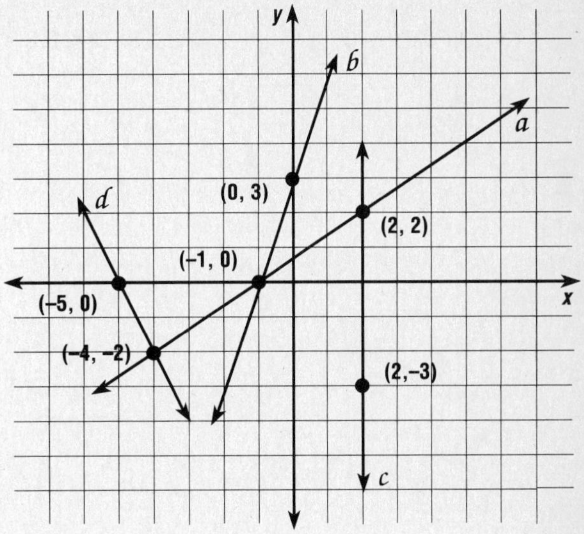

1. a

2. b

3. c

4. d

5. any line parallel to b

6. any line perpendicular to d

Graph the line that satisfies each description.

7. slope = –3, passes through $P(1, -2)$

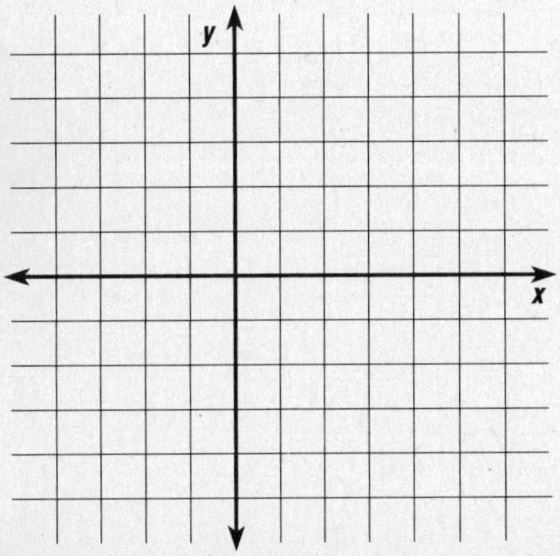

8. passes through $P(-1, 2)$ and is perpendicular to the line determined by $A(1, 4)$ and $B(2, 7)$

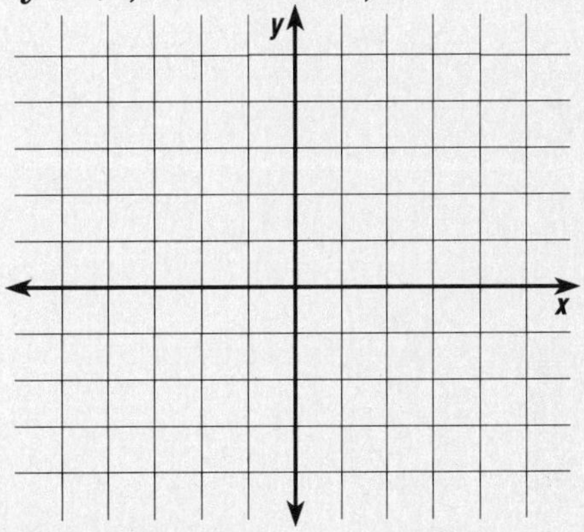

Determine the value of r so that a line through the points with the given coordinates has the given slope. Draw a sketch of each situation.

9. $(5, 3)$, $(r, 6)$; slope = 1

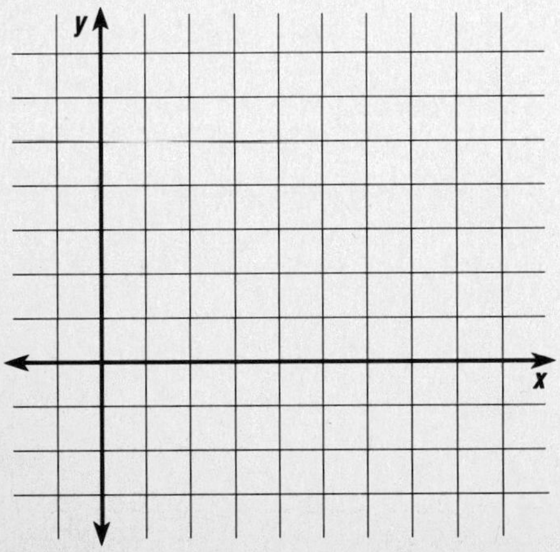

10. $(2, r)$, $(-2, 6)$; slope = $\frac{1}{2}$

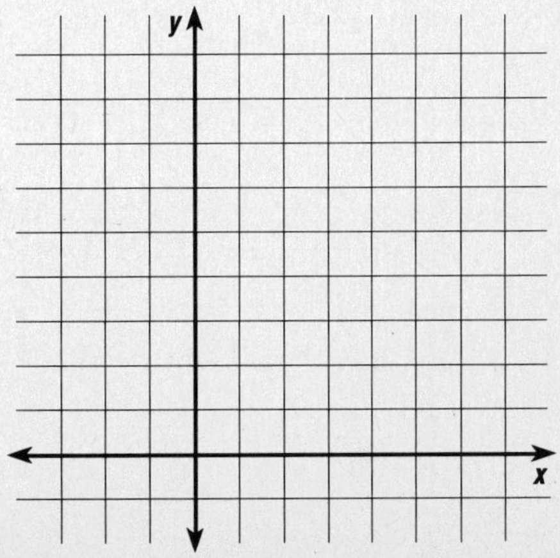

Glencoe Division, Macmillan/McGraw-Hill

NAME ____________________ DATE ____________

3-5

Practice Worksheet

Slopes of Lines

Determine the slope of each line named below.

1. *a* $\frac{2}{3}$
2. *b* 3
3. *c* undefined
4. *d* −2
5. any line parallel to *b* 3
6. any line perpendicular to *d* $\frac{1}{2}$

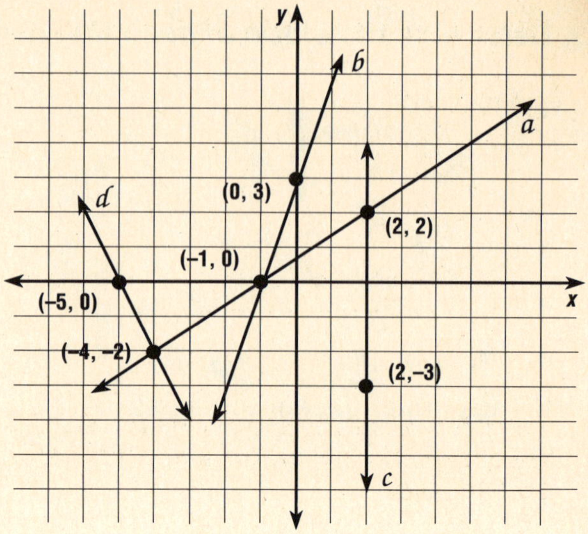

Graph the line that satisfies each description.

7. slope = −3, passes through $P(1, -2)$

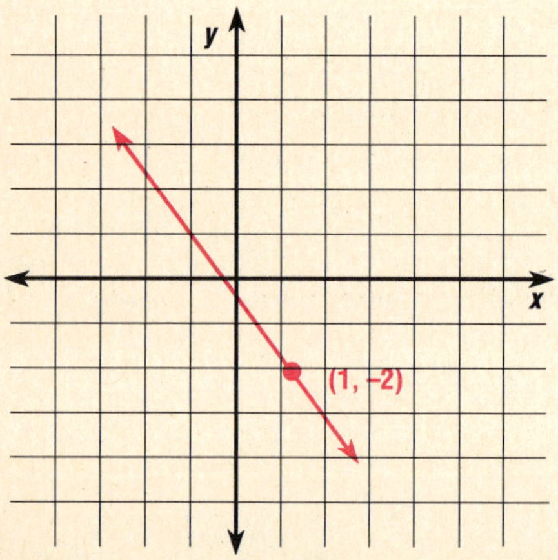

8. passes through $P(-1, 2)$ and is perpendicular to the line determined by $A(1, 4)$ and $B(2, 7)$

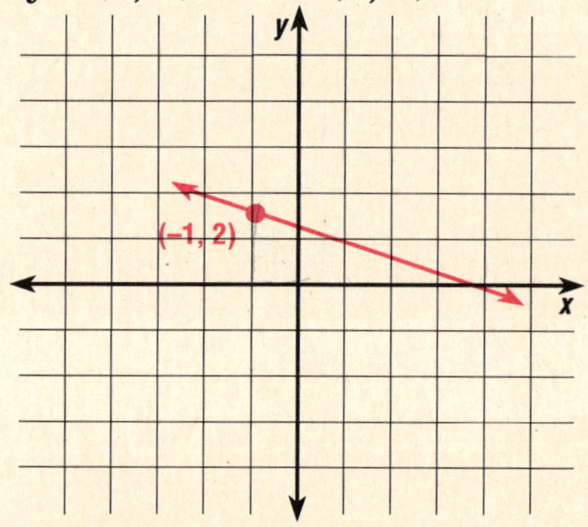

Determine the value of r so that a line through the points with the given coordinates has the given slope. Draw a sketch of each situation.

9. (5, 3), (*r*, 6); slope = 1 8

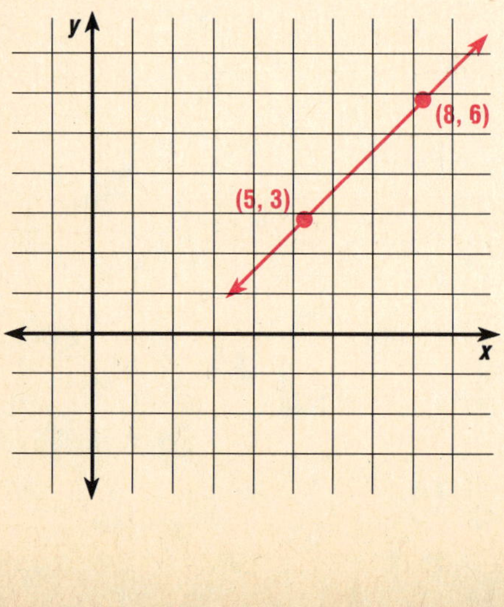

10. (2, *r*), (−2, 6); slope = $\frac{1}{2}$ 8

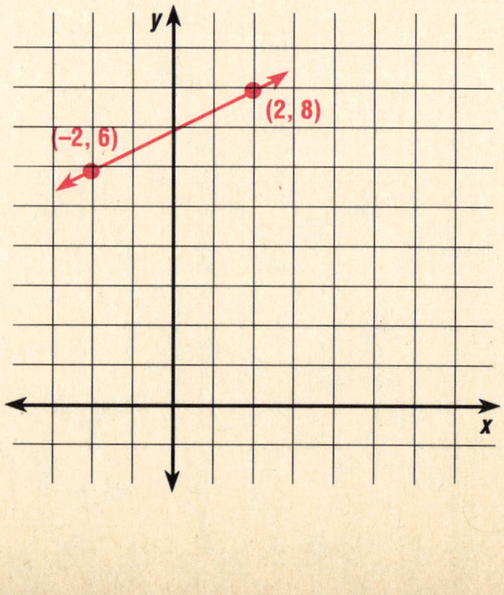

NAME ______________________________ DATE ______________

3-6

Practice Worksheet

Parallels and Distance

Draw in the segment that represents the distance indicated.

1. P to $\overleftrightarrow{RS}$

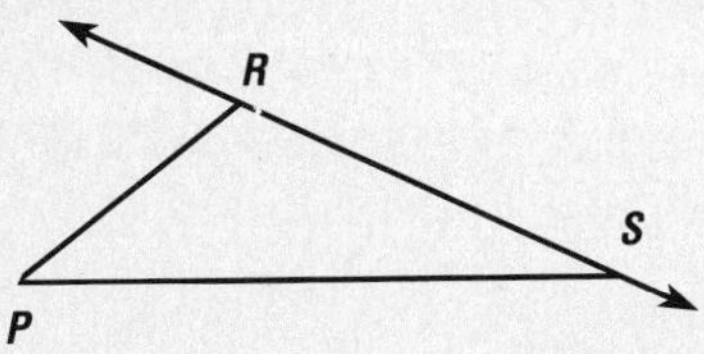

2. B to $\overline{AD}$

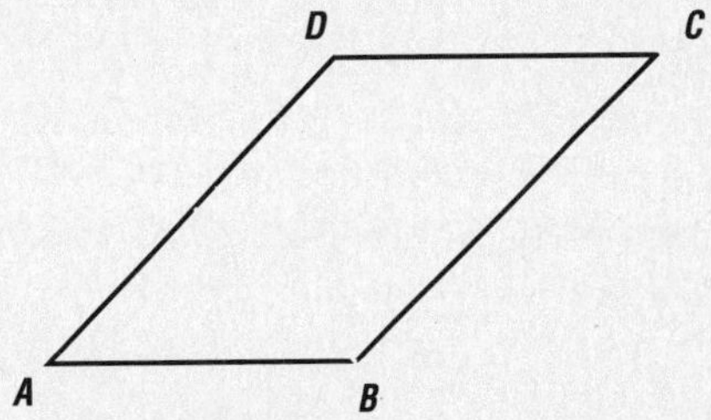

Graph each equation and plot the given point. Then construct a perpendicular segment and find the distance from the point to the line.

3. $y = x + 2$, $(2, -2)$

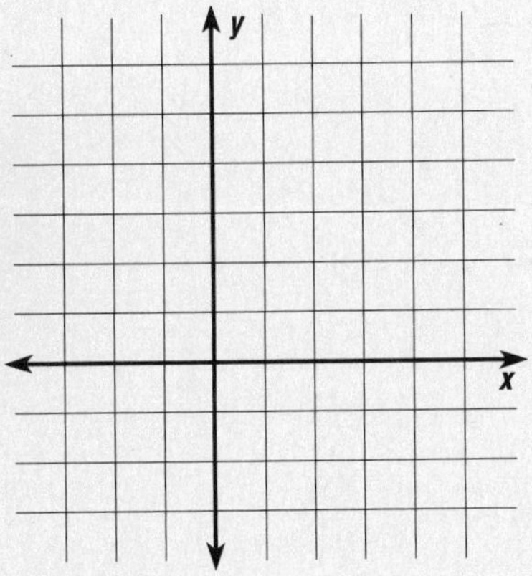

4. $x + y = 2$, $(3, 3)$

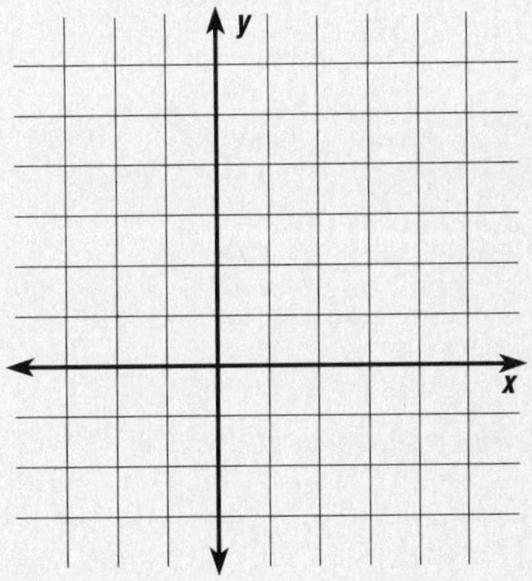

In the figure below, $\overline{BH} \perp \overline{AE}$, $\overline{CF} \perp \overline{AE}$, $\overline{BH} \perp \overline{BC}$, $\overline{BC} \perp \overline{CF}$, and $\overline{GD} \perp \overline{CE}$. Name the segment whose length represents the distance between the following points and lines.

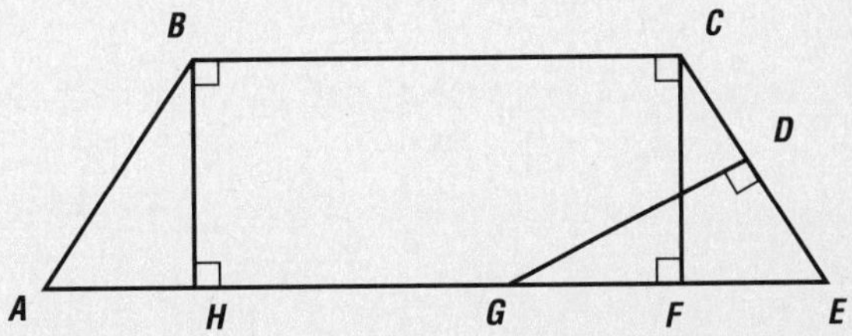

5. B to $\overline{AE}$

6. G to $\overline{CE}$

7. C to $\overline{BH}$

8. F to $\overline{BC}$

Glencoe Division, Macmillan/McGraw-Hill

3-6

NAME ______________________________ DATE ______________

Practice Worksheet

Parallels and Distance

Draw in the segment that represents the distance indicated.

1. P to $\overleftrightarrow{RS}$

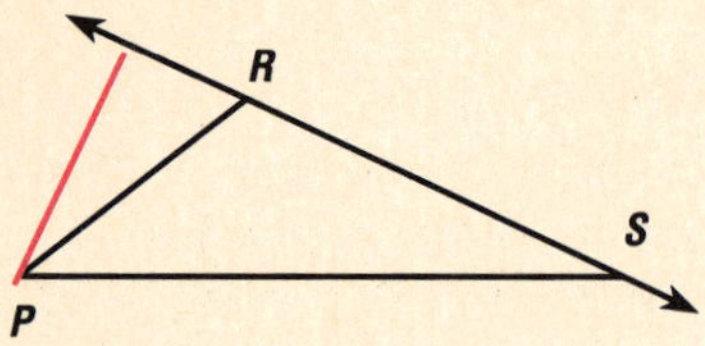

2. B to $\overline{AD}$

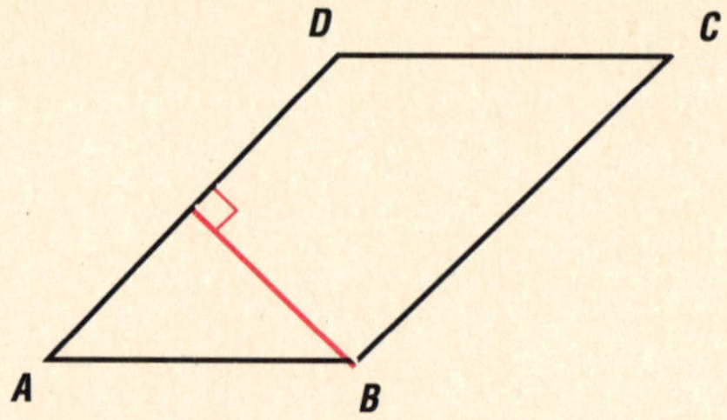

Graph each equation and plot the given point. Then construct a perpendicular segment and find the distance from the point to the line.

3. $y = x + 2$, $(2, -2)$ $3\sqrt{2}$

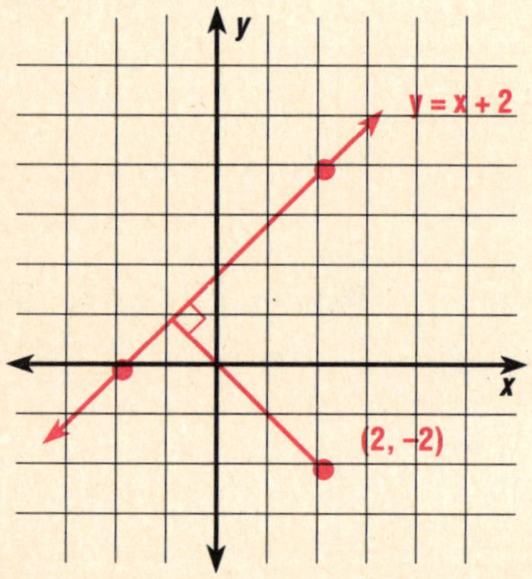

4. $x + y = 2$, $(3, 3)$ $2\sqrt{2}$

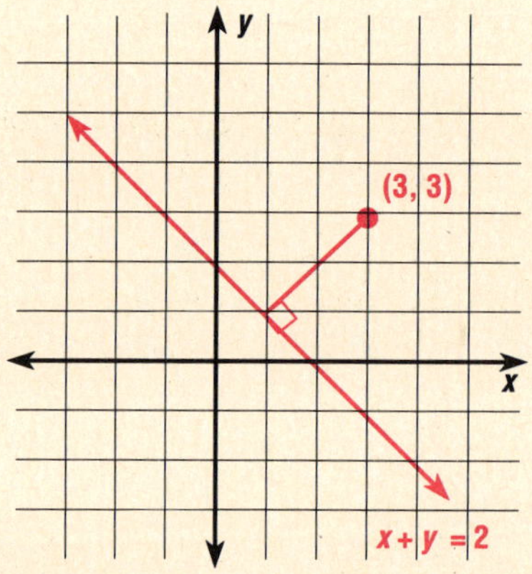

In the figure below, $\overline{BH} \perp \overline{AE}$, $\overline{CF} \perp \overline{AE}$, $\overline{BH} \perp \overline{BC}$, $\overline{BC} \perp \overline{CF}$, and $\overline{GD} \perp \overline{CE}$. Name the segment whose length represents the distance between the following points and lines.

5. B to $\overline{AE}$ $\overline{BH}$ or $\overline{CF}$

6. G to $\overline{CE}$ $\overline{GD}$

7. C to $\overline{BH}$ $\overline{CB}$ or $\overline{FH}$

8. F to $\overline{BC}$ $\overline{FC}$ or $\overline{BH}$

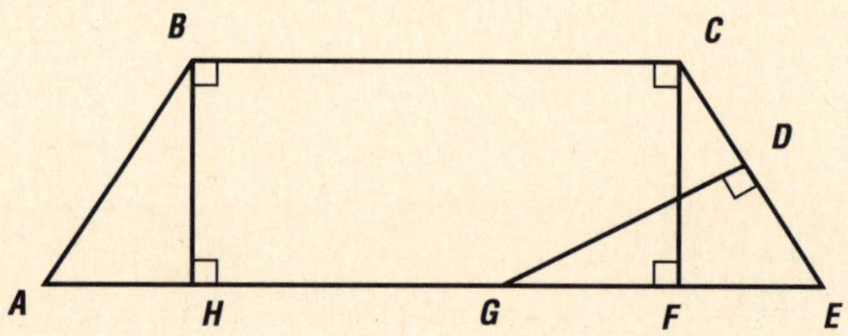

NAME ______________________ DATE ____________

4-1

Practice Worksheet

Classifying Triangles

For Exercises 1-7, refer to the figure at the right, below. Triangle ABC is isosceles with AB > AC and AB > BC. Also, $\overleftrightarrow{XY} \parallel \overline{AB}$. Name each of the following.

1. sides of the triangle

2. angles of the triangle

3. vertex angle

4. base angles

5. side opposite $\angle BCA$

6. congruent sides

7. angle opposite $\overline{AC}$

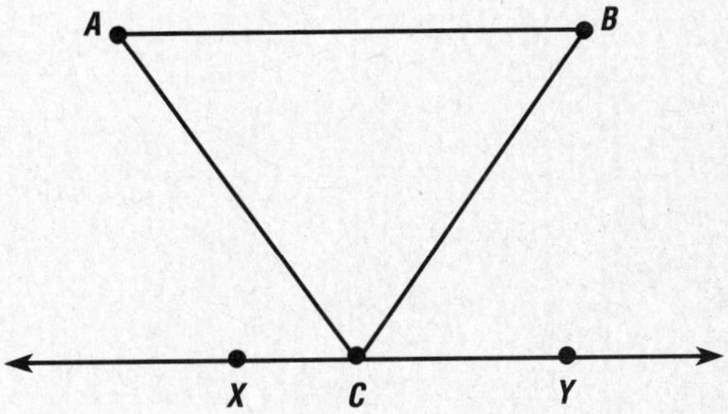

Use a ruler and protractor to draw triangles with the given parts. Classify each triangle in terms of its angles and in terms of its sides.

8. $\triangle BHE$, $BE = 1$ inch, $m\angle E = 60$, $HE = \frac{1}{2}$ inch

9. $\triangle QTR$, $m\angle T = 60$, $QT = TR = 4$ cm

10. Find the measures of the legs of isosceles triangle ABC if $AB = 2x + 4$, $BC = 3x - 1$, $AC = x + 1$, and the perimeter of $\triangle ABC$ is 34 units.

Glencoe Division, Macmillan/McGraw-Hill

NAME ______________________ DATE ____________

Practice Worksheet

Classifying Triangles

For Exercises 1-7, refer to the figure at the right, below. Triangle ABC is isosceles with AB > AC and AB > BC. Also, $\overleftrightarrow{XY} \parallel \overline{AB}$. Name each of the following.

1. sides of the triangle $\overline{AB}$, $\overline{BC}$, $\overline{AC}$
2. angles of the triangle $\angle A$, $\angle B$, $\angle ACB$
3. vertex angle $\angle ACB$
4. base angles $\angle A$, $\angle B$
5. side opposite $\angle BCA$ $\overline{AB}$
6. congruent sides $\overline{AC}$, $\overline{BC}$
7. angle opposite $\overline{AC}$ $\angle B$

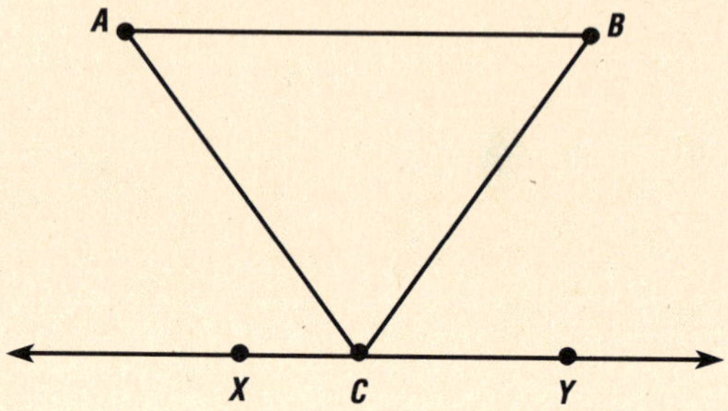

Use a ruler and protractor to draw triangles with the given parts. Classify each triangle in terms of its angles and in terms of its sides.

8. $\triangle BHE$, $BE = 1$ inch, $m\angle E = 60$, $HE = \frac{1}{2}$ inch
   right, scalene

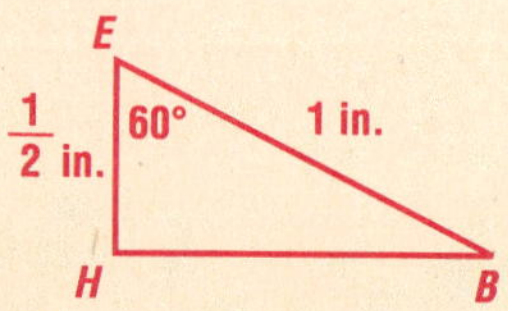

9. $\triangle QTR$, $m\angle T = 60$, $QT = TR = 4$ cm
   equiangular, equilateral

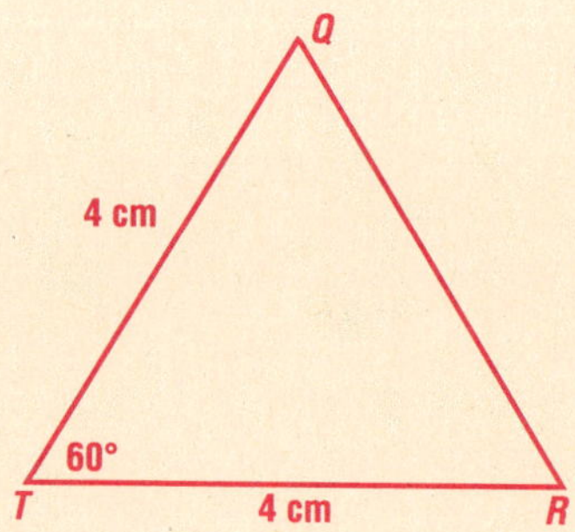

10. Find the measures of the legs of isosceles triangle ABC if $AB = 2x + 4$, $BC = 3x - 1$, $AC = x + 1$, and the perimeter of $\triangle ABC$ is 34 units. 14 units

NAME ______________________________ DATE ______________

4-2

Practice Worksheet

Angle Measures in Triangles

Find the value of x.

1.

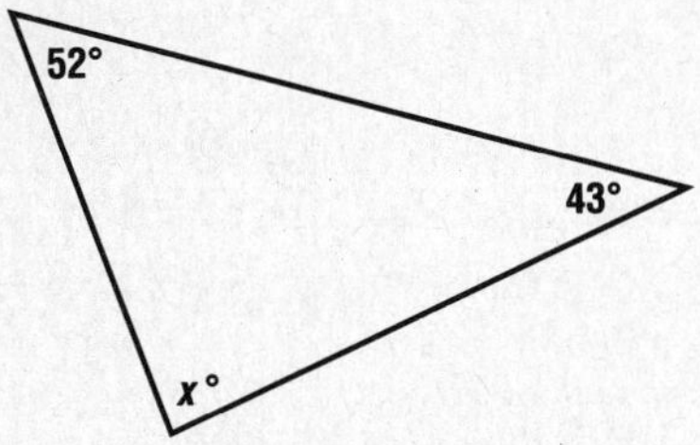

2.

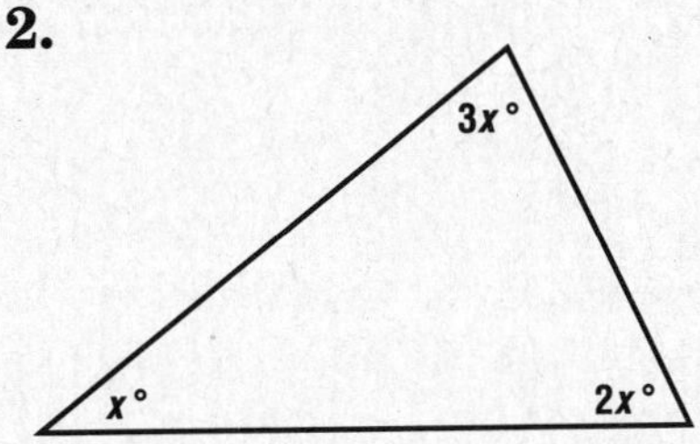

3.

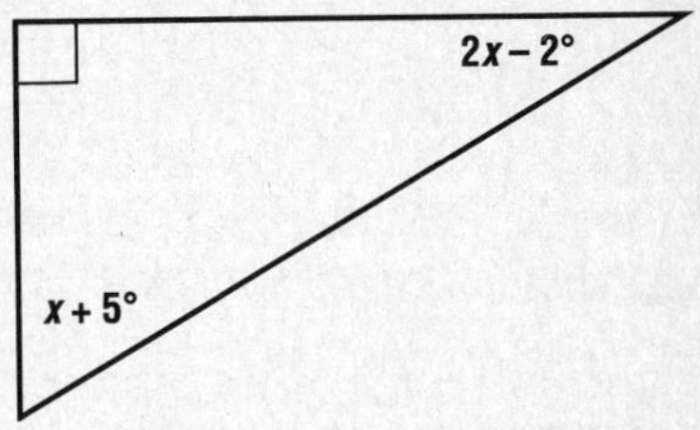

4.

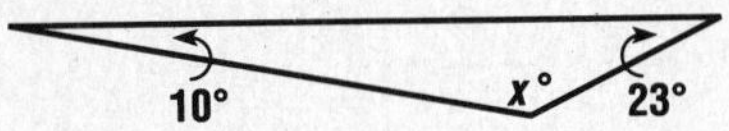

5.

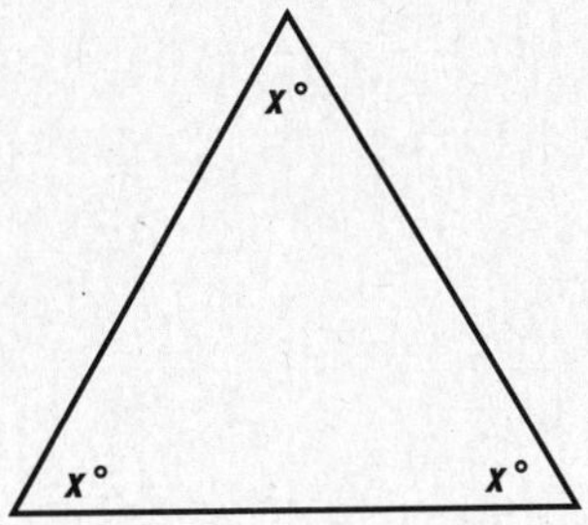

6.

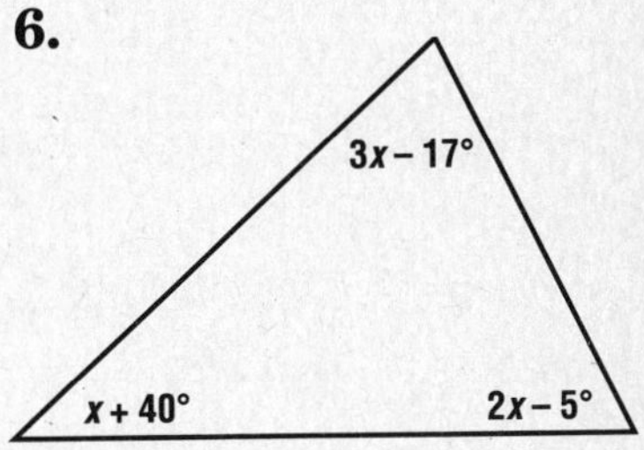

7.

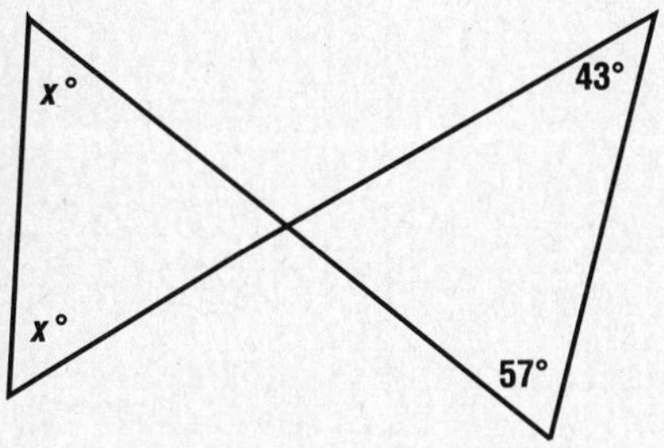

8.

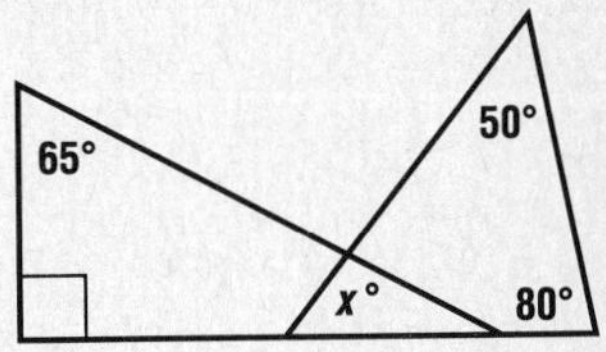

9.

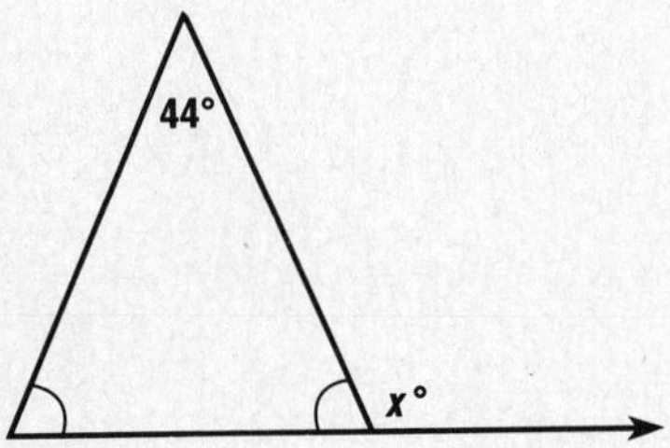

10.

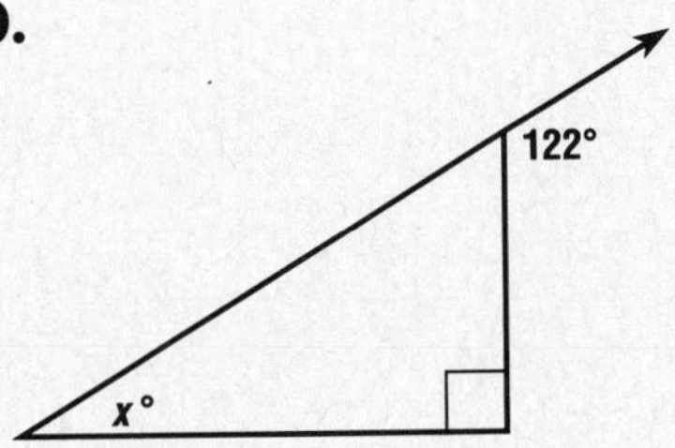

11.

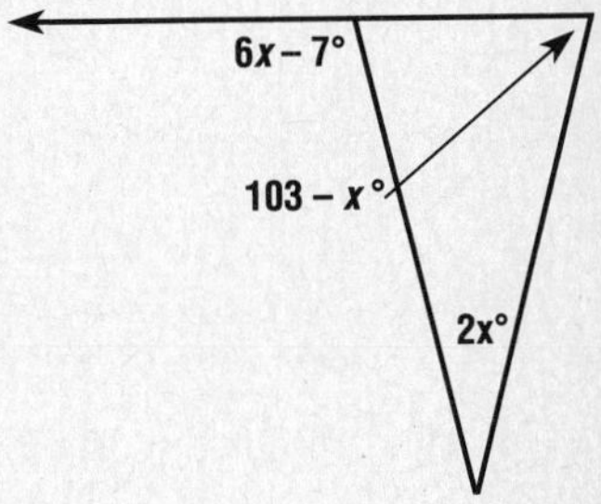

12.

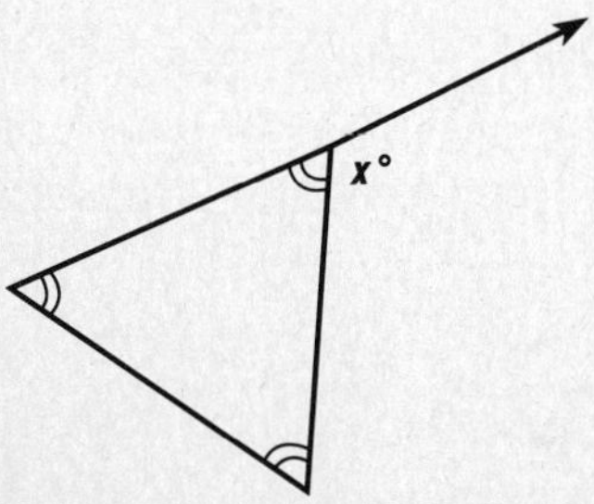

13.

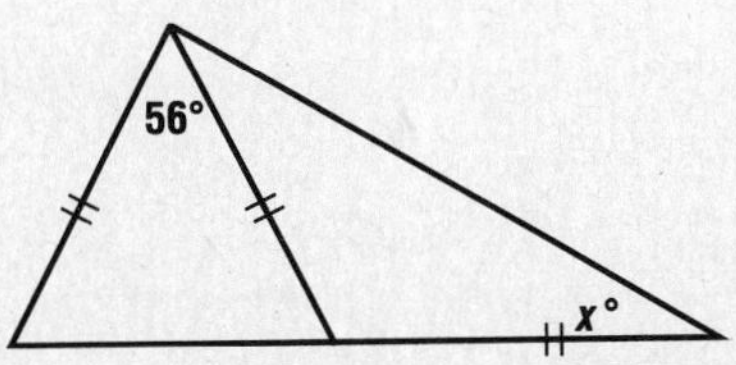

14.

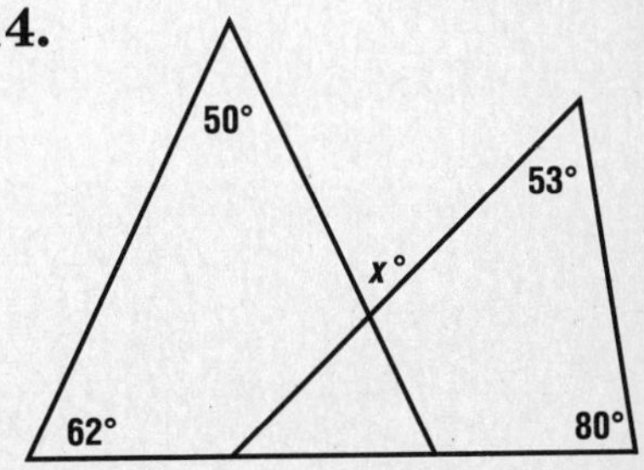

Glencoe Division, Macmillan/McGraw-Hill

4-2

NAME ________________________________ DATE ______________

Practice Worksheet

Angle Measures in Triangles

Find the value of x.

1.

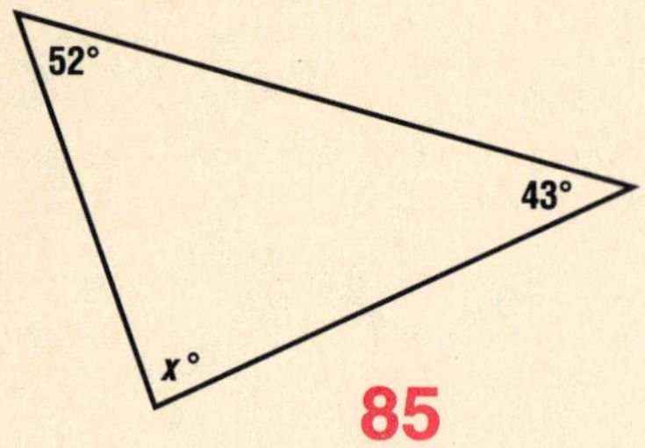

85

2.

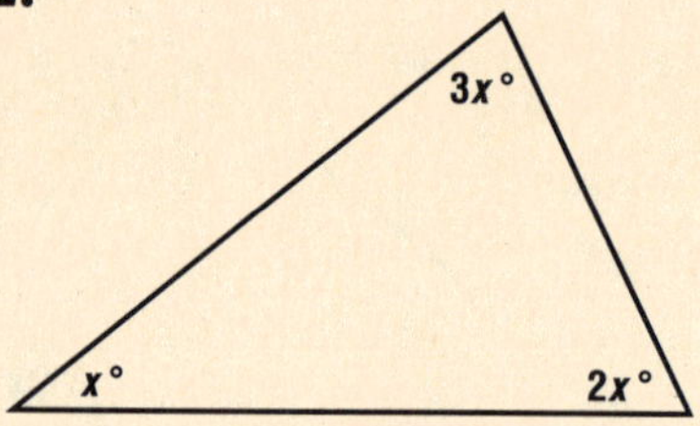

30

3.

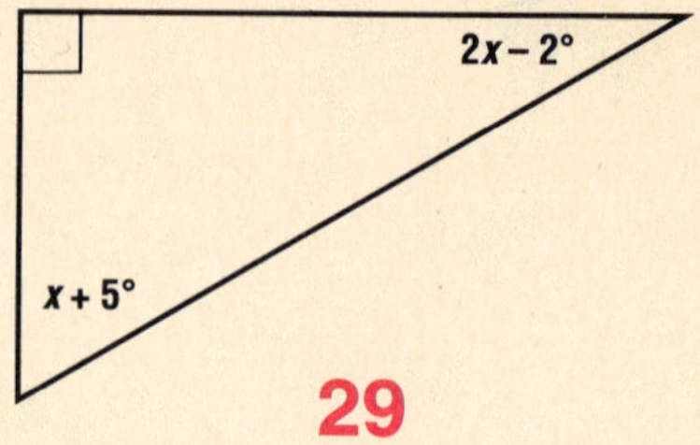

29

4.

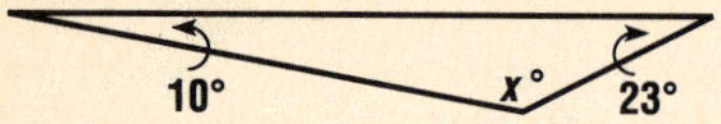

147

5.

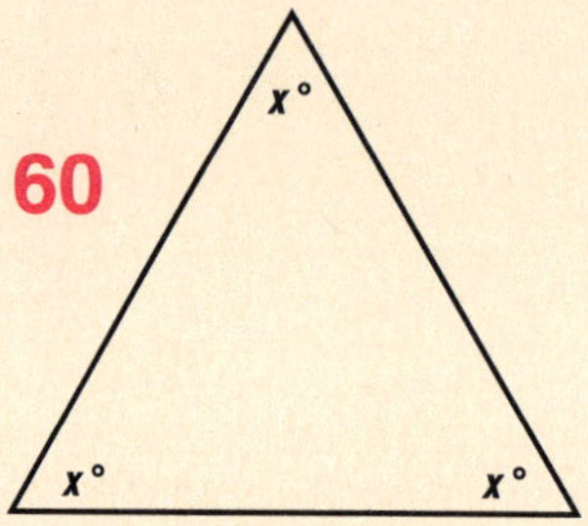

60

6.

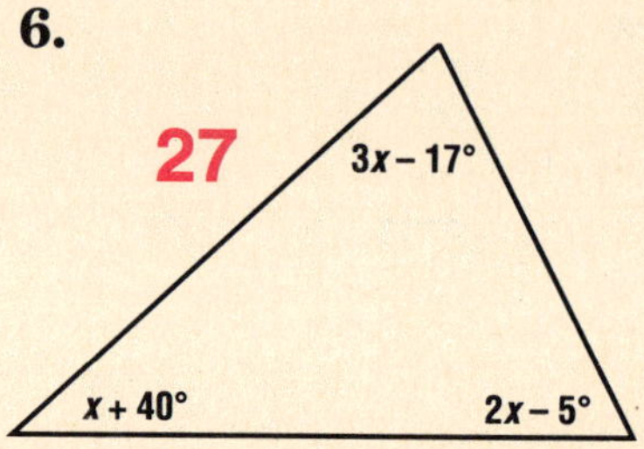

27

7.

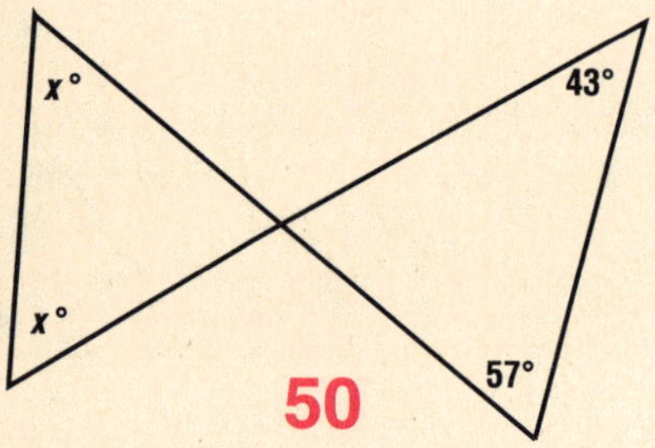

50

8.

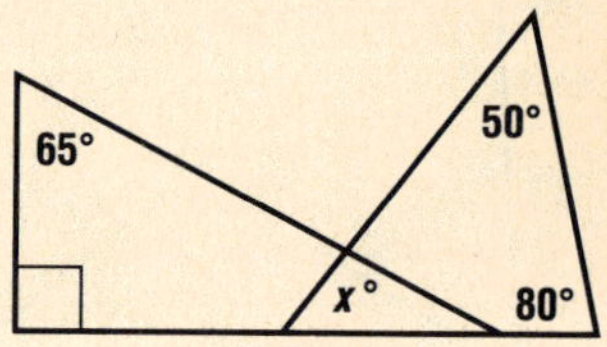

105

9.

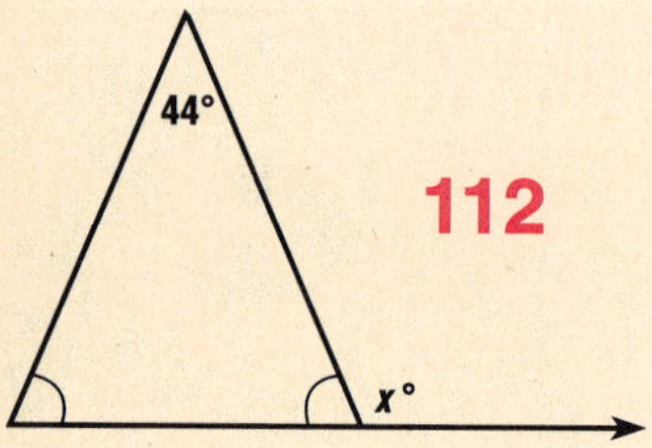

112

10.

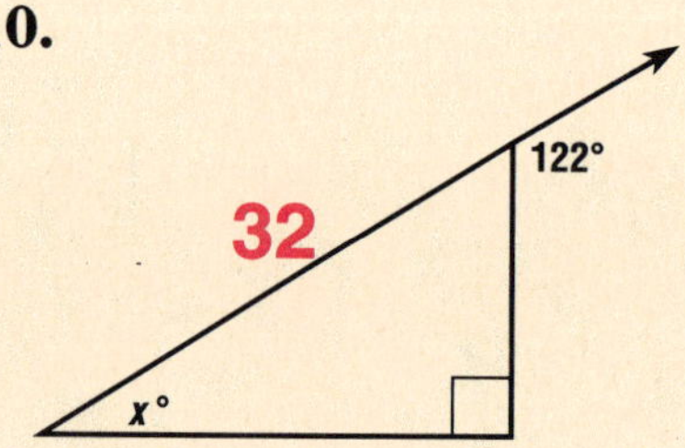

32

11.

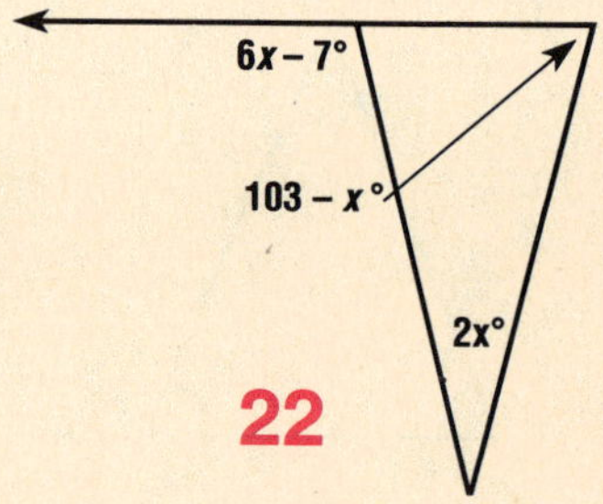

22

12.

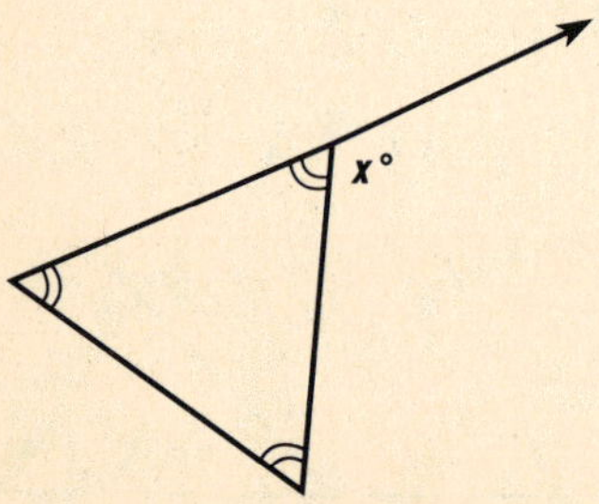

120

13.

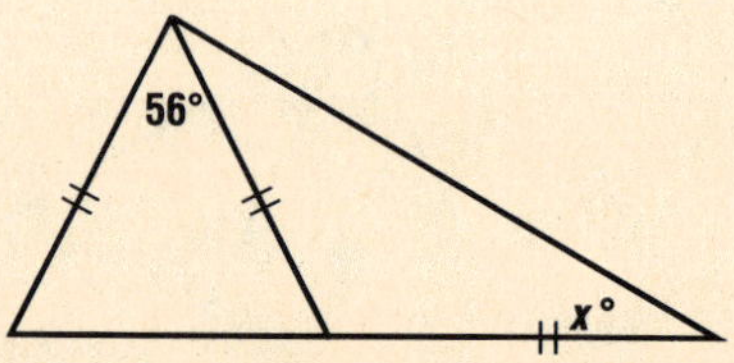

31

14.

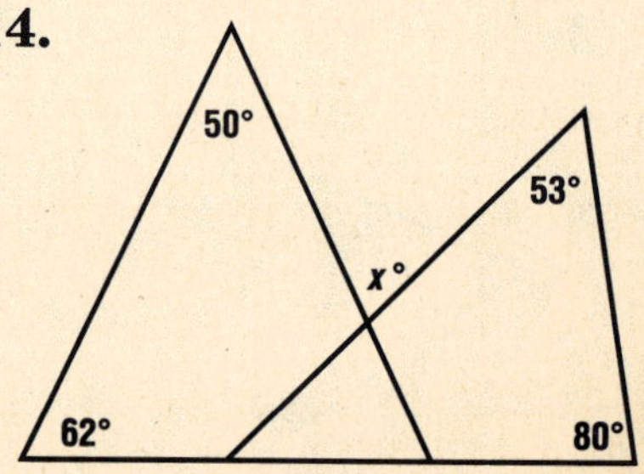

65

NAME ____________________ DATE ____________

4-3 Practice Worksheet

Congruent Triangles

Label the corresponding parts if △RST ≅ △ABC. Use the figures to complete each statement.

1. $\angle C \cong$ __?__

2. $\angle R \cong$ __?__

3. $\overline{AC} \cong$ __?__

4. $\overline{ST} \cong$ __?__

5. $\overline{RS} \cong$ __?__

6. $\angle B \cong$ __?__

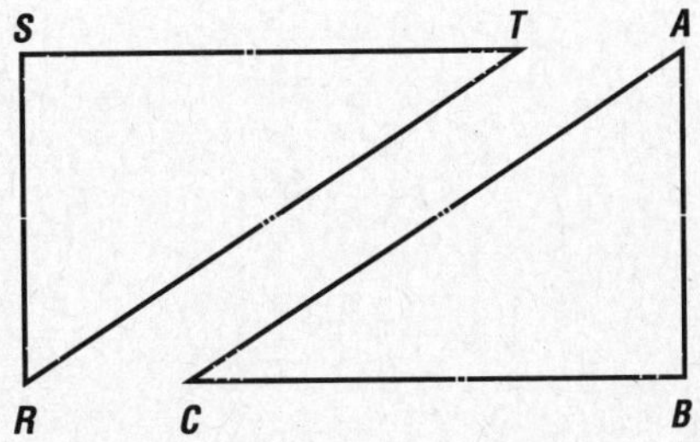

Write a congruence statement for the congruent triangles in each diagram.

7.

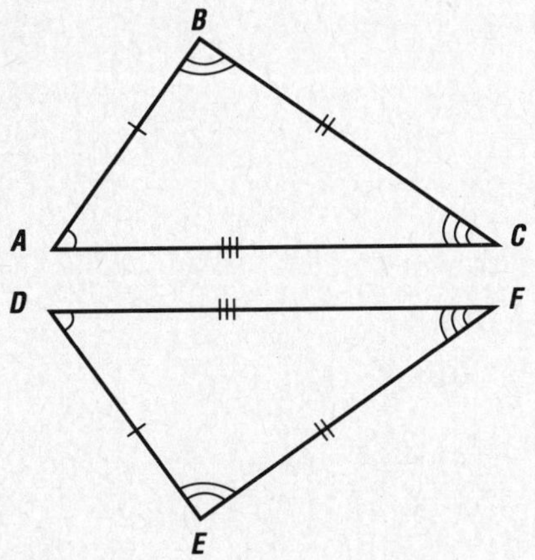

8.

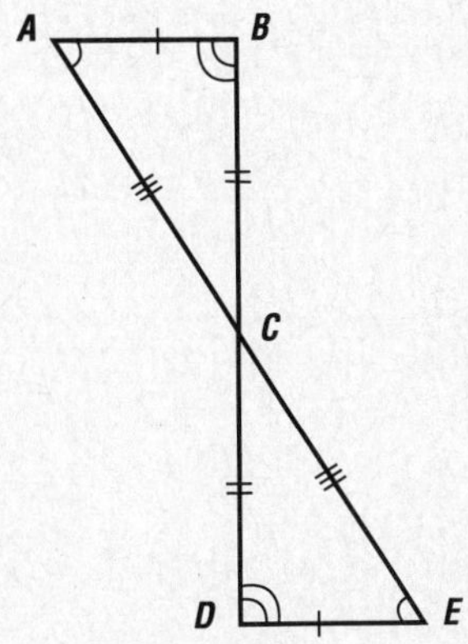

9. Given $\triangle ABC \cong \triangle DEF$, $AB = 15$, $BC = 20$, $AC = 25$, and $FE = 3x - 7$, find x.

10. Given $\triangle ABC \cong \triangle DEF$, $DE = 10$, $EF = 13$, $DF = 16$, and $AC = 4x - 8$, find x.

Glencoe Division, Macmillan/McGraw-Hill

4-3 NAME ______________________ DATE ____________

Practice Worksheet

Congruent Triangles

Label the corresponding parts if △RST ≅ △ABC. Use the figures to complete each statement.

1. $\angle C \cong$? $\angle T$

2. $\angle R \cong$? $\angle A$

3. $\overline{AC} \cong$? $\overline{RT}$

4. $\overline{ST} \cong$? $\overline{BC}$

5. $\overline{RS} \cong$? $\overline{AB}$

6. $\angle B \cong$? $\angle S$

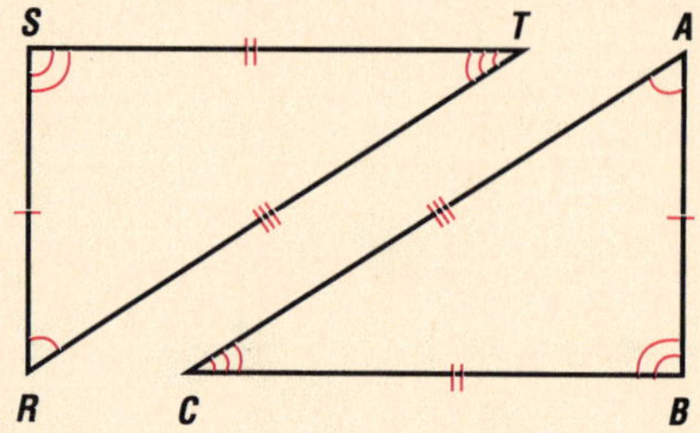

Write a congruence statement for the congruent triangles in each diagram.

7. 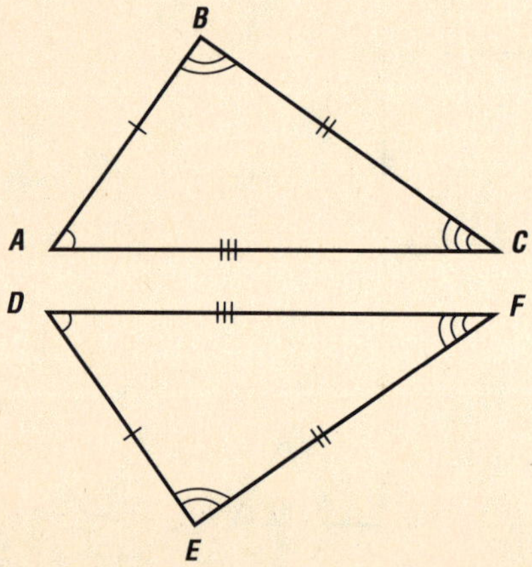

$\triangle ABC \cong \triangle DEF$

8. 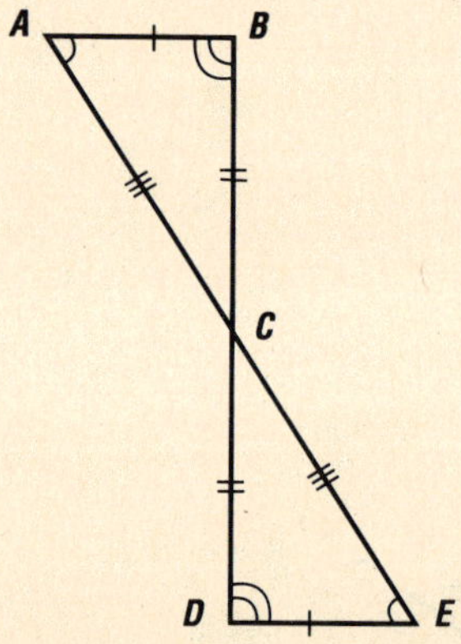

$\triangle ACB \cong \triangle ECD$

9. Given $\triangle ABC \cong \triangle DEF$, $AB = 15$, $BC = 20$, $AC = 25$, and $FE = 3x - 7$, find x. **9**

10. Given $\triangle ABC \cong \triangle DEF$, $DE = 10$, $EF = 13$, $DF = 16$, and $AC = 4x - 8$, find x. **6**

NAME ______________________________ DATE ______________

4-4 Practice Worksheet

Tests for Congruent Triangles

For each figure, mark all congruent parts. Then complete the prove statement and identify the postulate that can be used to prove the triangles congruent.

1.

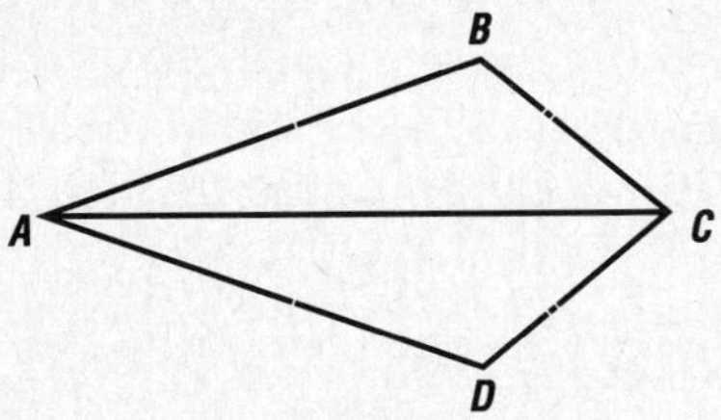

Given: $\overline{AB} \cong \overline{AD}$

$\overline{BC} \cong \overline{DC}$

Prove: $\triangle BCA \cong$ __?__

2.

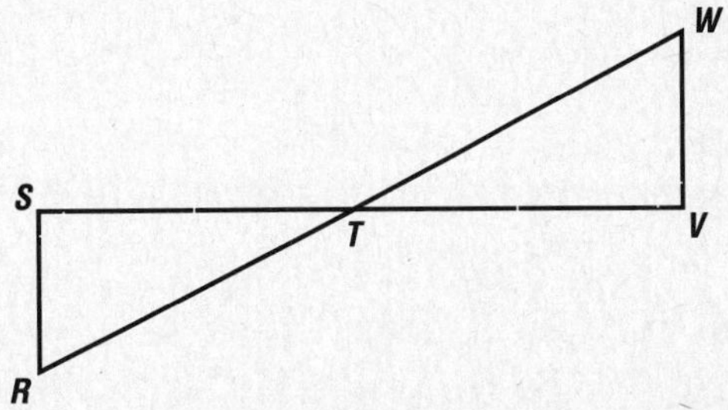

Given: $\angle S$ and $\angle V$ are right angles.

T bisects $\overline{SV}$.

Prove: $\triangle RST \cong$ __?__

Write a two-column proof.

3. Given: $\overline{BD} \perp \overline{AC}$

D bisects $\overline{AC}$.

Prove: $\overline{AB} \cong \overline{CB}$

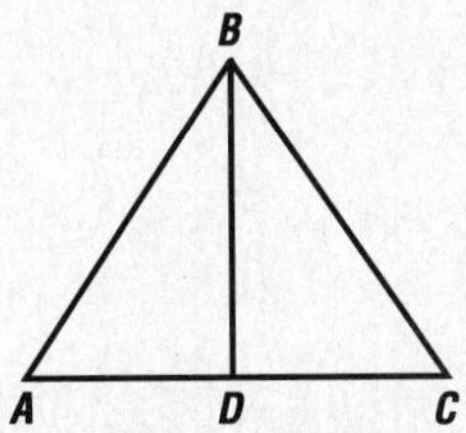

Statements	Reasons

4. Given: $\angle 1 \cong \angle 2$

$\angle 4 \cong \angle 5$

Prove: $\overline{BC} \cong \overline{DC}$

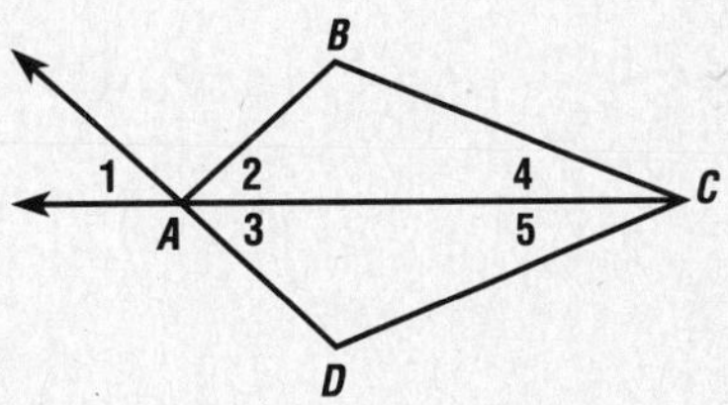

Statements	Reasons

Glencoe Division, Macmillan/McGraw-Hill

NAME ______________________ DATE ____________

4-4 Practice Worksheet

Tests for Congruent Triangles

For each figure, mark all congruent parts. Then complete the prove statement and identify the postulate that can be used to prove the triangles congruent.

1.

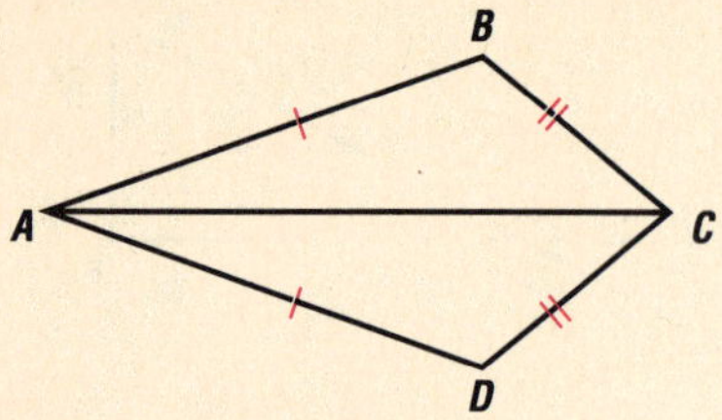

Given: $\overline{AB} \cong \overline{AD}$
$\overline{BC} \cong \overline{DC}$

Prove: $\triangle BCA \cong$ _?_ $\triangle DCA$; SSS

2.

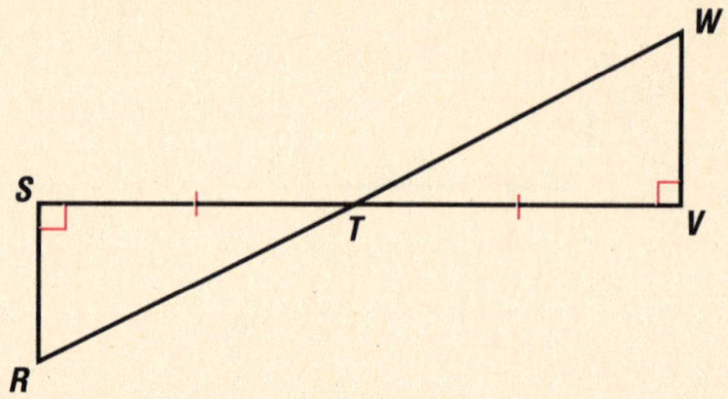

Given: $\angle S$ and $\angle V$ are right angles.
T bisects $\overline{SV}$.

Prove: $\triangle RST \cong$ _?_ $\triangle WVT$; ASA

Write a two-column proof.

3. Given: $\overline{BD} \perp \overline{AC}$
D bisects $\overline{AC}$.

Prove: $\overline{AB} \cong \overline{CB}$

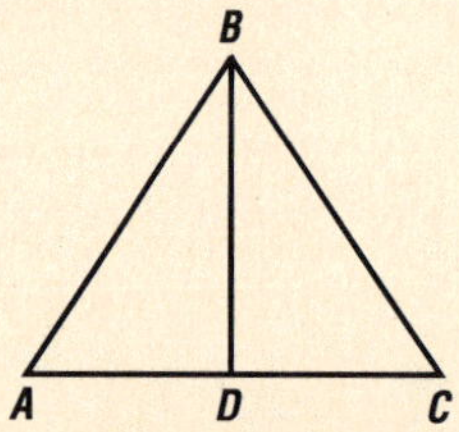

Statements	Reasons
a. D bisects $\overline{AC}$	a. Given
b. $\overline{AD} \cong \overline{CD}$	b. Definition of bisector
c. $\overline{BD} \perp \overline{AC}$	c. Given
d. $\angle ADB$ and $\angle CDB$ are right $\angle$s.	d. $\perp$ lines form 4 rt. $\angle$s.
e. $\angle ADB \cong \angle CDB$	e. All rt. $\angle$s are $\cong$.
f. $\overline{BD} \cong \overline{BD}$	f. Congruence of segments is reflexive.
g. $\triangle ABD \cong \triangle CBD$	g. SAS
h. $\overline{AB} \cong \overline{CB}$	h. CPCTC

4. Given: $\angle 1 \cong \angle 2$
$\angle 4 \cong \angle 5$

Prove: $\overline{BC} \cong \overline{DC}$

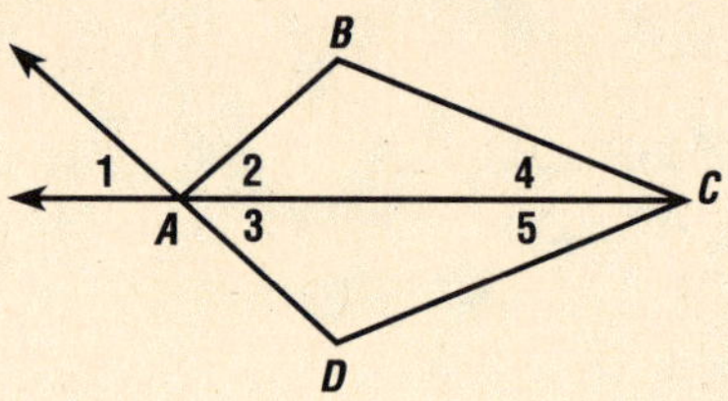

Statements	Reasons
a. $\angle 1 \cong \angle 2$	a. Given
b. $\angle 1 \cong \angle 3$	b. Vert. $\angle$s are $\cong$.
c. $\angle 2 \cong \angle 3$	c. Congruence of $\angle$s is transitive.
d. $\overline{AC} \cong \overline{AC}$	d. Congruence of segments is reflexive.
e. $\angle 4 \cong \angle 5$	e. Given
f. $\triangle ABC \cong \triangle ADC$	f. ASA
g. $\overline{BC} \cong \overline{DC}$	g. CPCTC

NAME ______________________________ DATE ______________

4-5 Practice Worksheet

Another Test for Congruent Triangles

Draw and label triangles MNO and XYZ. Indicate the additional pairs of corresponding parts that would have to be proved congruent in order to use the given postulate or theorem to prove the triangles congruent.

1. $\angle N \cong \angle Y$ and $\overline{NO} \cong \overline{YZ}$ by ASA

2. $\angle O \cong \angle Z$ and $\angle M \cong \angle X$ by AAS

3. $\angle O \cong \angle Z$ and $\overline{MO} \cong \overline{XZ}$ by AAS

4. $\angle N \cong \angle Y$ and $\angle M \cong \angle X$ by ASA

5. The steps in the following proof are *not* in logical order. Rearrange them in a correct sequence and give the reasons.

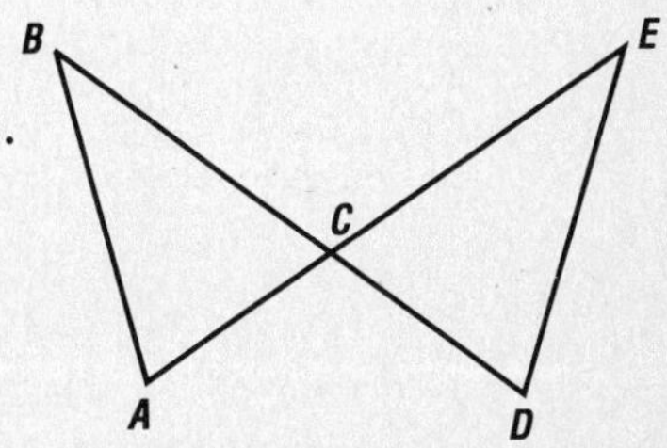

Given: $\angle A \cong \angle D$
$\overline{AB} \cong \overline{DE}$
Prove: $\overline{CA} \cong \overline{CD}$

Statements	Reasons
a. $\triangle CAB \cong \triangle CDE$	**a.** ______________
b. $\angle A \cong \angle D$	**b.** ______________
c. $\angle BCA \cong \angle ECD$	**c.** ______________
d. $\overline{CA} \cong \overline{CD}$	**d.** ______________
e. $\overline{BA} \cong \overline{ED}$	**e.** ______________

6. Write a two-column proof.

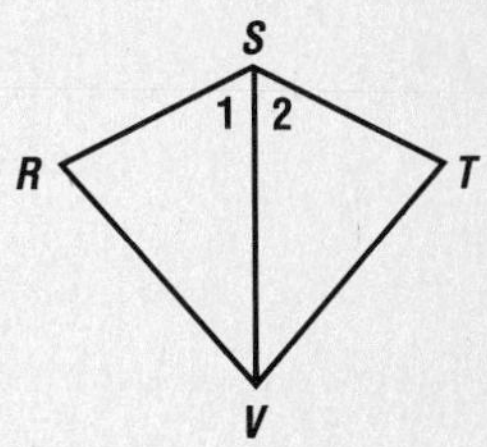

Given: $\angle R$ and $\angle T$ are right angles.
$\angle 1 \cong \angle 2$
Prove: $\overline{RV} \cong \overline{TV}$

NAME ______________________ DATE ____________

Practice Worksheet

Another Test for Congruent Triangles

Draw and label triangles MNO and XYZ. Indicate the additional pairs of corresponding parts that would have to be proved congruent in order to use the given postulate or theorem to prove the triangles congruent.

1. $\angle N \cong \angle Y$ and $\overline{NO} \cong \overline{YZ}$ by ASA

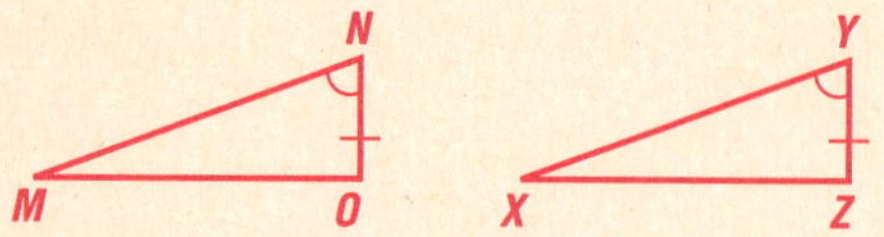

$\angle O \cong \angle Z$

2. $\angle O \cong \angle Z$ and $\angle M \cong \angle X$ by AAS

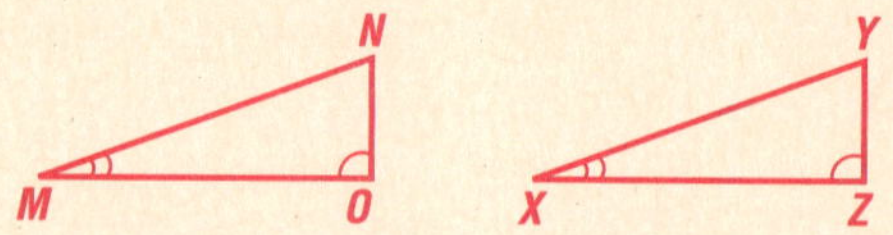

$\angle \overline{NO} \cong \overline{YZ}$ or $\overline{MN} \cong \overline{XY}$

3. $\angle O \cong \angle Z$ and $\overline{MO} \cong \overline{XZ}$ by AAS

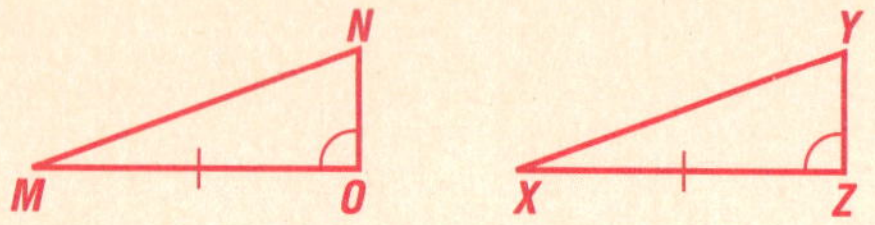

$\angle N \cong \angle Y$

4. $\angle N \cong \angle Y$ and $\angle M \cong \angle X$ by ASA

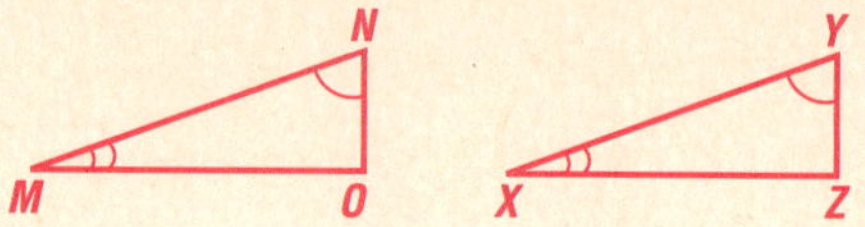

$\overline{MN} \cong \overline{XY}$

5. The steps in the following proof are *not* in logical order. Rearrange them in a correct sequence and give the reasons.

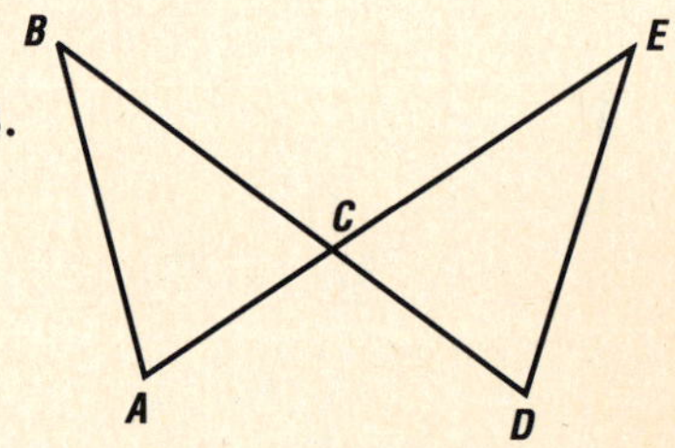

Given: $\angle A \cong \angle D$
$\overline{AB} \cong \overline{DE}$
Prove: $\overline{CA} \cong \overline{CD}$

One order of steps: b, c, e, a, d

Statements	Reasons
a. $\triangle CAB \cong \triangle CDE$	**a.** AAS
b. $\angle A \cong \angle D$	**b.** Given
c. $\angle BCA \cong \angle ECD$	**c.** Vertical $\angle$s are $\cong$.
d. $\overline{CA} \cong \overline{CD}$	**d.** CPCTC
e. $\overline{BA} \cong \overline{ED}$	**e.** Given

6. Write a two-column proof.

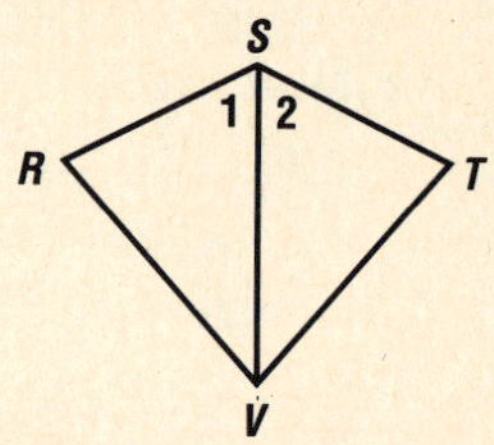

Given: $\angle R$ and $\angle T$ are right angles.
$\angle 1 \cong \angle 2$
Prove: $\overline{RV} \cong \overline{TV}$

Statements	Reasons
a. $\angle R$ and $\angle T$ are right angles.	a. Given
b. $\angle R \cong \angle T$	b. All rt. $\angle$s are $\cong$.
c. $\angle 1 \cong \angle 2$	c. Given
d. $\overline{SV} \cong \overline{SV}$	d. Congruence of segments is reflexive.
e. $\triangle SRV \cong \triangle STV$	e. AAS
f. $\overline{RV} \cong \overline{TV}$	f. CPCTC

NAME ______________________ DATE ______________

4-6

Practice Worksheet

Problem-Solving Strategy: Identify Subgoals

Solve. Use any strategy.

1. How many whole numbers less than 1,000 have digits whose sum is 9?

2. Find the least positive integer having remainders of 1, 2, and 3 when divided by 3, 5, and 7, respectively.

3. How many zeros would you have at the end of your answer if you found the product of the first twenty whole numbers?

4. Identify the subgoals you would need to accomplish to complete the proof. Then complete the proof.

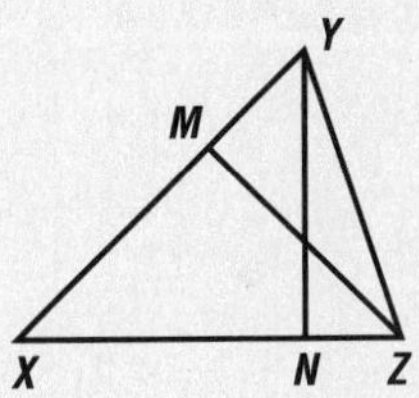

Given: $\overline{YN} \perp \overline{XZ}$, $\overline{ZM} \perp \overline{XY}$

$\overline{XZ} \cong \overline{XY}$

Prove: $\overline{XM} \cong \overline{XN}$

Statements	Reasons

NAME ______________________ DATE ____________

4-6

Practice Worksheet

Problem-Solving Strategy: Identify Subgoals

Solve. Use any strategy.

1. How many whole numbers less than 1,000 have digits whose sum is 9? **55**

2. Find the least positive integer having remainders of 1, 2, and 3 when divided by 3, 5, and 7, respectively. **52**

3. How many zeros would you have at the end of your answer if you found the product of the first twenty whole numbers? **4**

4. Identify the subgoals you would need to accomplish to complete the proof. Then complete the proof.

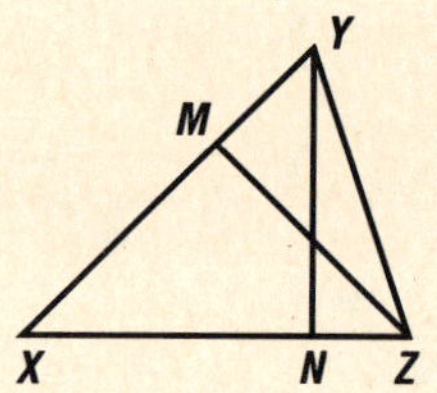

Given: $\overline{YN} \perp \overline{XZ}$, $\overline{ZM} \perp \overline{XY}$

$\overline{XZ} \cong \overline{XY}$

Prove: $\overline{XM} \cong \overline{XN}$

Subgoals: $\triangle XZM \cong \triangle XYN$, $\angle XMZ \cong \angle XNY$

Statements	Reasons
a. $\overline{YN} \perp \overline{XZ}$, $\overline{ZM} \perp \overline{XY}$	a. Given
b. $\angle XMZ$ and $\angle XNY$ are rt. $\angle$s.	b. $\perp$ lines form 4 rt. $\angle$s.
c. $\angle XMZ \cong \angle XNY$	c. All rt. $\angle$s are $\cong$.
d. $\angle X \cong \angle X$	d. Congruence of $\angle$s is reflexive.
e. $\overline{XZ} \cong \overline{XY}$	e. Given
f. $\triangle XZM \cong \triangle XYN$	f. AAS
g. $\overline{XM} \cong \overline{XN}$	g. CPCTC

Glencoe Division, Macmillan/McGraw-Hill

NAME ________________________________ DATE ______________

4-7 Practice Worksheet

Isosceles Triangles

Find the value of x.

1.

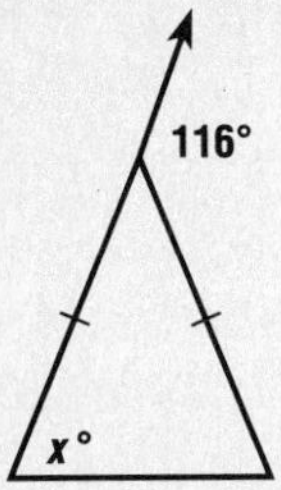

2.

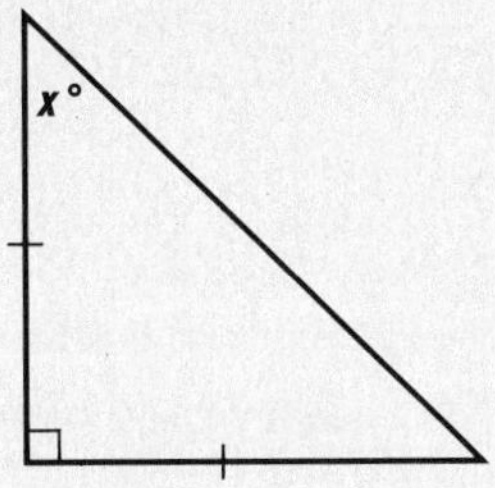

3.

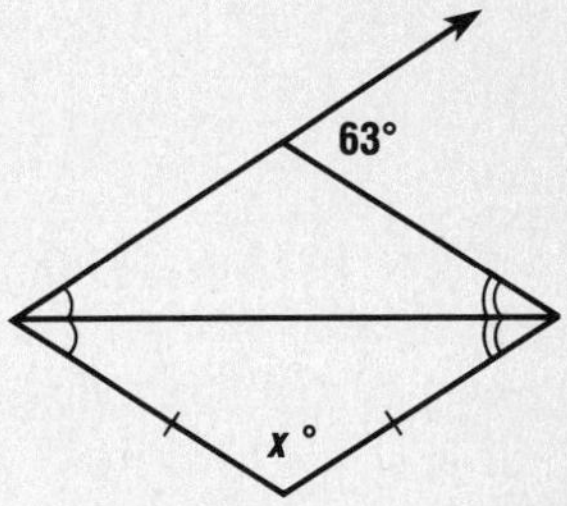

4.

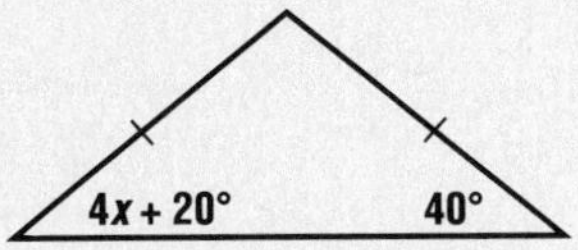

5.

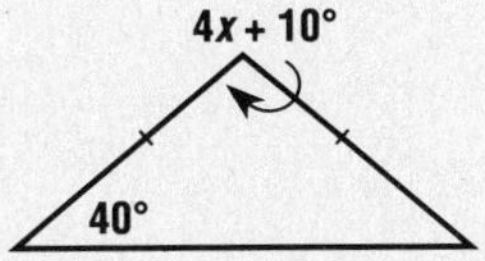

6.

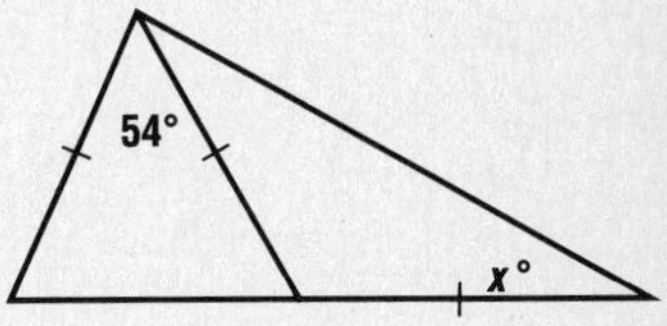

Write a two-column proof for each.

7. Given: $\angle 3 \cong \angle 4$

Prove: $\overline{AB} \cong \overline{BC}$

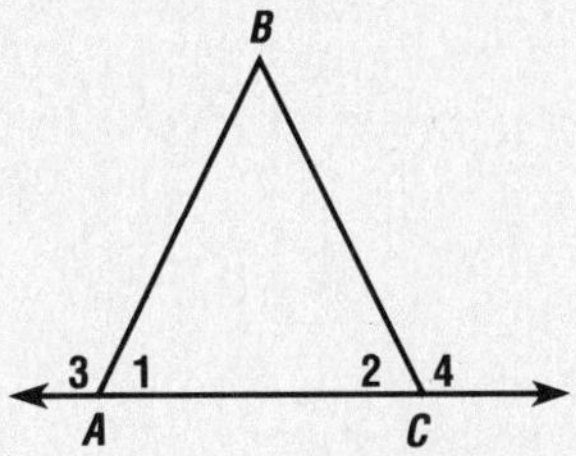

Statements	Reasons

8. Given: $\overline{AR} \cong \overline{AQ}$

$\overline{RS} \cong \overline{QT}$

Prove: $\overline{AS} \cong \overline{AT}$

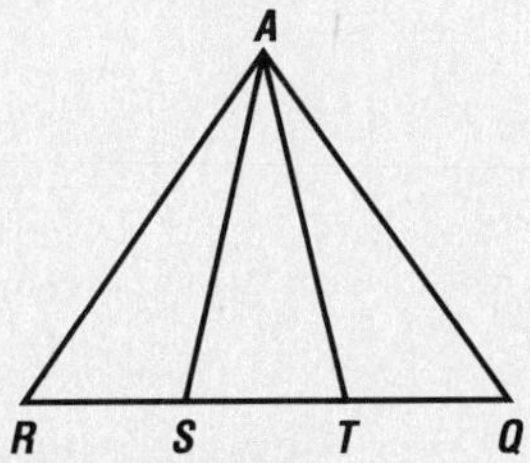

NAME ______________________________ DATE ____________

4-7

Practice Worksheet

Isosceles Triangles

Find the value of x.

1.

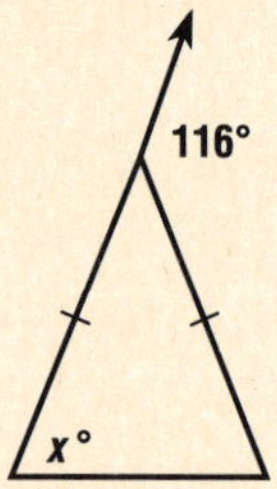

58

2.

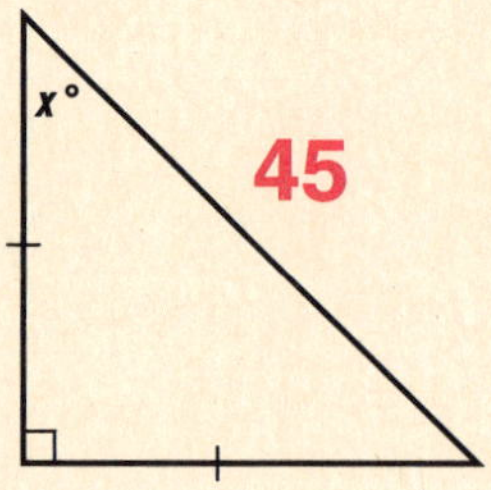

45

3.

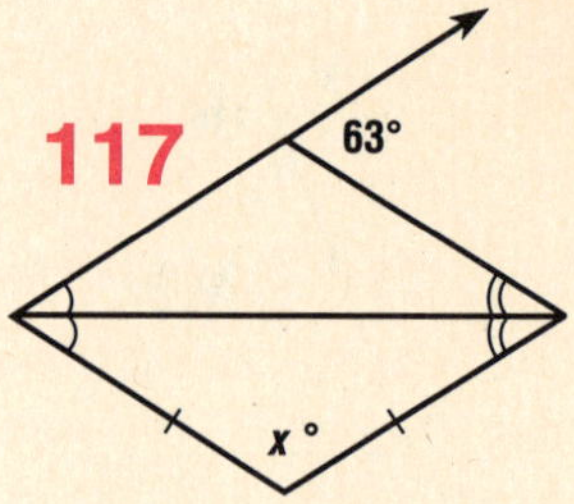

117

4.

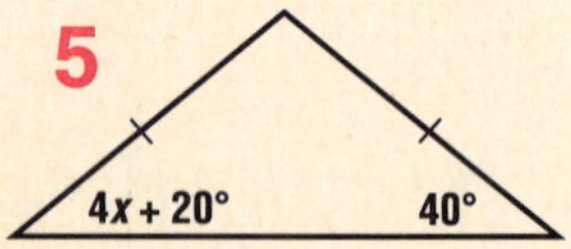

5

5.

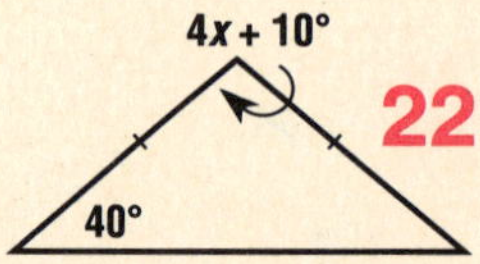

22.5

6.

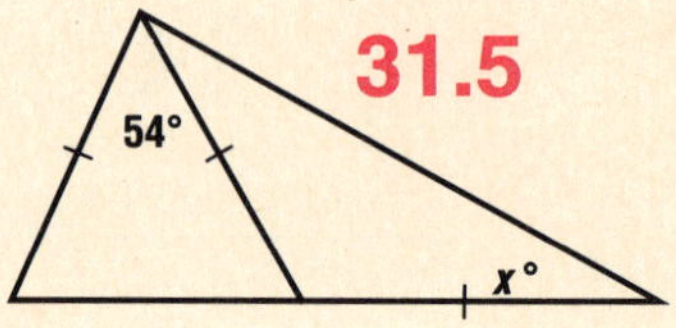

31.5

Write a two-column proof for each.

7. Given: $\angle 3 \cong \angle 4$

Prove: $\overline{AB} \cong \overline{BC}$

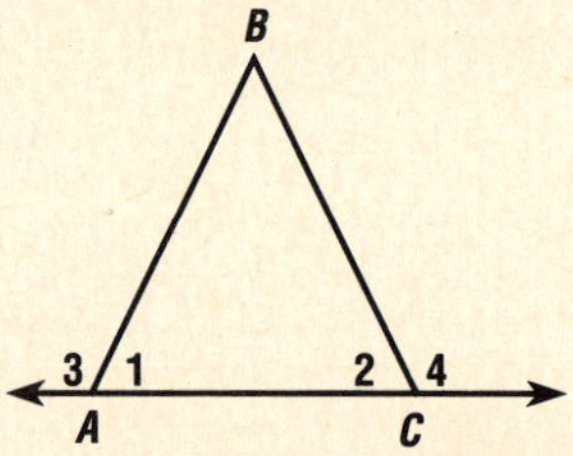

Statements	Reasons
a. $\angle 3 \cong \angle 4$	a. Given
b. $\angle 3$ and $\angle 1$ form a linear pair. $\angle 4$ and $\angle 2$ form a linear pair.	b. Definition of linear pair.
c. $\angle 3$ and $\angle 1$ are supplementary. $\angle 4$ and $\angle 2$ are supplementary.	c. If 2 $\angle$s form a linear pair, they are supp.
d. $\angle 1 \cong \angle 2$	d. $\angle$s supp. to $\cong$ $\angle$s are $\cong$.
e. $\overline{AB} \cong \overline{BC}$	e. If 2 $\angle$s of a $\triangle$ are $\cong$, the sides opp. the $\angle$s are $\cong$.

8. Given: $\overline{AR} \cong \overline{AQ}$

$\overline{RS} \cong \overline{QT}$

Prove: $\overline{AS} \cong \overline{AT}$

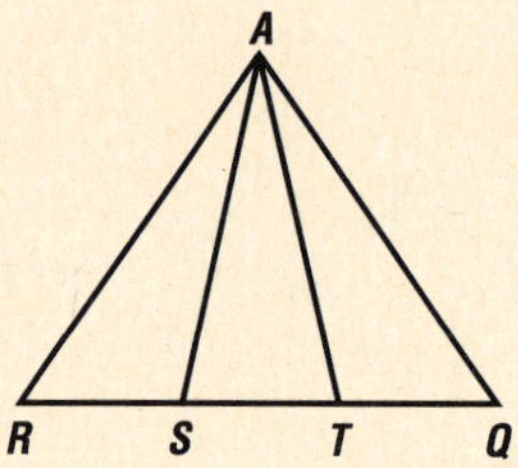

Statements	Reasons
a. $\overline{AR} \cong \overline{AQ}$	a. Given
b. $\angle R \cong \angle Q$	b. If 2 sides of a $\triangle$ are $\cong$, the $\angle$s opp. the sides are $\cong$.
c. $\overline{RS} \cong \overline{QT}$	c. Given
d. $\triangle ARS \cong \triangle AQT$	d. SAS
e. $\overline{AS} \cong \overline{AT}$	e. CPCTC

Glencoe Division, Macmillan/McGraw-Hill

NAME ______________________ DATE ______________

Practice Worksheet

Special Segments in Triangles

1. Find AB if $\overline{BD}$ is a median of $\triangle ABC$.

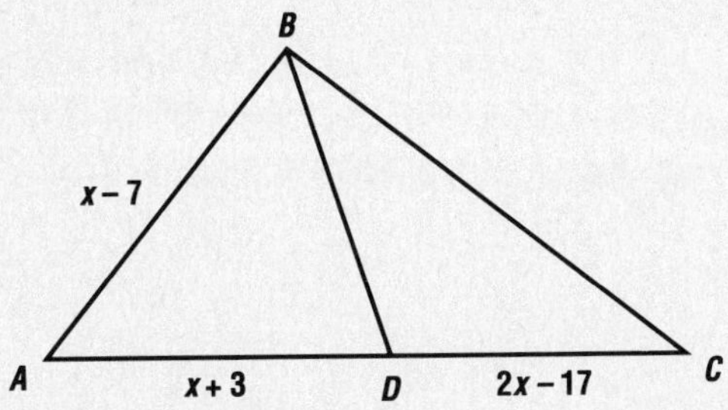

2. Find BC if $\overline{AD}$ is an altitude of $\triangle ABC$.

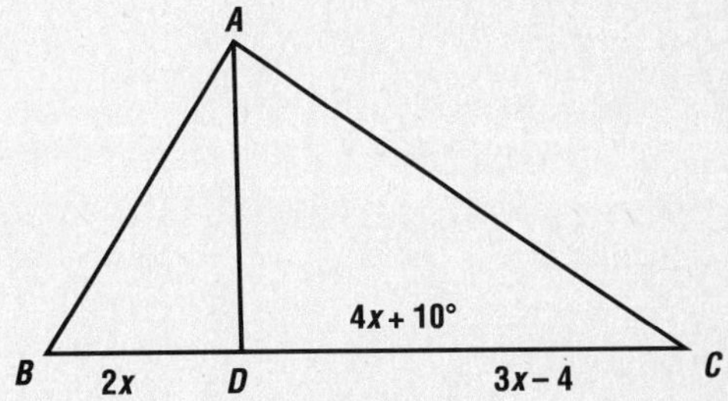

3. Find $m\angle ABC$ if $\overline{BD}$ is an angle bisector of $\triangle ABC$.

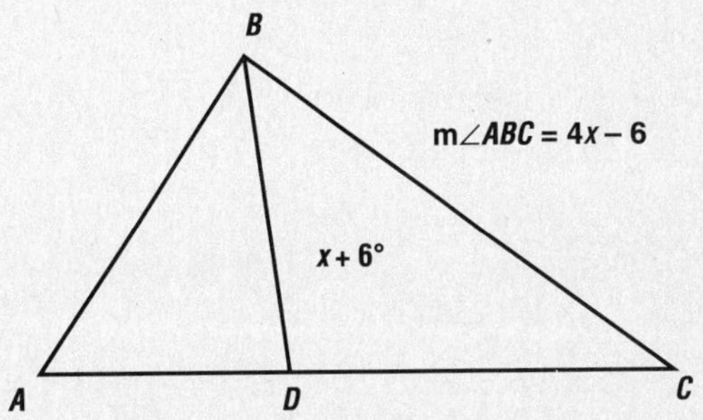

Answer each question if A(2, 5), B(12, −1), and C(−6, 8) are the vertices of △ABC.

4. What are the coordinates K if $\overline{CK}$ is a median of $\triangle ABC$?

5. What is the slope of the perpendicular bisector of $\overline{AB}$? What is the slope of $\overline{CL}$ if $\overline{CL}$ is the altitude from point C?

6. Point N on $\overleftrightarrow{BC}$ has coordinates $\left(\frac{8}{5}, \frac{21}{5}\right)$. Is $\overline{NA}$ an altitude of $\triangle ABC$? Explain your answer.

Glencoe Division, Macmillan/McGraw-Hill

NAME ______________________ DATE ____________

5-1

Practice Worksheet

Special Segments in Triangles

1. Find AB if $\overline{BD}$ is a median of $\triangle ABC$.
13

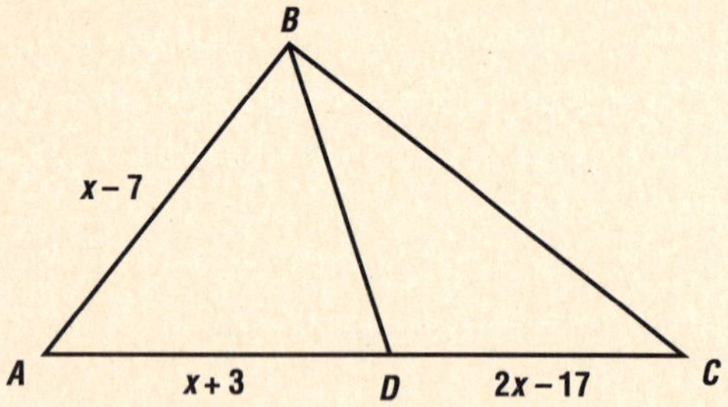

2. Find BC if $\overline{AD}$ is an altitude of $\triangle ABC$.
96

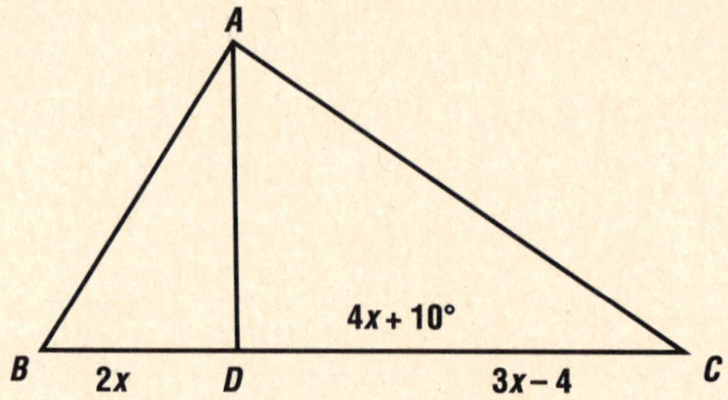

3. Find $m\angle ABC$ if $\overline{BD}$ is an angle bisector of $\triangle ABC$. **15°**

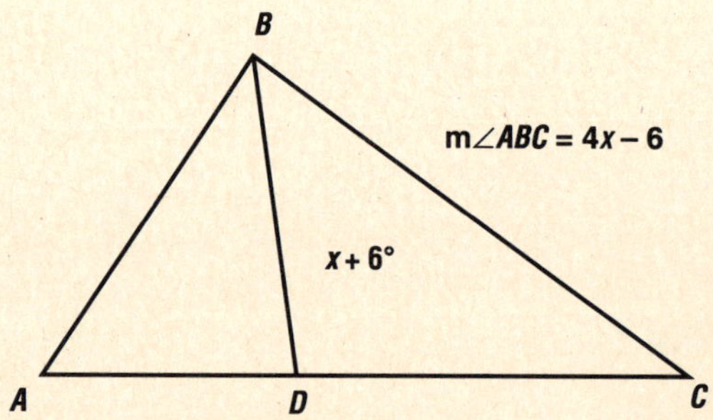

Answer each question if A(2, 5), B(12, −1), and C(−6, 8) are the vertices of △ABC.

4. What are the coordinates K if $\overline{CK}$ is a median of $\triangle ABC$? **(7, 2)**

5. What is the slope of the perpendicular bisector of $\overline{AB}$? What is the slope of $\overline{CL}$ if $\overline{CL}$ is the altitude from point C?
$\frac{5}{3}$, $\frac{5}{3}$

6. Point N on $\overleftrightarrow{BC}$ has coordinates $\left(\frac{8}{5}, \frac{21}{5}\right)$. Is $\overline{NA}$ an altitude of $\triangle ABC$? Explain your answer.
yes; *N* is on the line that contains $\overline{BC}$, and the product of the slope of $\overline{BC}$, $-\frac{1}{2}$, and the slope of $\overline{AN}$, 2, is −1.

NAME ________________________________ DATE ____________

5-2

Practice Worksheet

Right Triangles

For each figure, find the values of x and y so that △DEF ≅ △PQR by the indicated theorem or postulate.

1. *HA*

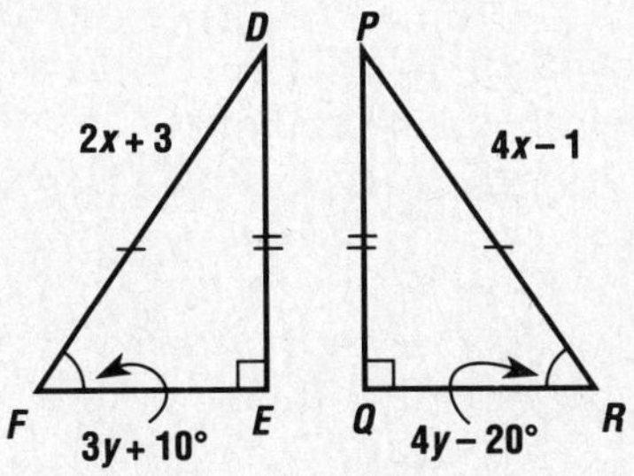

2. *LL*

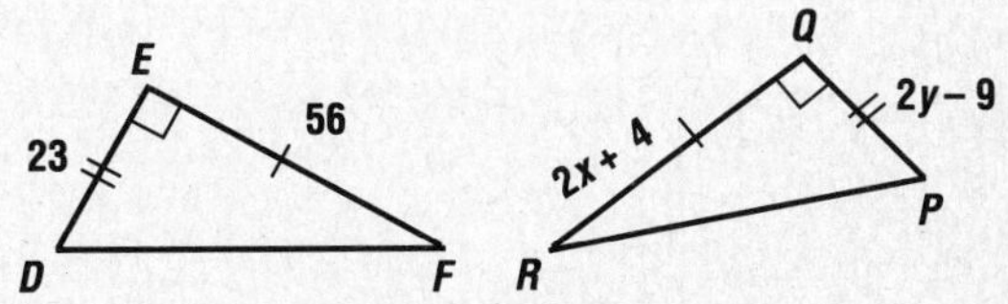

3. *LA*

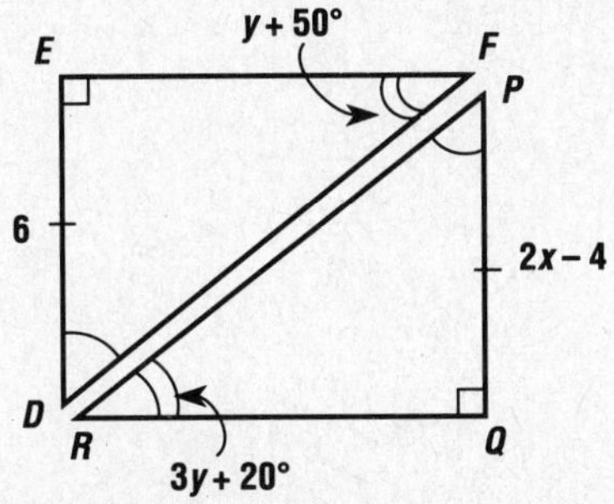

4. *HL*

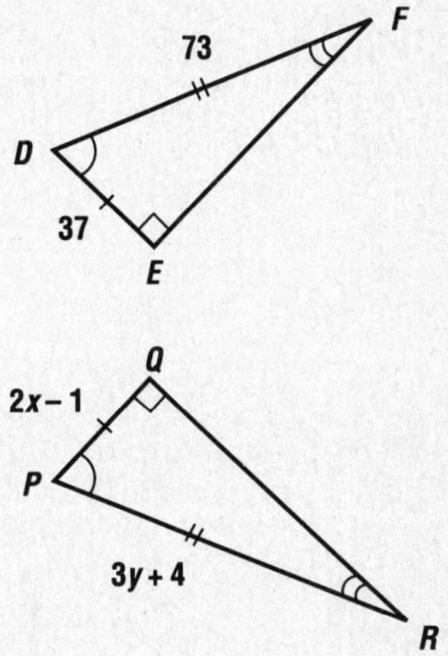

5. Write a two-column proof.

Given: $\overline{AB}$ bisects $\angle DAC$.
$\angle C$ and $\angle D$ are right angles.

Prove: $\overline{BC} \cong \overline{BD}$

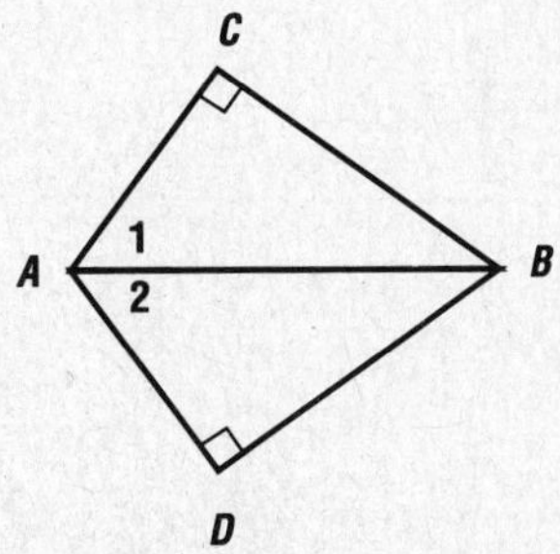

5-2 Practice Worksheet

NAME ______________________ DATE ____________

Right Triangles

For each figure, find the values of x and y so that $\triangle DEF \cong \triangle PQR$ by the indicated theorem or postulate.

1. *HA* **2, 30**

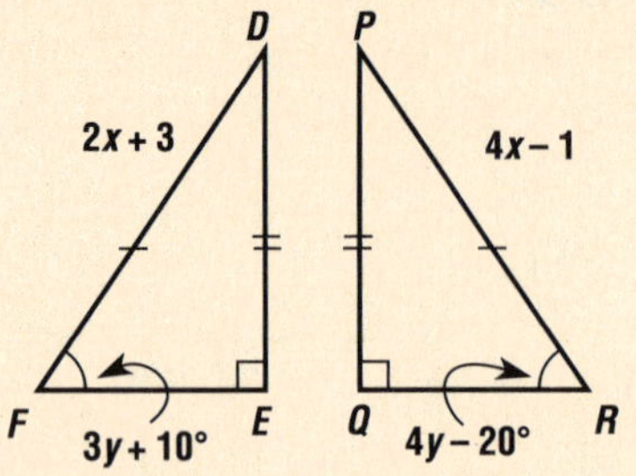

2. *LL* **26, 16**

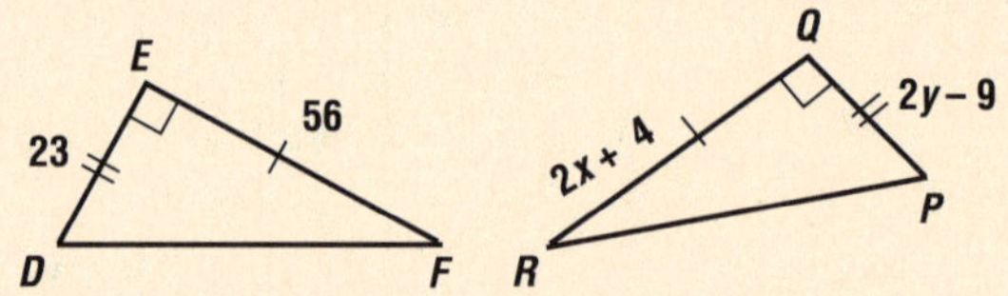

3. *LA* **5, 15**

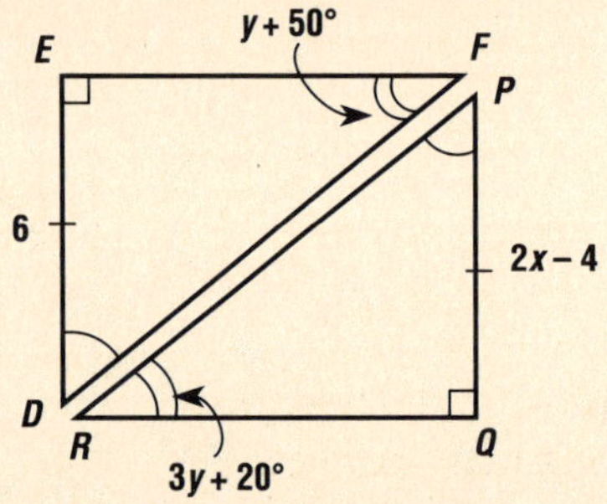

4. *HL* **19, 23**

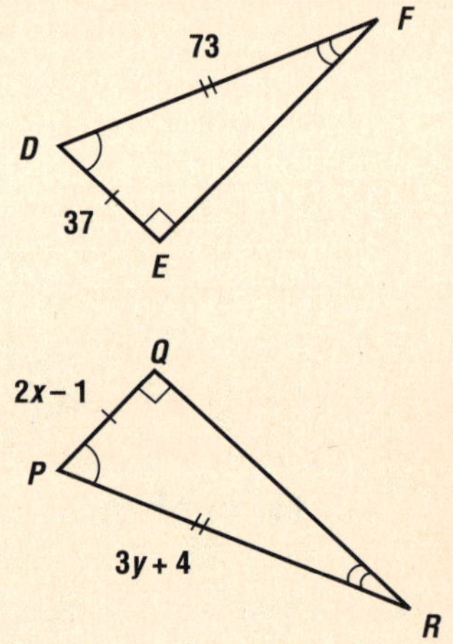

5. Write a two-column proof.

Given: $\overline{AB}$ bisects $\angle DAC$.
$\angle C$ and $\angle D$ are right angles.

Prove: $\overline{BC} \cong \overline{BD}$

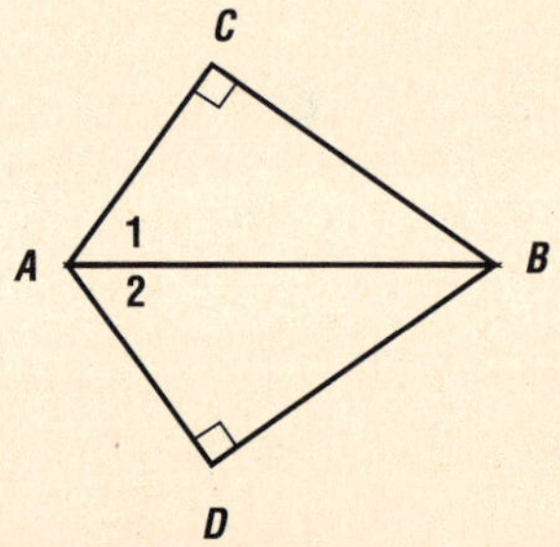

Statements	Reasons
a. $\overline{AB}$ bisects $\angle DAC$.	a. Given
b. $\angle 1 \cong \angle 2$	b. Definition of bisector
c. $\angle C$ and $\angle D$ are right angles.	c. Given
d. $\triangle ACB$ and $\triangle ADB$ are right triangles.	d. Definition of right triangle
e. $\overline{AB} \cong \overline{AB}$	e. Congruence of segments is reflexive.
f. $\triangle ACB \cong \triangle ADB$	f. HA
g. $\overline{BC} \cong \overline{BD}$	g. CPCTC

5-3

NAME ______________________ DATE ______________

Practice Worksheet

Problem-Solving Strategy: Work Backward

Solve. Use any strategy.

1. A can of apple juice in a vending machine costs $0.75. Exact change must be used. If the machine will accept quarters, dimes, and nickels, how many different combinations of coins must the machine be programmed to accept?

2. Increase a number by 63, then divide the result by 4, then add 39, and multiply by 3. If the final result is 186, with what number did you start?

3. Katie sews costumes for plays. One type of sequins she buys comes 36 to a package. Another type comes 48 to a package. If she buys 600 sequins for the costumes, how many of each type of package did she buy?

4. Joe spent half of the money in his wallet on a table for his computer printer. He then spent $8.12 for a printer ribbon. Then he spent half of what he had left on supplies for his office. He then had $14.50 remaining. How much money did Joe have to start with?

5. For a tug of war game, several parallel lines were drawn on the ground. The first team to pull the lead member of the opposing team across all the lines will win. The lines are numbered, with team A at line 1 and team B at the last line. Ten seconds after the start, the lead member of team A is at the middle line. Team A is then pulled forward 3 lines. Team A pulls team B for 5 lines. Then team B pulls team A forward 7 lines, then another 4 lines and wins. How many lines were there?

Glencoe Division, Macmillan/McGraw-Hill

5-3

NAME ______________________ DATE ____________

Practice Worksheet

Problem-Solving Strategy: Work Backward

Solve. Use any strategy.

1. A can of apple juice in a vending machine costs $0.75. Exact change must be used. If the machine will accept quarters, dimes, and nickels, how many different combinations of coins must the machine be programmed to accept?
18 combinations

2. Increase a number by 63, then divide the result by 4, then add 39, and multiply by 3. If the final result is 186, with what number did you start? **29**

3. Katie sews costumes for plays. One type of sequins she buys comes 36 to a package. Another type comes 48 to a package. If she buys 600 sequins for the costumes, how many of each type of package did she buy? **2 of 36, 11 of 48; 6 of 36, 8 of 48; 10 of 36, 5 of 48; or 14 of 36, 2 of 48**

4. Joe spent half of the money in his wallet on a table for his computer printer. He then spent $8.12 for a printer ribbon. Then he spent half of what he had left on supplies for his office. He then had $14.50 remaining. How much money did Joe have to start with? **$74.24**

5. For a tug of war game, several parallel lines were drawn on the ground. The first team to pull the lead member of the opposing team across all the lines will win. The lines are numbered, with team A at line 1 and team B at the last line. Ten seconds after the start, the lead member of team A is at the middle line. Team A is then pulled forward 3 lines. Team A pulls team B for 5 lines. Then team B pulls team A forward 7 lines, then another 4 lines and wins. How many lines were there? **19 lines**

5-4

NAME ______________________________ DATE ______________

Practice Worksheet

Inequalities and Indirect Proof

Use the figure to complete each statement with either $<$ or $>$.

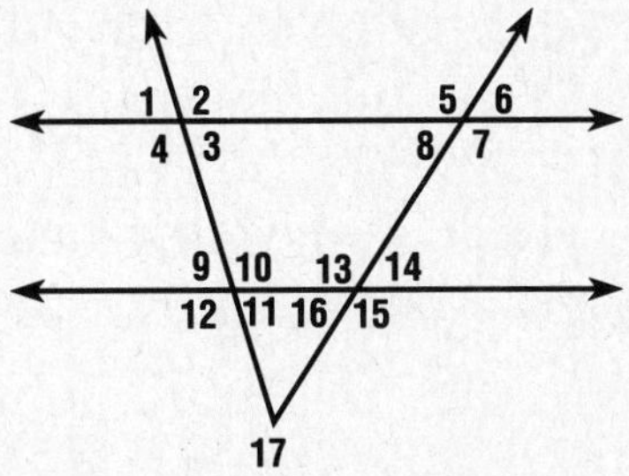

1. $\angle 4$ __?__ $\angle 8$

2. $\angle 13$ __?__ $\angle 11$

3. $\angle 17$ __?__ $\angle 2$

4. If $m\angle 5 = m\angle 7$, then $m\angle 11$ __?__ $\angle 7$.

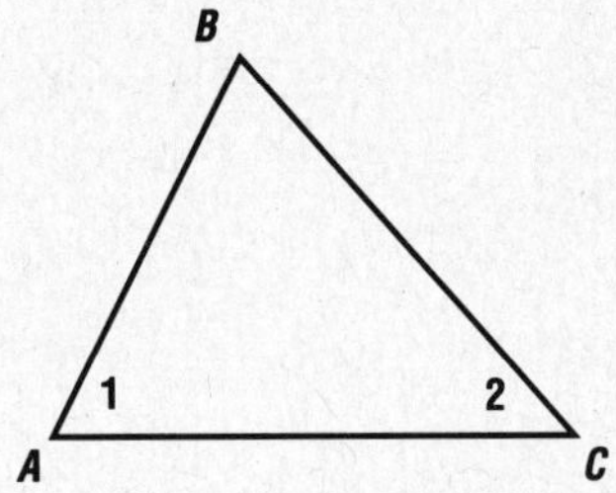

Write an indirect proof for each.

5. Given: $m\angle 1 \neq m\angle 2$

Prove: $\triangle ABC$ is not an isosceles triangle with vertex B.

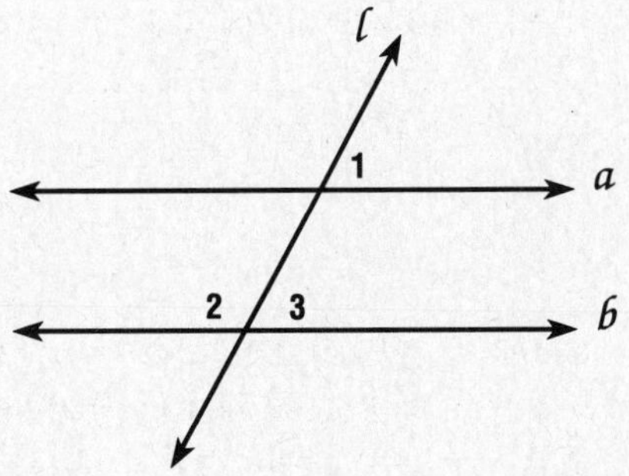

6. Given: $m\angle 1 + m\angle 2 \neq 180$

Prove: $a \nparallel b$

Glencoe Division, Macmillan/McGraw-Hill

NAME ______________________ DATE ____________

5-4

Practice Worksheet

Inequalities and Indirect Proof

Use the figure to complete each statement with either $<$ or $>$.

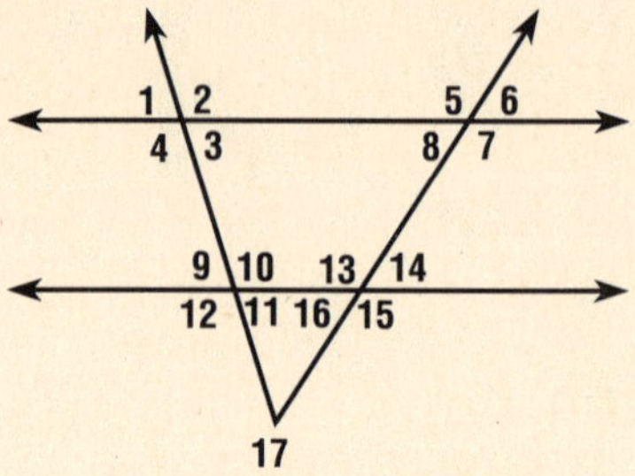

1. $\angle 4$ __?__ $\angle 8$ $>$

2. $\angle 13$ __?__ $\angle 11$ $>$

3. $\angle 17$ __?__ $\angle 2$ $<$

4. If $m\angle 5 = m\angle 7$, then $m\angle 11$ __?__ $\angle 7$. $<$

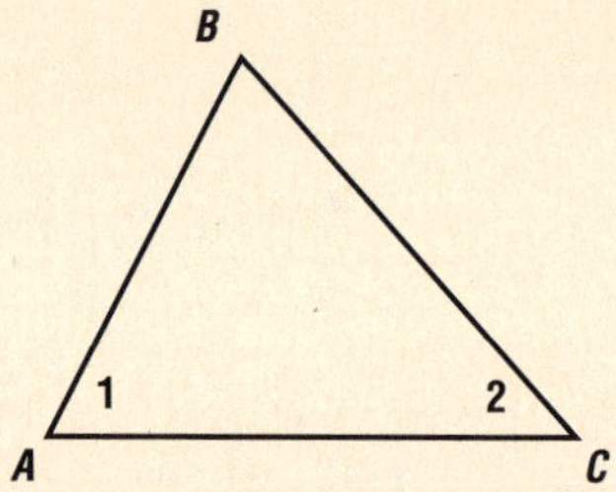

Write an indirect proof for each.

5. Given: $m\angle 1 \neq m\angle 2$

   Prove: $\triangle ABC$ is not an isosceles triangle with vertex B.

   Assume $\triangle ABC$ is isosceles, with vertex B. Then $AB = BC$ by the definition of isosceles triangle. This implies that $m\angle 1 = m\angle 2$, since angles opposite congruent sides of a triangle are congruent. This contradicts the given information. Therefore, $\triangle ABC$ cannot be an isosceles triangle with vertex B.

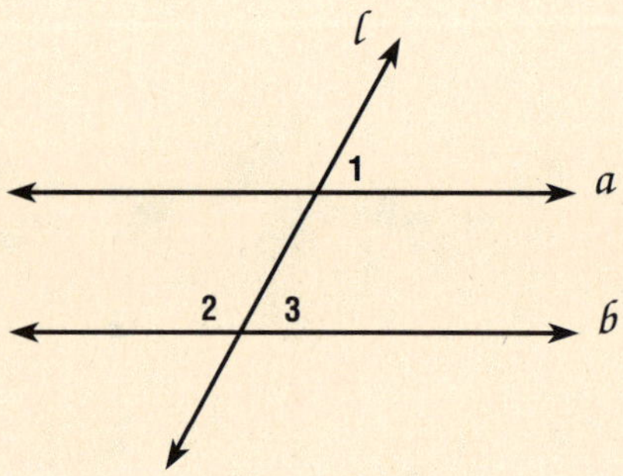

6. Given: $m\angle 1 + m\angle 2 \neq 180$

   Prove: $a \nparallel b$

   Assume $a \parallel b$. Then $m\angle 1 = m\angle 3$, since they are corresponding angles formed by parallel lines and a transversal. $\angle 2$ and $\angle 3$ are a linear pair, so $m\angle 2 + m\angle 3 = 180$. By substitution, $m\angle 2 + m\angle 1 = 180$. This contradicts given information. Therefore, $a \nparallel b$.

NAME ________________________________ DATE ______________

5-5 Practice Worksheet

Inequalities for Sides and Angles of a Triangle

For Exercises 1 to 4, refer to the figure on the right. This figure is not drawn to scale.

1. Name the shortest and the longest segments in $\triangle BCD$.

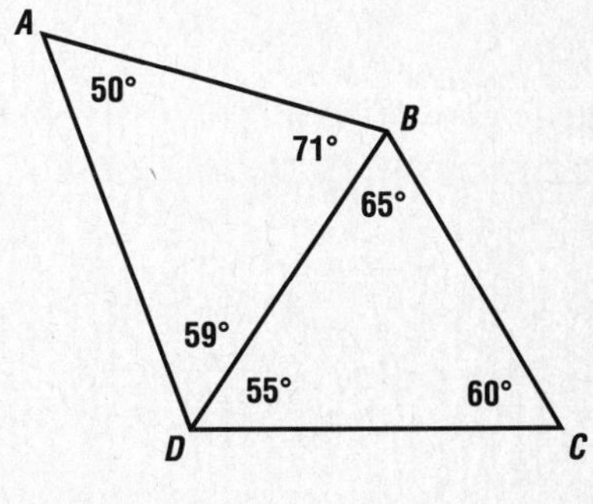

2. Name the shortest and the longest segments in $\triangle ABD$.

3. Find the shortest segment in the figure.

4. How many of the segments in the figure are longer than $\overline{BD}$?

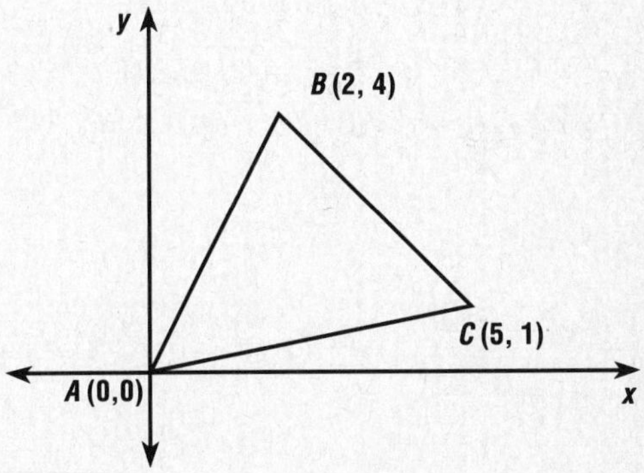

5. List the angles in order from least to greatest.

6. List the sides of $\triangle MNO$ in order from longest to shortest if $m\angle M = 4x + 20$, $m\angle N = 2x + 10$, and $m\angle 0 = 3x - 20$.

7. List the angles of $\triangle KLM$ in order from least to greatest if $KL = x - 4$, $LM = x + 4$, $KM = 2x - 1$, and the perimeter of $\triangle KLM$ is 27.

Glencoe Division, Macmillan/McGraw-Hill

NAME ______________________ DATE ____________

5-5

Practice Worksheet

Inequalities for Sides and Angles of a Triangle

For Exercises 1 to 4, refer to the figure on the right. This figure is not drawn to scale.

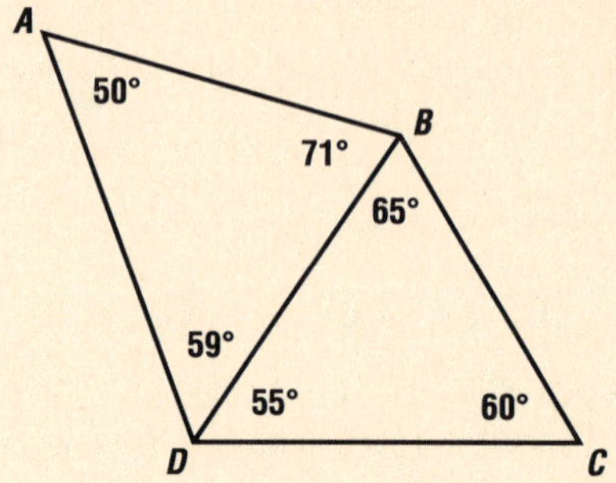

1. Name the shortest and the longest segments in $\triangle BCD$.
$\overline{BC}, \overline{DC}$

2. Name the shortest and the longest segments in $\triangle ABD$.
$\overline{BD}, \overline{AD}$

3. Find the shortest segment in the figure. $\overline{BC}$

4. How many of the segments in the figure are longer than $\overline{BD}$? **3**

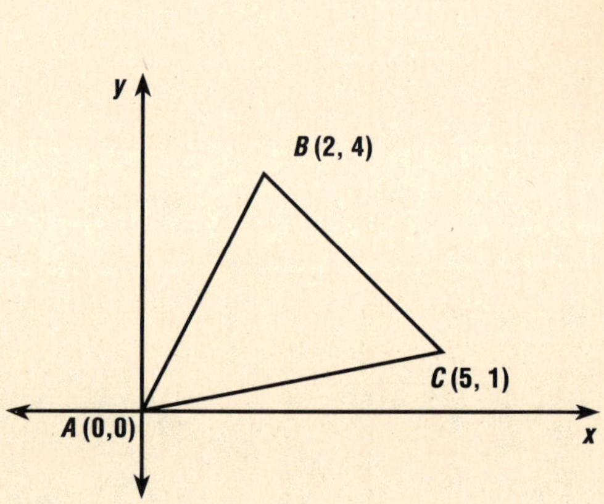

5. List the angles in order from least to greatest.
$\angle A, \angle C, \angle B$

6. List the sides of $\triangle MNO$ in order from longest to shortest if $m\angle M = 4x + 20$, $m\angle N = 2x + 10$, and $m\angle 0 = 3x - 20$.
$\overline{NO}, \overline{MO}, \overline{MN}$

7. List the angles of $\triangle KLM$ in order from least to greatest if $KL = x - 4$, $LM = x + 4$, $KM = 2x - 1$, and the perimeter of $\triangle KLM$ is 27. $\angle M, \angle K, \angle L$

NAME ______________________ DATE ______________

5-6 Practice Worksheet

The Triangle Inequality

Determine whether it is possible to have a triangle with sides of the given measure. Write yes or no.

1. 3, 3, 3

2. 2, 3, 4

3. 1, 2, 3

4. 8.9, 9.3, 18.3

5. 16.5, 20.5, 38.5

6. 19, 19, 0.5

Is it possible to have a triangle with the given vertices? Write yes or no. Explain your answer. Use your calculator as needed.

7. $A(-2, -2)$, $B(-1, 1)$, $C(1, 4)$

8. $A(-4, 2)$, $B(-2, 1)$, $C(2, -1)$

9. $A(2, 5)$, $B(-3, 5)$, $C(6, -1)$

10. $A(3, -6)$, $B(1, 2)$, $C(-2, 10)$

Two sides of a triangle are 21 and 24 inches in length. Determine whether each measurement can be the length of the third side.

11. 3 inches

12. 40 inches

13. 56 inches

If the sides of a triangle have the following lengths, find all possible values for x.

14. $AB = 2x + 5$, $BC = 3x - 2$, $AC = 4x - 8$

15. $PQ = 3x$, $QR = 4x - 7$, $PR = 2x + 9$

Glencoe Division, Macmillan/McGraw-Hill

NAME ________________________________ DATE ______________

5-6 Practice Worksheet

The Triangle Inequality

Determine whether it is possible to have a triangle with sides of the given measure. Write yes or no.

1. 3, 3, 3 **yes**

2. 2, 3, 4 **yes**

3. 1, 2, 3 **no**

4. 8.9, 9.3, 18.3 **no**

5. 16.5, 20.5, 38.5 **no**

6. 19, 19, 0.5 **yes**

Is it possible to have a triangle with the given vertices? Write yes or no. Explain your answer. Use your calculator as needed.

7. $A(-2, -2)$, $B(-1, 1)$, $C(1, 4)$
yes; $AB = \sqrt{10}$, $BC = \sqrt{13}$, $AC = 3\sqrt{5}$ and all triangle inequalities are satisfied.

8. $A(-4, 2)$, $B(-2, 1)$, $C(2, -1)$
no; $AB = \sqrt{5}$, $BC = 2\sqrt{5}$, $AC = 3\sqrt{5}$
$AB + BC = AC$

9. $A(2, 5)$, $B(-3, 5)$, $C(6, -1)$
yes; $AB = 5$, $BC = 3\sqrt{13}$, $AC = 2\sqrt{13}$ and all triangle inequalities are satisfied.

10. $A(3, -6)$, $B(1, 2)$, $C(-2, 10)$
yes; $AB = 2\sqrt{17}$, $BC = \sqrt{73}$, $AC = \sqrt{281}$ and all triangle inequalities are satisfied.

Two sides of a triangle are 21 and 24 inches in length. Determine whether each measurement can be the length of the third side.

11. 3 inches **no**

12. 40 inches **yes**

13. 56 inches **no**

If the sides of a triangle have the following lengths, find all possible values for x.

14. $AB = 2x + 5$, $BC = 3x - 2$, $AC = 4x - 8$

$\{x \mid x > 3\}$

15. $PQ = 3x$, $QR = 4x - 7$, $PR = 2x + 9$

$\{x \mid x > \frac{16}{5}\}$

NAME ____________________ DATE ____________

5-7

Practice Worksheet

Inequalities Involving Two Triangles

Write an inequality relating each pair of measures.

1. m∠*PRQ*, m∠*PRS*

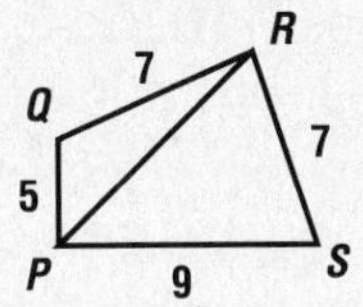

2. m∠*ABE*, m∠*DBC*

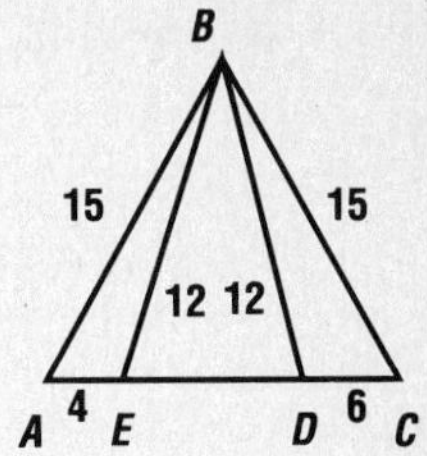

For each figure, write an inequality or pair of inequalities to describe the possible values of x.

3.

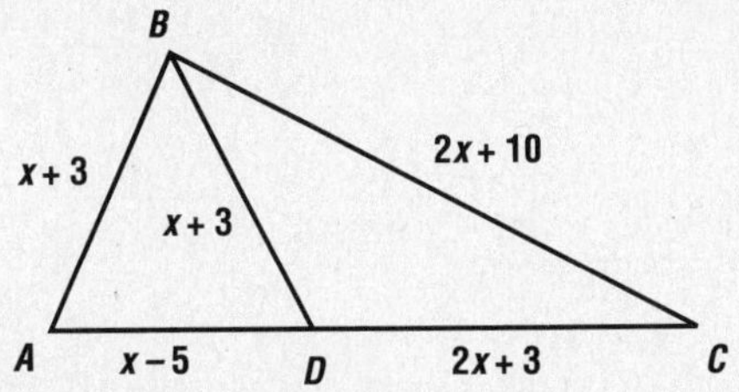

4.

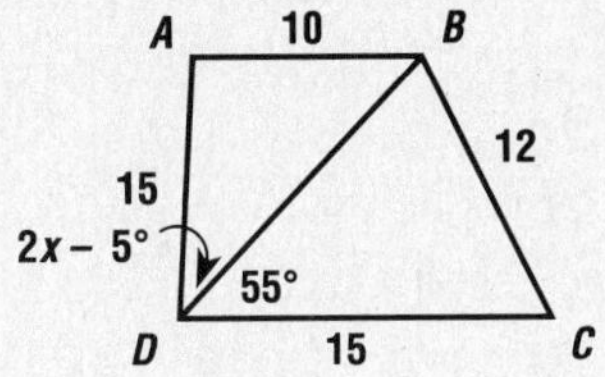

Write a two-column proof for each.

5. Given: $\overline{AD} \cong \overline{EC}$

m∠ADC > m∠ECD

Prove: $AC > ED$

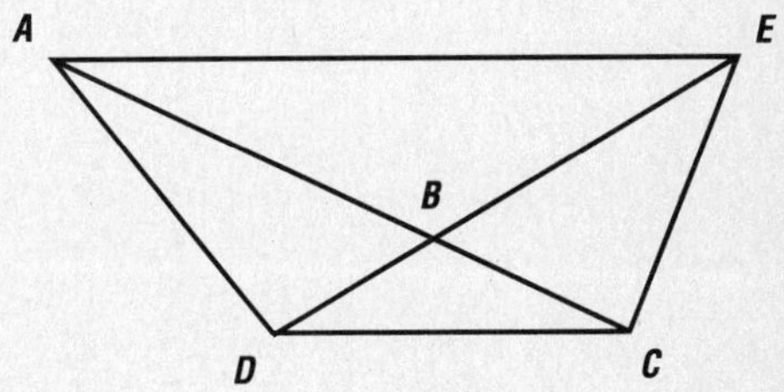

6. Given: *D* is the midpoint of $\overline{AC}$.

$BC > AB$

Prove: m∠1 > m∠2

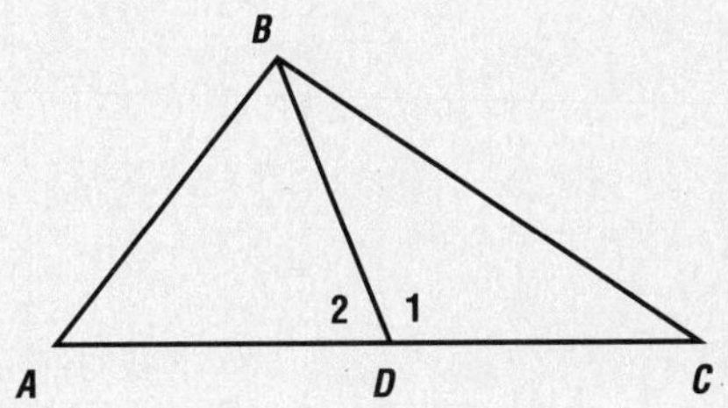

Glencoe Division, Macmillan/McGraw-Hill

NAME ________________________________ DATE ______________

5-7

Practice Worksheet

Inequalities Involving Two Triangles

Write an inequality relating each pair of measures.

1. m∠*PRQ*, m∠*PRS*

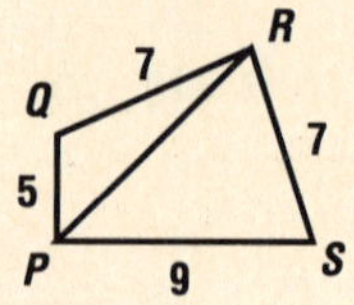

m∠*PRQ* < m∠*PRS*

2. m∠*ABE*, m∠*DBC*

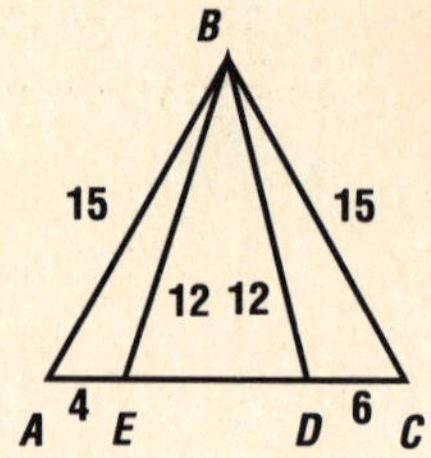

m∠*ABE* < m∠*DBC*

For each figure, write an inequality or pair of inequalities to describe the possible values of x.

3.

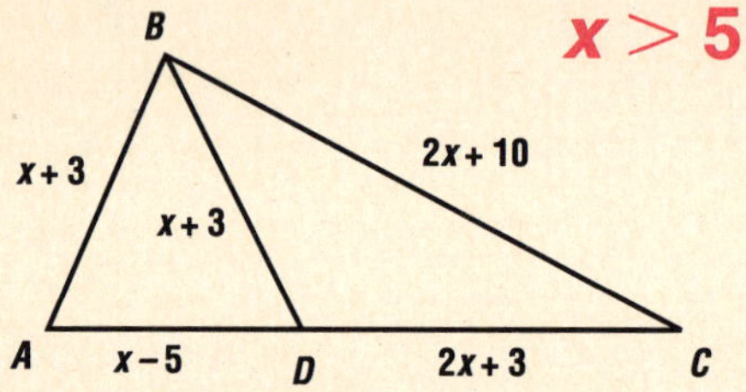

$x > 5$

4.

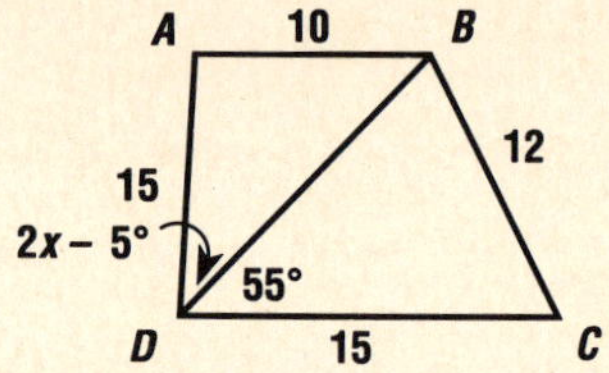

$x < 30$

Write a two-column proof for each.

5. Given: $\overline{AD} \cong \overline{EC}$
m∠ADC > m∠ECD
Prove: $AC > ED$

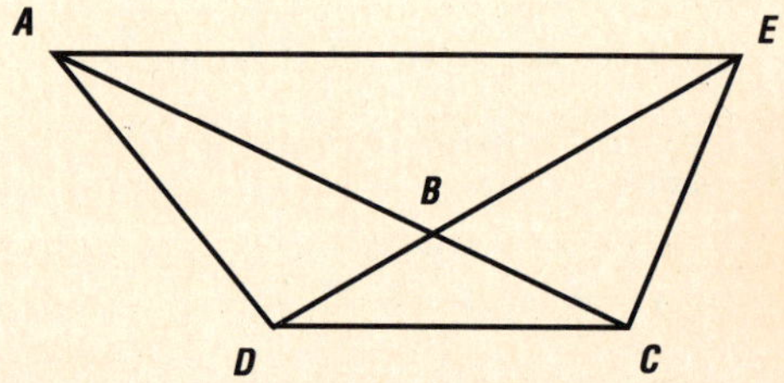

Statements	Reasons
a. $\overline{AD} \cong \overline{EC}$	a. Given
b. $\overline{DC} \cong \overline{DC}$	b. Congruence of segments is reflexive.
c. m∠*ADC* > m∠*ECD*	c. Given
d. $AC > ED$	d. SAS Inequality

6. Given: *D* is the midpoint of $\overline{AC}$.
$BC > AB$
Prove: m∠1 > m∠2

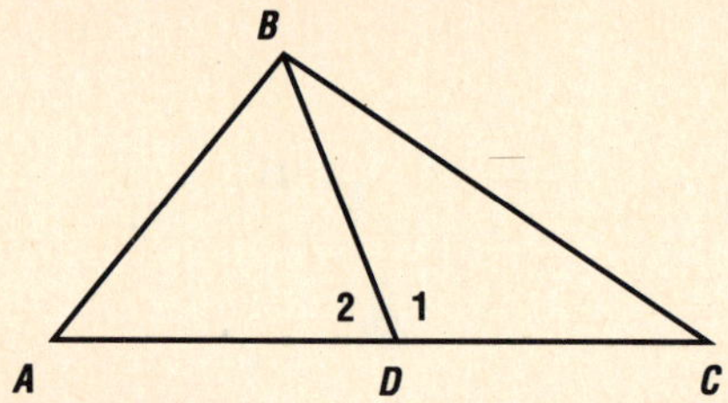

Statements	Reasons
a. *D* is the midpoint of $\overline{AC}$.	a. Given
b. $\overline{AD} \cong \overline{DC}$	b. Definition of midpoint
c. $\overline{BD} \cong \overline{BD}$	c. Congruence of segments is reflexive.
d. $\overline{BC} > \overline{AB}$	d. Given
e. m∠1 > m∠2	e. SSS Inequality

Glencoe Division, Macmillan/McGraw-Hill

NAME ______________________ DATE ____________

6-1

Practice Worksheet

Parallelograms

The figure at the right is a parallelogram. Use this figure and the information given to solve each problem.

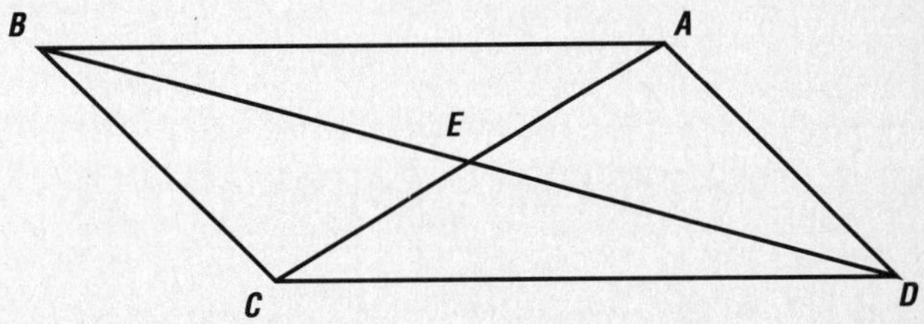

1. If $m\angle BCD = 125$, find $m\angle BAD$.
2. If $m\angle BAC = 45$, find $m\angle ACD$.
3. If $m\angle BEA = 135$, find $m\angle AED$.
4. If $m\angle ABC = 50$, find $m\angle BCD$.
5. If $AB = 5x - 3$ and $CD = 2x + 9$, find AB.
6. If $m\angle DAB = 2x - 10$ and $m\angle ADC = 3x$, find $m\angle DAB$.
7. If $m\angle BAD = 3x - 12$ and $m\angle BCD = x + 40$, find $m\angle BAD$.

Write a two-column proof.

8. Given: $ABCD$ is a parallelogram, $\overline{BE} \cong \overline{AD}$

   Prove: $\angle 1 \cong \angle C$

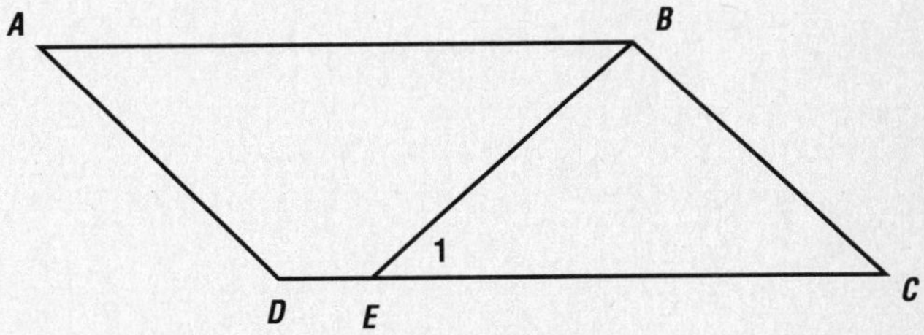

Statements	Reasons

NAME ______________________ DATE __________

6-1

Practice Worksheet

Parallelograms

The figure at the right is a parallelogram. Use this figure and the information given to solve each problem.

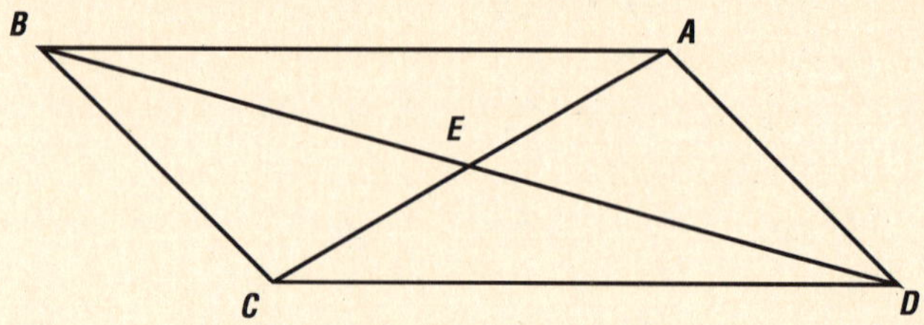

1. If m$\angle BCD$ = 125, find m$\angle BAD$. **125**
2. If m$\angle BAC$ = 45, find m$\angle ACD$. **45**
3. If m$\angle BEA$ = 135, find m$\angle AED$. **45**
4. If m$\angle ABC$ = 50, find m$\angle BCD$. **130**
5. If $AB = 5x - 3$ and $CD = 2x + 9$, find AB. **17**
6. If m$\angle DAB = 2x - 10$ and m$\angle ADC = 3x$, find m$\angle DAB$. **66**
7. If m$\angle BAD = 3x - 12$ and m$\angle BCD = x + 40$, find m$\angle BAD$. **66**

Write a two-column proof.

8. Given: $ABCD$ is a parallelogram, $\overline{BE} \cong \overline{AD}$
   Prove: $\angle 1 \cong \angle C$

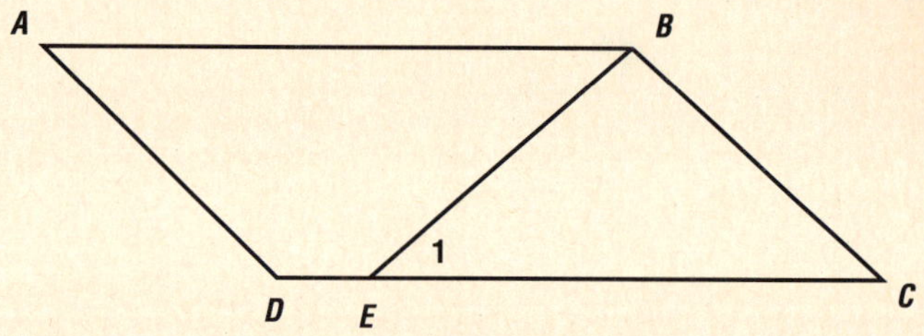

Statements	Reasons
a. *ABCD* is a parallelogram.	a. Given
b. $\overline{AD} \cong \overline{BC}$	b. Opposite sides of a parrallelogram are congruent.
c. $\overline{BE} \cong \overline{AD}$	c. Given
d. $\overline{BC} \cong \overline{BE}$	d. Congruence of segments is transitive.
e. $\angle 1 \cong \angle C$	e. If two sides of a triangle are congruent, then the angles opposite those sides are congruent.

NAME ________________ DATE ________

6-2

Practice Worksheet

Problem-Solving Strategy: Look for a Pattern

Solve. Use any strategy.

1. Two hundred people signed up for a ping pong singles tournament. After two people play, the looser is eliminated from the competition. How many matches will there be before a winner can be determined?

2. How many squares are there in the figure shown at the right? (Hint: There are more than 16.)

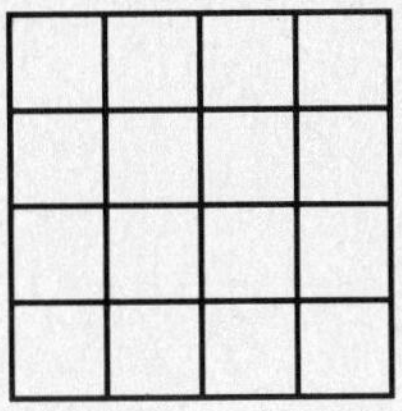

3. Find the sum of the first 20 terms of the following sequence: 1, 1, 2, 3, 5, 8, 13, . . .

4. Tanya likes to burn candles. She has found that, once a candle has burned, she can melt 3 candle stubs, add a new wick, and have one more candle to burn. How many candles total can she burn from a box of 15 candles?

5. In a college senior class $\frac{2}{7}$ of the young men are engaged to $\frac{1}{5}$ of the women. What fraction of the senior class consists of women.

NAME ______________________ DATE ____________

6-2 Practice Worksheet

Problem-Solving Strategy: Look for a Pattern

Solve. Use any strategy.

1. Two hundred people signed up for a ping pong singles tournament. After two people play, the looser is eliminated from the competition. How many matches will there be before a winner can be determined? **199**

2. How many squares are there in the figure shown at the right? (Hint: There are more than 16.) **30**

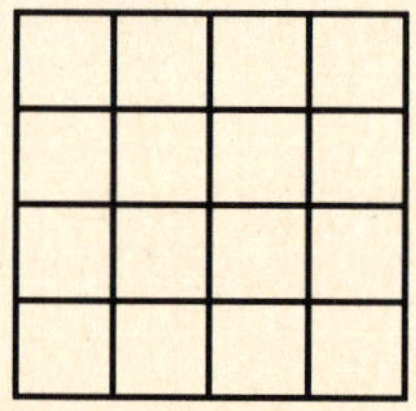

3. Find the sum of the first 20 terms of the following sequence: 1, 1, 2, 3, 5, 8, 13, . . . **17,710**

4. Tanya likes to burn candles. She has found that, once a candle has burned, she can melt 3 candle stubs, add a new wick, and have one more candle to burn. How many candles total can she burn from a box of 15 candles? **22**

5. In a college senior class $\frac{2}{7}$ of the young men are engaged to $\frac{1}{5}$ of the women. What fraction of the senior class consists of women.

   $\frac{10}{17}$

Glencoe Division, Macmillan/McGraw-Hill

NAME ____________________ DATE __________

6-3

Practice Worksheet

Tests for Parallelograms

What values must x and y have in order for each quadrilateral to be a parallelogram?

1.

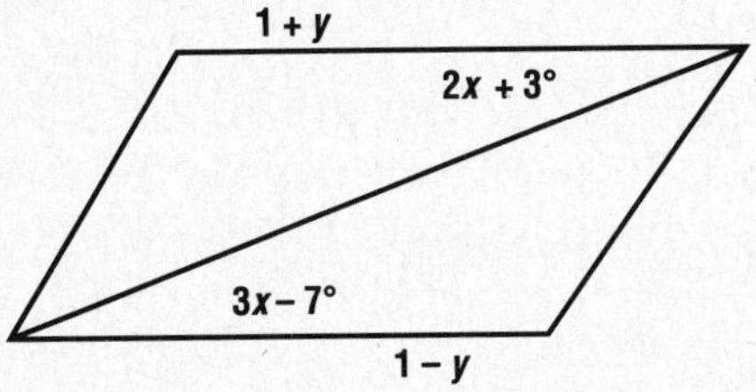

2.

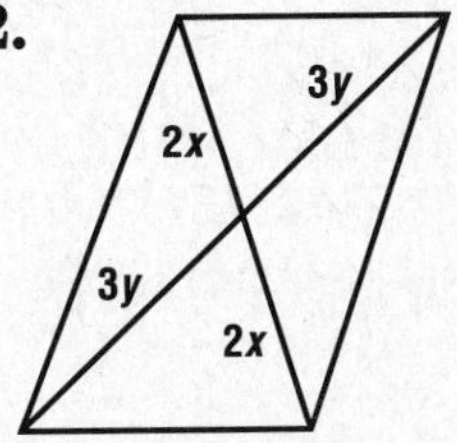

3. Refer to the figure at the right. $\overline{YZ}$ bisects $\angle XYK$ and $\overline{LK}$ bisects $\angle ZLM$. Also, $\angle 1 \cong \angle 2$. Is $YZLK$ a parallelogram? Explain.

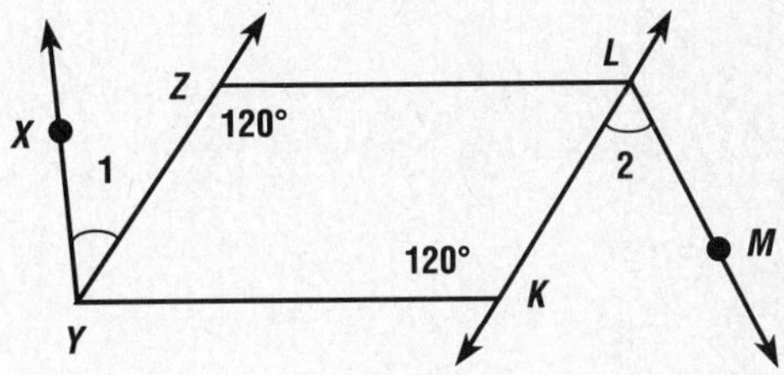

4. Refer to the figure at the right. $\overline{AC} = \overline{WV}$ and $\overline{BD} = \overline{WV}$. Also, $\overline{AB} = \overline{XY}$ and $\overline{CD} = \overline{XY}$. Is $ABCD$ a parallelogram? Explain.

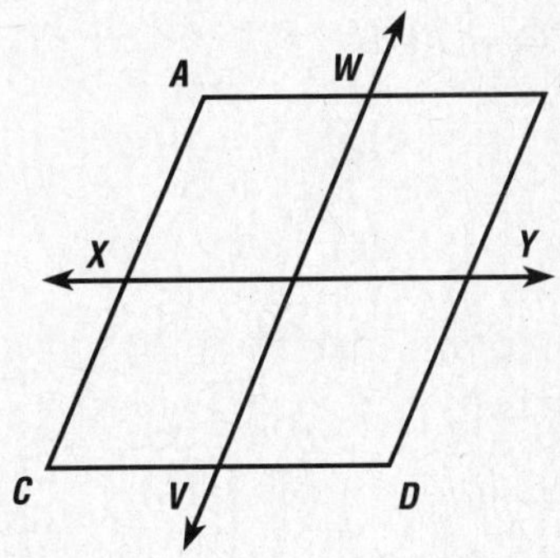

Determine whether ABCD is a parallelogram given each set of vertices. Explain.

5. $A(2, 5)$, $B(5, 9)$, $C(3, -1)$, $D(6, 3)$

6. $A(-1, 6)$, $B(2, -3)$, $C(5, 9)$, $D(2, 7)$

6-3

NAME ______________________ DATE ____________

Practice Worksheet

Tests for Parallelograms

What values must x and y have in order for each quadrilateral to be a parallelogram?

1.

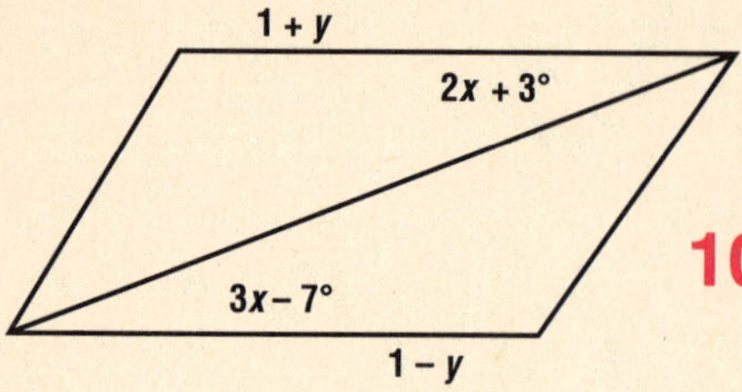

10, 0

2.

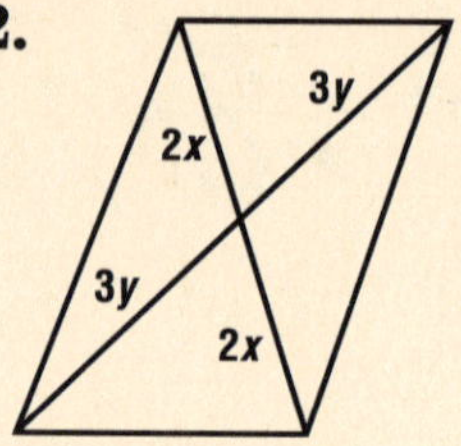

any values for x and y that are greater than 0

3. Refer to the figure at the right. $\overline{YZ}$ bisects $\angle XYK$ and $\overline{LK}$ bisects $\angle ZLM$. Also, $\angle 1 \cong \angle 2$. Is $YZLK$ a parallelogram? Explain.

yes; Opposite angles are congruent.

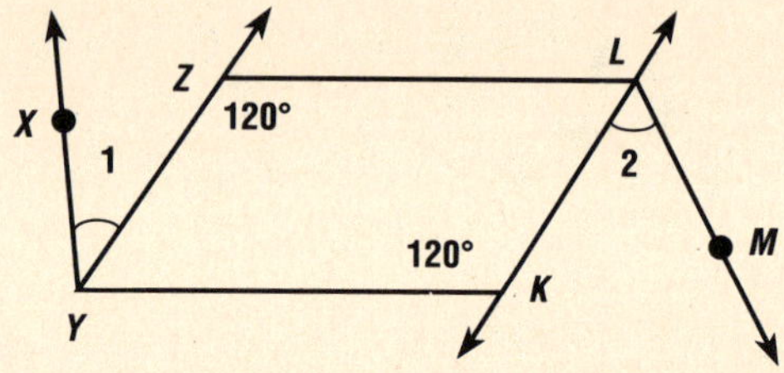

4. Refer to the figure at the right. $\overline{AC} = \overline{WV}$ and $\overline{BD} = \overline{WV}$. Also, $\overline{AB} = \overline{XY}$ and $\overline{CD} = \overline{XY}$. Is $ABCD$ a parallelogram? Explain.

yes; Both pairs of opposite sides are congruent.

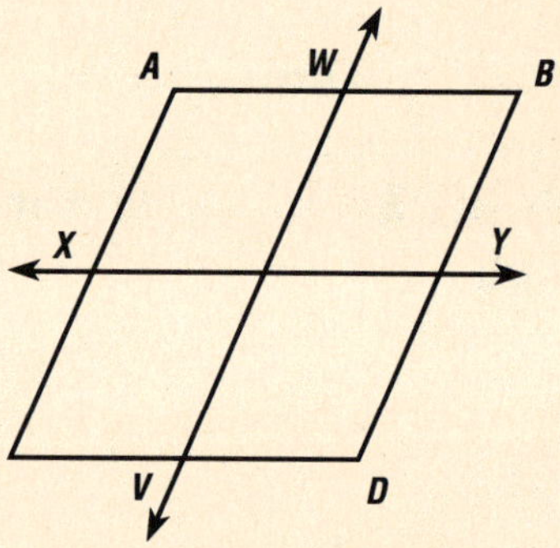

Determine whether ABCD is a parallelogram given each set of vertices. Explain.

5. $A(2, 5)$, $B(5, 9)$, $C(3, -1)$, $D(6, 3)$

yes; Opposite sides have equal lengths and are therefore congruent.

6. $A(-1, 6)$, $B(2, -3)$, $C(5, 9)$, $D(2, 7)$

no; Slope $\overline{AD} = \frac{1}{3}$ and slope $\overline{BC} = 4$, so these two opposite sides are not parallel.

6-4

NAME ______________________ DATE ______________

Practice Worksheet

Rectangles

Use rectangles ABCD and the given information to solve each problem.

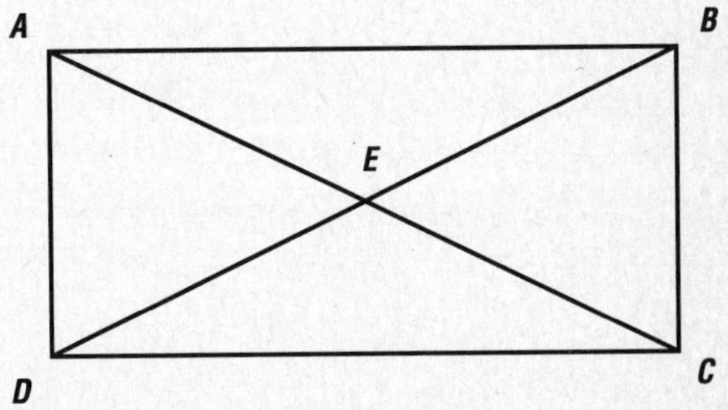

1. If $AC = 4x - 60$ and $BD = 30 - x$, find BD.

2. If $AC = 4x - 60$ and $AE = x + 5$, find EC.

3. If $m\angle BAC = 4x + 5$ and $m\angle CAD = 5x - 14$, find $m\angle CAD$.

4. If $AE = 2x + 3$ and $BE = 12 - x$, find BD.

5. $m\angle BAC = 3x + 5$ and $m\angle ACD = 40 - 2x$. Find $m\angle AED$.

Determine whether PQRS is a rectangle. Explain.

6. $P(2, 3)$, $Q(5, 9)$, $R(11, 6)$, $S(8,0)$

7. $P(-1, 4)$, $Q(3, 6)$, $R(9, -3)$, $S(5, -5)$

8. $P(1, 3)$, $Q(4, 7)$, $R(6, 2)$, $S(2, 4)$

9. $P(-1, -3)$, $Q(-4, 6)$, $R(8, 10)$, $S(11, 1)$

10. $P(-1, -2)$, $Q(5, 2)$, $R(13, -10)$, $S(7, -14)$

Glencoe Division, Macmillan/McGraw-Hill

NAME ______________________ DATE ____________

6-4

Practice Worksheet

Rectangles

Use rectangles ABCD and the given information to solve each problem.

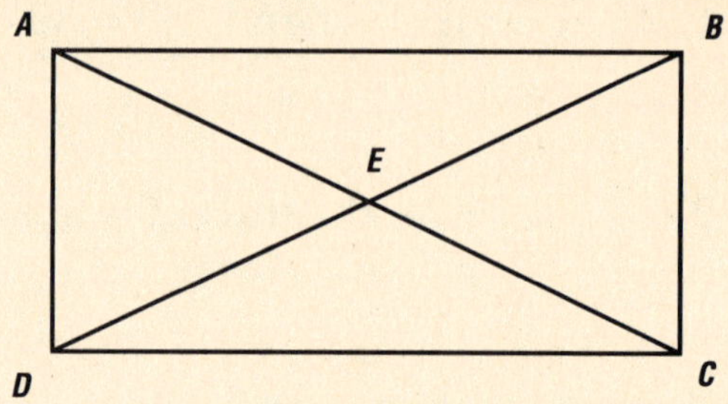

1. If $AC = 4x - 60$ and $BD = 30 - x$, find BD.
   12

2. If $AC = 4x - 60$ and $AE = x + 5$, find EC.
   40

3. If $m\angle BAC = 4x + 5$ and $m\angle CAD = 5x - 14$, find $m\angle CAD$.
   41

4. If $AE = 2x + 3$ and $BE = 12 - x$, find BD.
   18

5. $m\angle BAC = 3x + 5$ and $m\angle ACD = 40 - 2x$. Find $m\angle AED$.
   52

Determine whether PQRS is a rectangle. Explain.

6. $P(2, 3)$, $Q(5, 9)$, $R(11, 6)$, $S(8,0)$
   yes; opposite sides parallel and all right angles

7. $P(-1, 4)$, $Q(3, 6)$, $R(9, -3)$, $S(5, -5)$
   no; not all right angles

8. $P(1, 3)$, $Q(4, 7)$, $R(6, 2)$, $S(2, 4)$
   no; opposite sides not parallel

9. $P(-1, -3)$, $Q(-4, 6)$, $R(8, 10)$, $S(11, 1)$
   yes; opposite sides parallel and all right angles

10. $P(-1, -2)$, $Q(5, 2)$, $R(13, -10)$, $S(7, -14)$
    yes; opposite sides parallel and all right angles

NAME ______________________ DATE ______________

6-5

Practice Worksheet

Squares and Rhombi

Use square ABCD and the given information to solve each problem.

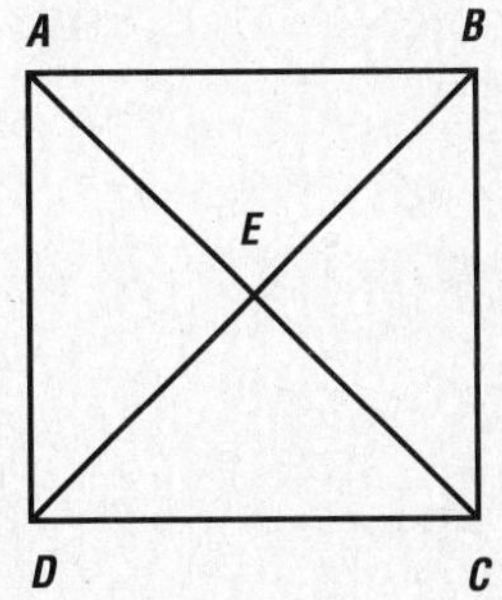

1. If m∠AEB = $3x$, find x.

2. If m∠BAC = $9x$, find x.

3. If $AB = 2x + 4$ and $CD = 3x - 5$, find BC.

4. If m∠DAC = y and m∠BAC = $3x$, find x.

5. If $AB = x^2 - 15$ and $BC = 2x$, find x.

Use rhombus ABCD and the given information to find each measure.

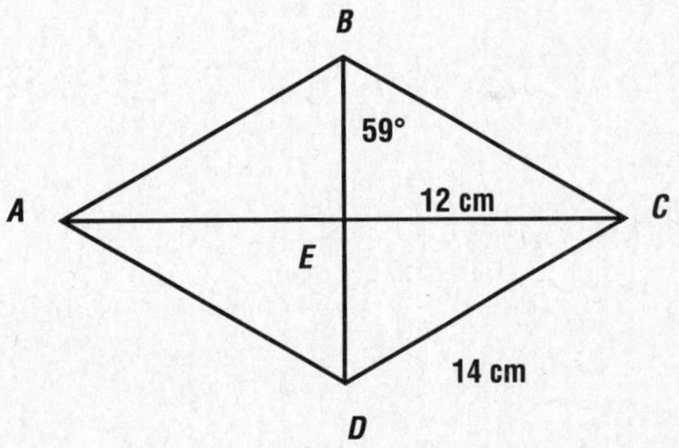

6. m∠BCE

7. m∠BEC

8. AC

9. m∠ABD

10. AD

Determine whether EFGH is a parallelogram, rectangle, rhombus, or square. List all that apply.

11. $E(0, -3)$, $F(-3, 0)$, $G(0, 3)$, $H(3, 0)$

12. $E(2, 1)$, $F(3, 4)$, $G(7, 2)$, $H(6, -1)$

NAME ______________________________ DATE ______________

6-5

Practice Worksheet

Squares and Rhombi

Use square ABCD and the given information to solve each problem.

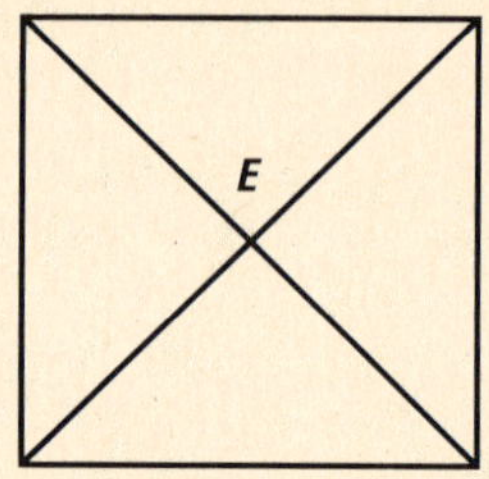

1. If m$\angle AEB = 3x$, find x. **30**

2. If m$\angle BAC = 9x$, find x. **5**

3. If $AB = 2x + 4$ and $CD = 3x - 5$, find BC. **22**

4. If m$\angle DAC = y$ and m$\angle BAC = 3x$, find x. **15**

5. If $AB = x^2 - 15$ and $BC = 2x$, find x. **5**

Use rhombus ABCD and the given information to find each measure.

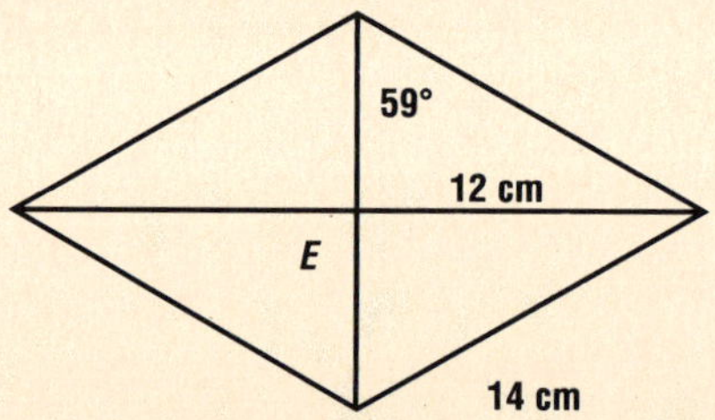

6. m$\angle BCE$ **31**

7. m$\angle BEC$ **90**

8. AC **24 cm**

9. m$\angle ABD$ **56**

10. AD **14 cm**

Determine whether EFGH is a parallelogram, rectangle, rhombus, or square. List all that apply.

11. $E(0, -3)$, $F(-3, 0)$, $G(0, 3)$, $H(3, 0)$
    parallelogram, rectangle, rhombus, square

12. $E(2, 1)$, $F(3, 4)$, $G(7, 2)$, $H(6, -1)$
    parallelogram

NAME ______________________ DATE ____________

6-6 Practice Worksheet

Trapezoids

For Exercises 1-7, refer to the figure. MATH is an isosceles trapezoid with bases $\overline{MA}$ and $\overline{TH}$. Use the figure and the given information to solve each problem.

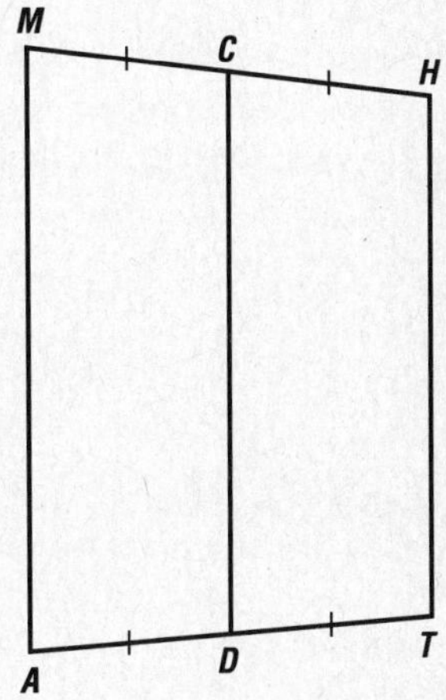

1. If $\overline{MA} = 34$ and $\overline{HT} = 20$, find $\overline{CD}$.

2. If $\overline{HT} = 17.6$ and $\overline{CD} = 28.6$, find $\overline{MA}$.

3. If $\overline{MA} = 23.9$ and $\overline{CD} = 16.4$, find $\overline{HT}$.

4. If $\overline{CD} = x + 12$ and $\overline{MA} + \overline{HT} = 4x + 3$, find x.

5. If $m\angle TAM = 63$, find $m\angle HMA$.

6. If $m\angle HCD = 52$, find $m\angle TDC$.

7. If $m\angle DCM = 2x$, find $m\angle CMA$ in terms of x.

8. If the measure of the median of an isosceles trapezoid is 5.5, what are the possible integral measures for the bases?

9. $\overline{VW}$ is the median of a trapezoid that has bases $\overline{MN}$ and $\overline{PO}$, with V on $\overline{OM}$ and W on $\overline{PN}$. If the vertices of the trapezoid are $M(2, 6)$, $N(4, 6)$ $P(10, 0)$, and $O(0, 0)$, find the coordinates of V and W.

10. $\overline{VW}$ is the median of a trapezoid that has bases $\overline{MN}$ and $\overline{PO}$, with V on $\overline{PM}$ and W on $\overline{ON}$. If four of the points are $M(5, 10)$, $N(9, 10)$, $V(3, 7)$, and $W(11, 7)$, find the coordinates of P and O.

NAME ______________________________ DATE ____________

6-6 Practice Worksheet

Trapezoids

For Exercises 1-7, refer to the figure. MATH is an isosceles trapezoid with bases $\overline{MA}$ and $\overline{TH}$. Use the figure and the given information to solve each problem.

1. If $\overline{MA} = 34$ and $\overline{HT} = 20$, find $\overline{CD}$. **27**

2. If $\overline{HT} = 17.6$ and $\overline{CD} = 28.6$, find $\overline{MA}$. **39.6**

3. If $\overline{MA} = 23.9$ and $\overline{CD} = 16.4$, find $\overline{HT}$. **8.9**

4. If $\overline{CD} = x + 12$ and $\overline{MA} + \overline{HT} = 4x + 3$, find x. **10.5**

5. If $m\angle TAM = 63$, find $m\angle HMA$. **63**

6. If $m\angle HCD = 52$, find $m\angle TDC$. **52**

7. If $m\angle DCM = 2x$, find $m\angle CMA$ in terms of x. **$180 - 2x$**

8. If the measure of the median of an isosceles trapezoid is 5.5, what are the possible integral measures for the bases?
1, 10; 2, 9; 3, 8; 4, 7; 5, 6

9. $\overline{VW}$ is the median of a trapezoid that has bases $\overline{MN}$ and $\overline{PO}$, with V on $\overline{OM}$ and W on $\overline{PN}$. If the vertices of the trapezoid are $M(2, 6)$, $N(4, 6)$ $P(10, 0)$, and $O(0, 0)$, find the coordinates of V and W. **$V(1, 3)$, $W(7, 3)$**

10. $\overline{VW}$ is the median of a trapezoid that has bases $\overline{MN}$ and $\overline{PO}$, with V on $\overline{PM}$ and W on $\overline{ON}$. If four of the points are $M(5, 10)$, $N(9, 10)$, $V(3, 7)$, and $W(11, 7)$, find the coordinates of P and O. **$P(1, 4)$, $O(13, 4)$**

NAME ______________________ DATE ______________

7-1

Practice Worksheet

Properties of Proportions

Solve each proportion using cross products.

1. $\frac{3}{5} = \frac{x}{15}$

2. $\frac{20 - x}{x} = \frac{6}{4}$

3. $\frac{x + 1}{5} = \frac{x - 1}{2}$

4. $\frac{x}{x - 3} = \frac{x + 4}{x}$

5. $\frac{x + 1}{6} = \frac{x - 1}{x}$

6. $\frac{1}{x} = \frac{6}{x + 9}$

7. $\frac{x}{x + 8} = \frac{2}{3}$

8. $\frac{4}{12} = \frac{x + 2}{2x + 5}$

In the figure at the right, $\frac{AC}{CD} = \frac{CE}{CB}$. Use proportions to complete the table.

	AC	BC	AB	CE	ED	DC
9.	10	4		8		
10.	12			10		9

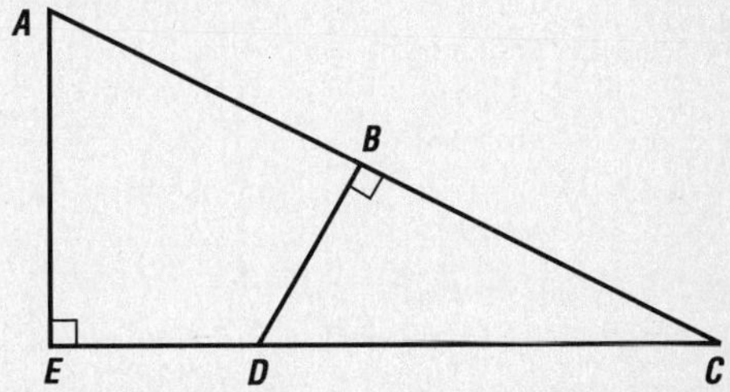

Glencoe Division, Macmillan/McGraw-Hill

NAME ______________________ DATE ____________

Practice Worksheet

Properties of Proportions

Solve each proportion using cross products.

1. $\frac{3}{5} = \frac{x}{15}$ 9

2. $\frac{20 - x}{x} = \frac{6}{4}$ 8

3. $\frac{x + 1}{5} = \frac{x - 1}{2}$ $\frac{7}{3}$

4. $\frac{x}{x - 3} = \frac{x + 4}{x}$ 12

5. $\frac{x + 1}{6} = \frac{x - 1}{x}$ 2, 3

6. $\frac{1}{x} = \frac{6}{x + 9}$ $\frac{9}{5}$

7. $\frac{x}{x + 8} = \frac{2}{3}$ 16

8. $\frac{4}{12} = \frac{x + 2}{2x + 5}$ −1

In the figure at the right, $\frac{AC}{CD} = \frac{CE}{CB}$. Use proportions to complete the table.

	AC	BC	AB	CE	ED	DC
9.	10	4	6	8	3	5
10.	12	7.5	4.5	10	1	9

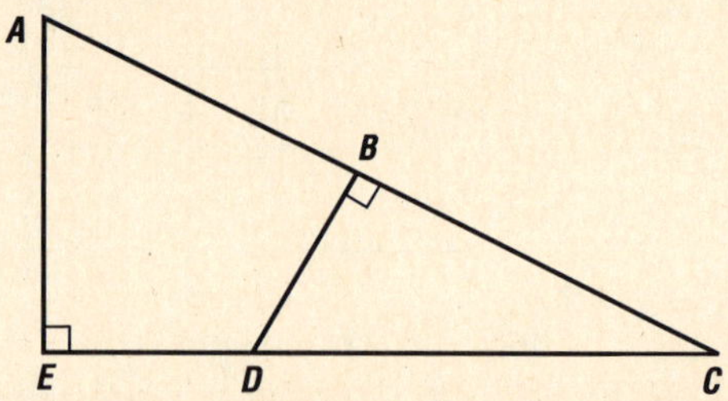

NAME ______________________ DATE ____________

7-2

Practice Worksheet

Applications of Proportions

Points M and N are midpoints of their respective segments.

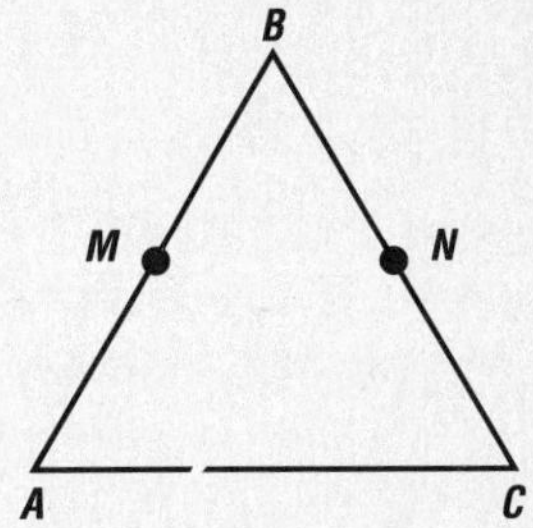

1. Find the ratio AM to MB.

2. Find the ratio AM to AB.

3. If $\triangle ABC$ is an equilateral triangle, find the ratio of AM to BN.

Suppose the measures of corresponding sides of $\triangle ABC$ and $\triangle DEF$ are proportional.

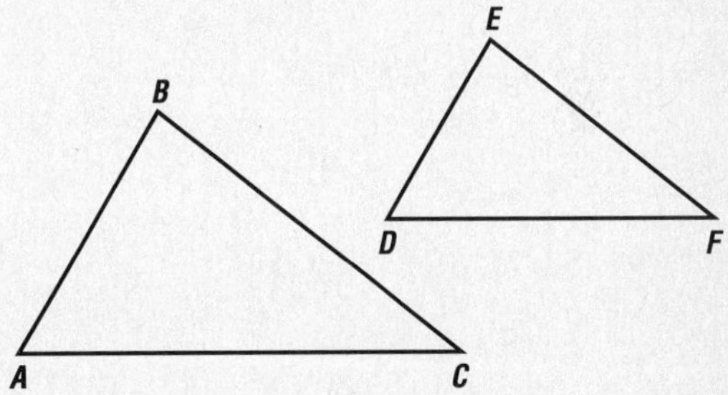

4. If $BC = 20$, $DE = 9$, and $EF = 15$, find AB.

5. If $BC = 24$, $EF = 9$, $AC = y + 30$, and $DF = y$, find AC.

6. If $AB = 5x + 3$, $BC = x + 1$, $DE = 4x$ and $EF = x$, find EF.

Use a proportion to solve each problem.

7. The ratio of seniors to juniors in the Math Club is 2:3. If there are 21 juniors, how many seniors are in the club?

8. A 15-foot building casts a 9-foot shadow. How tall is a building that casts a 30-foot shadow at the same time?

9. A photo that is 3 inches wide and 5 inches high was enlarged so that it is 12 inches wide. How high is the enlargement?

10. Philip has been eating 2 hamburgers every 5 days. At that rate, how many hamburgers will he eat in 30 days.

Glencoe Division, Macmillan/McGraw-Hill

7-2 NAME ______________________ DATE ______________

Practice Worksheet

Applications of Proportions

Points M and N are midpoints of their respective segments.

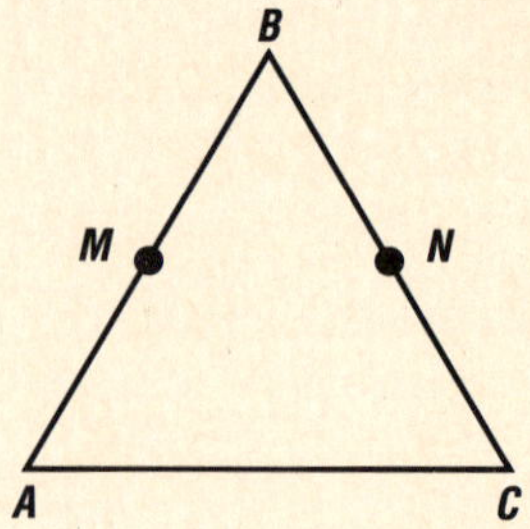

1. Find the ratio AM to MB. $\frac{1}{1}$

2. Find the ratio AM to AB. $\frac{1}{2}$

3. If $\triangle ABC$ is an equilateral triangle, find the ratio of AM to BN. $\frac{1}{1}$

Suppose the measures of corresponding sides of $\triangle ABC$ and $\triangle DEF$ are proportional.

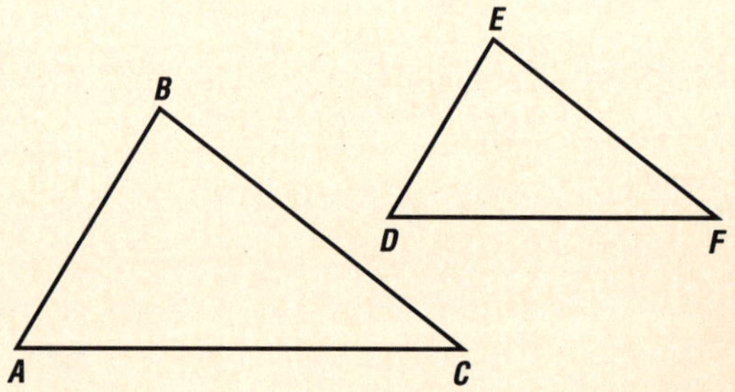

4. If $BC = 20$, $DE = 9$, and $EF = 15$, find AB. **12**

5. If $BC = 24$, $EF = 9$, $AC = y + 30$, and $DF = y$, find AC. **48**

6. If $AB = 5x + 3$, $BC = x + 1$, $DE = 4x$ and $EF = x$, find EF. **1**

Use a proportion to solve each problem.

7. The ratio of seniors to juniors in the Math Club is 2:3. If there are 21 juniors, how many seniors are in the club? **14 seniors**

8. A 15-foot building casts a 9-foot shadow. How tall is a building that casts a 30-foot shadow at the same time? **50 ft**

9. A photo that is 3 inches wide and 5 inches high was enlarged so that it is 12 inches wide. How high is the enlargement? **20 in.**

10. Philip has been eating 2 hamburgers every 5 days. At that rate, how many hamburgers will he eat in 30 days. **12 hamburgers**

Glencoe Division, Macmillan/McGraw-Hill

NAME ____________ DATE ____________

7-3 Practice Worksheet

Similar Polygons

In the figure at the right, $\triangle ABC$ is similar to $\triangle DEF$.

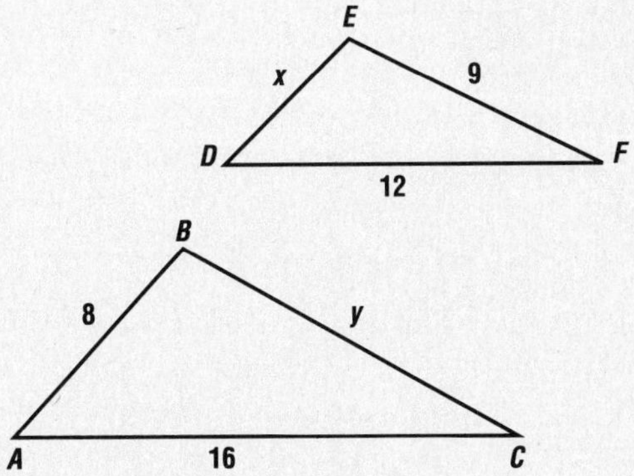

1. Write three equal ratios to show corresponding sides are proportional.

2. Find the value of x.

3. Find the value of y.

4. Find the ratio $\frac{m\angle A}{m\angle D}$.

In the figure at the right, quadrilateral $ABCD$ is similar to quadrilateral $EFGH$.

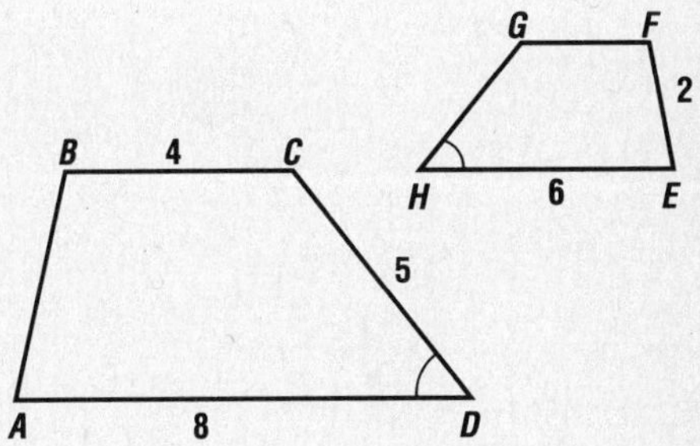

5. Write four equal ratios to show corresponding sides are proportional.

6. Find AB.

7. Find HG.

8. Find FG.

9. The sum of the measures of $\angle A$ and $\angle C$ equals the sum of the measures of which two angles of quadrilateral $EFGH$?

NAME ______________________ DATE ____________

7-3

Practice Worksheet

Similar Polygons

In the figure at the right, $\triangle ABC$ is similar to $\triangle DEF$.

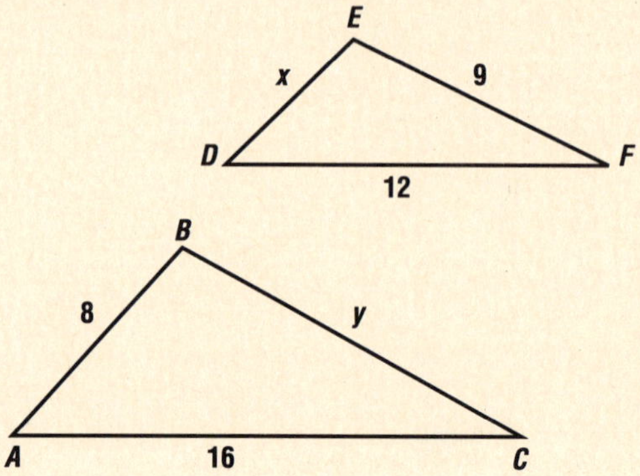

1. Write three equal ratios to show corresponding sides are proportional.
$\frac{AB}{DE} = \frac{AC}{DF} = \frac{BC}{EF}$

2. Find the value of x. 6

3. Find the value of y. 12

4. Find the ratio $\frac{m\angle A}{m\angle D}$. $\frac{1}{1}$

In the figure at the right, quadrilateral $ABCD$ is similar to quadrilateral $EFGH$.

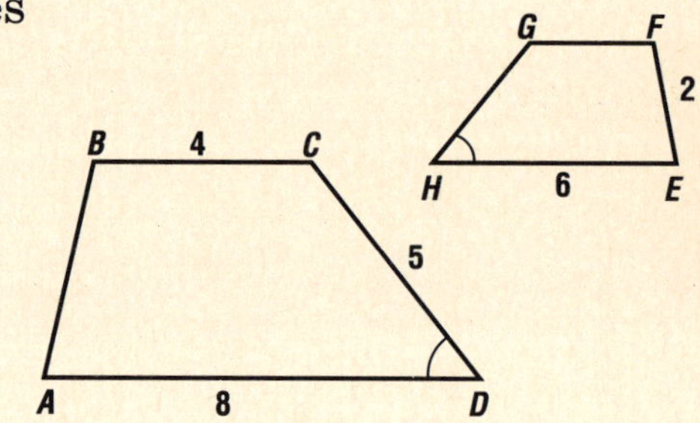

5. Write four equal ratios to show corresponding sides are proportional. $\frac{AB}{EF} = \frac{BC}{FG} = \frac{CD}{GH} = \frac{DA}{HE}$

6. Find AB. $2\frac{2}{3}$

7. Find HG. $3\frac{3}{4}$

8. Find FG. 3

9. The sum of the measures of $\angle A$ and $\angle C$ equals the sum of the measures of which two angles of quadrilateral $EFGH$? $\angle E$ and $\angle G$

NAME ______________________ DATE ____________

7-4

Practice Worksheet

Similar Triangles

Identify the similar triangles in each figure. Explain why they are similar and find the missing measures x and y.

1.

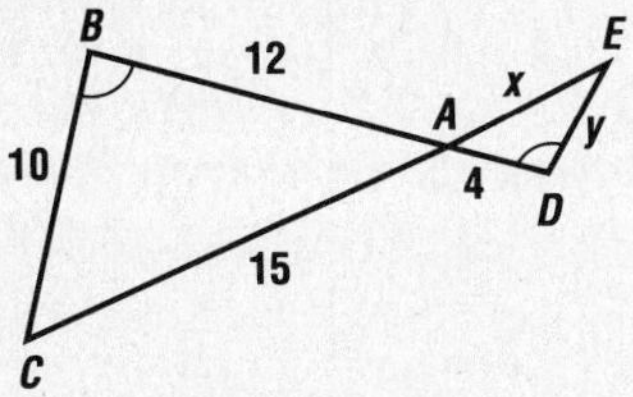

2.

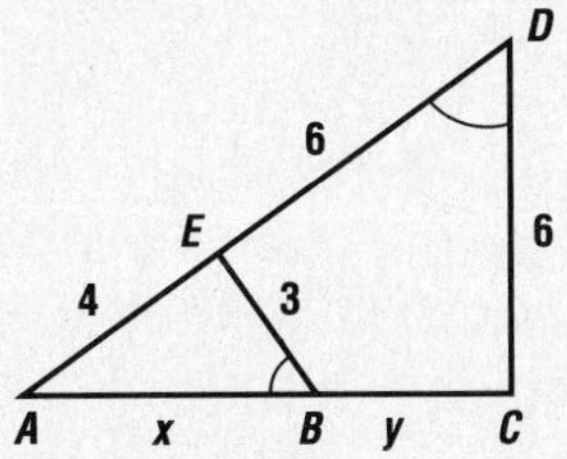

3.

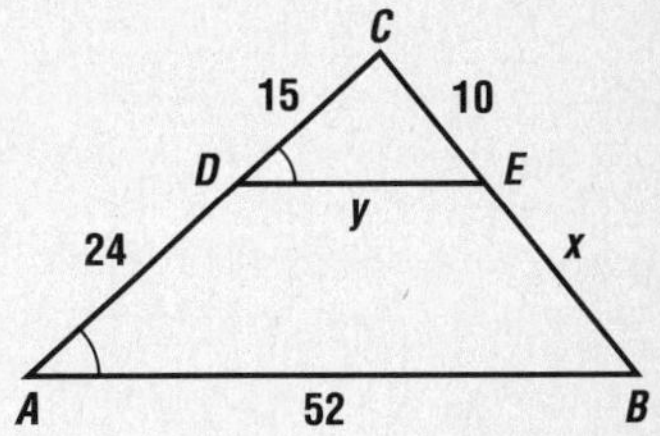

Write a two-column proof for each.

4. Given: $\overline{AB} \parallel \overline{EF}$
$\overline{AC} \parallel \overline{DF}$
Prove: $\triangle ABC \sim \triangle FED$

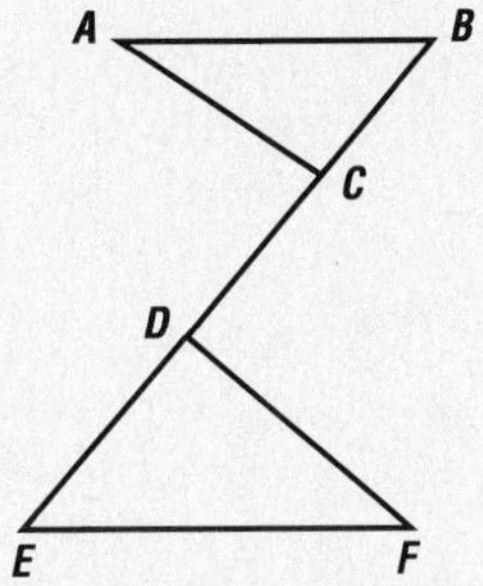

Statements	Reasons

5. Given: $\overline{AB} \perp \overline{BD}$
$\overline{EC} \perp \overline{BD}$
Prove: $\triangle BDA \sim \triangle CDE$

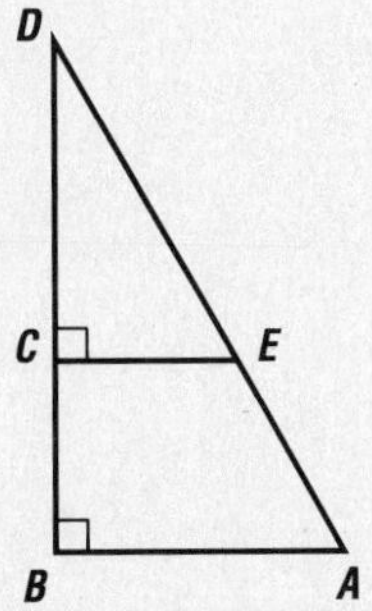

Statements	Reasons

Glencoe Division, Macmillan/McGraw-Hill

NAME ______________________________ DATE ______________

7-4

Practice Worksheet

Similar Triangles

Identify the similar triangles in each figure. Explain why they are similar and find the missing measures x and y.

1.

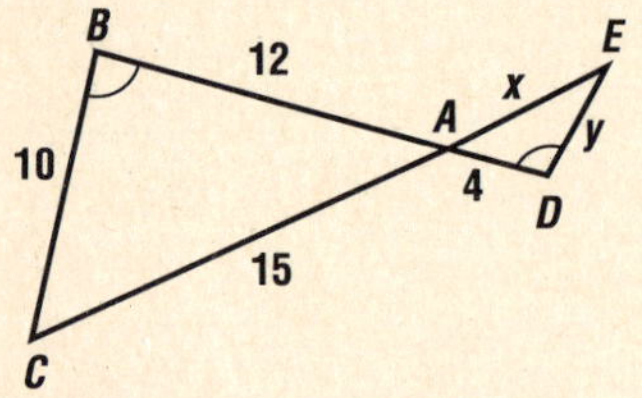

2.

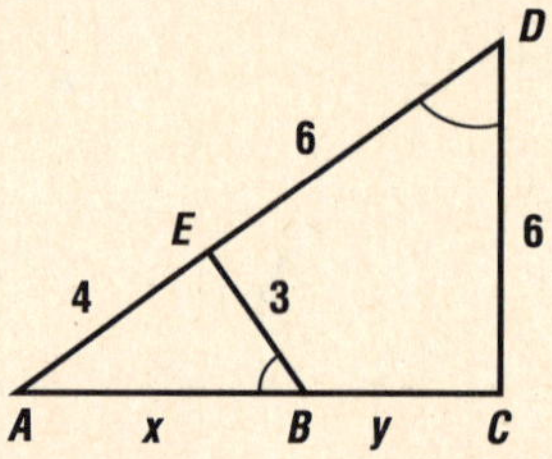

3.

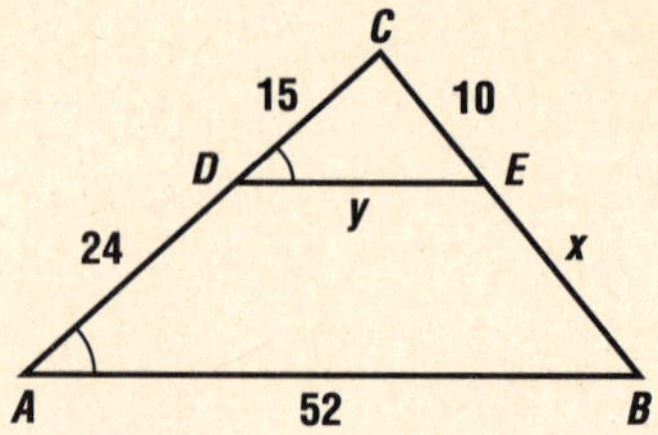

1. △ABC ~ △ADE; AA; 5, $3\frac{1}{3}$

2. △ABE ~ △ADC; AA; 5, 3

3. △ABC ~ △DEC; AA; 14, 20

Write a two-column proof for each.

4. Given: $\overline{AB} \parallel \overline{EF}$

$\overline{AC} \parallel \overline{DF}$

Prove: $\triangle ABC \sim \triangle FED$

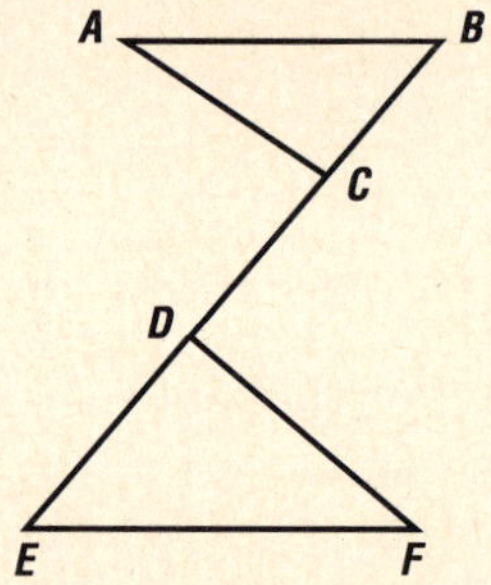

Statements	Reasons
a. $\overline{AB} \parallel \overline{EF}$	a. Given
b. $\angle B \cong \angle E$	b. If two ∥ lines are cut by a transversal, alt. int. ∠s are ≅.
c. $\overline{AC} \parallel \overline{DF}$	c. Given
d. $\angle ACB \cong \angle FDE$	d. If two ∥ lines are cut by a transversal, alt. ext. ∠s are ≅.
e. $\triangle ABC \sim \triangle FED$	e. AA Similarity

5. Given: $\overline{AB} \perp \overline{BD}$

$\overline{EC} \perp \overline{BD}$

Prove: $\triangle BDA \sim \triangle CDE$

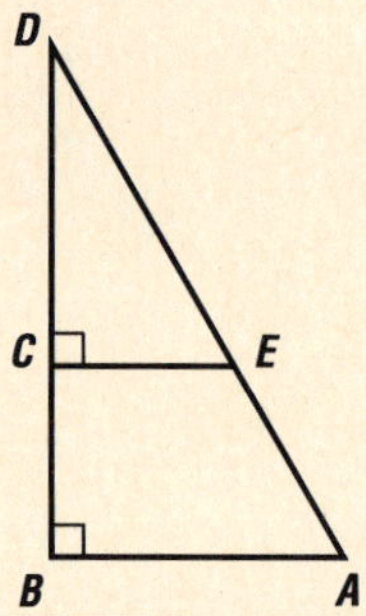

Statements	Reasons
a. $\overline{AB} \perp \overline{BD}$, $\overline{EC} \perp \overline{BD}$	a. Given
b. ∠ABD, ∠ECD are rt. ∠s.	b. ⊥ lines form four rt. ∠s.
c. $\angle ABD \cong \angle ECD$	c. All rt. ∠s are ≅.
d. $\angle D \cong \angle D$	d. Congruence of ∠s is reflexive.
e. $\triangle BDA \sim \triangle CDF$	e. AA Similarity

NAME ________________ DATE ________

7-5 **Practice Worksheet**

Proportional Parts

In △XYZ, determine whether it is always true that $\overline{AB} \parallel \overline{YZ}$ under the given conditions.

1. $XA = 6$
$AY = 4$
$XB = 8$
$BZ = 5$

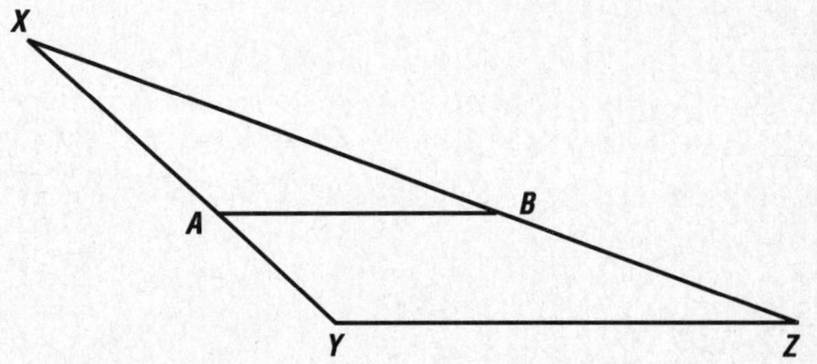

2. $XB = 3$
$BZ = 2$
$AB = 6$
$YZ = 10$

Using the figure at the right, determine the values of x and y if $\overline{JG} \parallel \overline{RQ}$.

3. $PJ = 6$
$JG = 5$
$PG = 4$
$GQ = x$
$RQ = x + 6$
$JR = y$

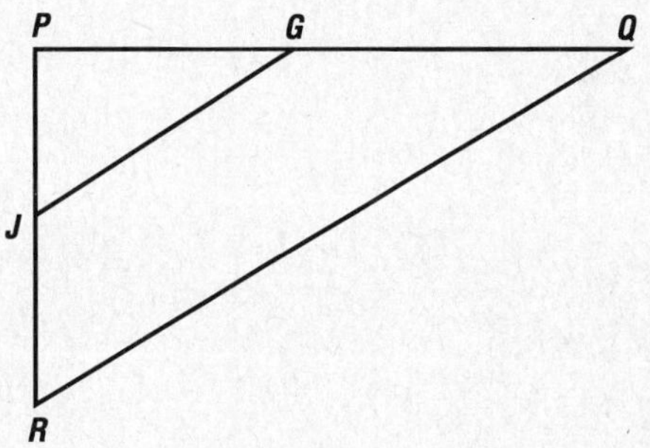

4. $RQ = 10$
$JG = 8$
$PJ = 8x - 5$
$JR = x$
$PG = 3y + 2$
$QG = y$

5. In the figure at the right, $\overleftrightarrow{YA} \parallel \overleftrightarrow{OE} \parallel \overleftrightarrow{BR}$. Find the values of x and y if $YO = 4$, $ER = 16$, and $AR = 24$.

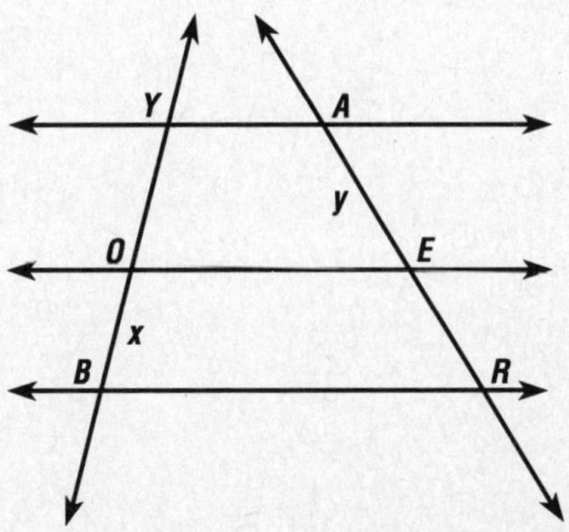

Glencoe Division, Macmillan/McGraw-Hill

NAME ________________________________ DATE ______________

7-5

Practice Worksheet

Proportional Parts

In △XYZ, determine whether it is always true that $\overline{AB} \parallel \overline{YZ}$ under the given conditions.

3/5 = 6/10

1. $XA = 6$
$AY = 4$
$XB = 8$
$BZ = 5$ **no**

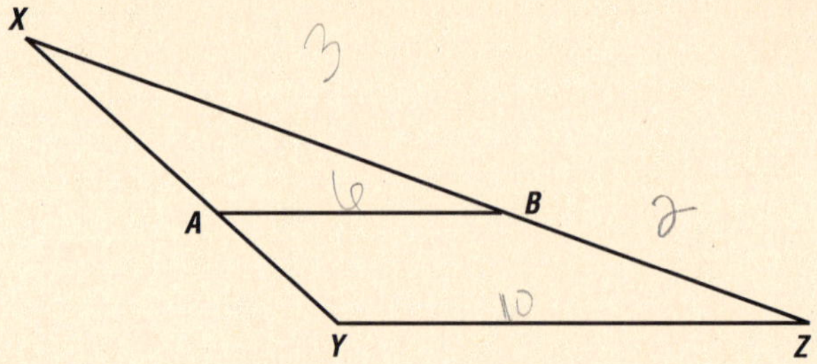

2. $XB = 3$
$BZ = 2$
$AB = 6$
$YZ = 10$ **no**

Using the figure at the right, determine the values of x and y if $\overline{JG} \parallel \overline{RQ}$.

3. $PJ = 6$
$JG = 5$
$PG = 4$
$GQ = x$
$RQ = x + 6$
$JR = y$ **4, 6**

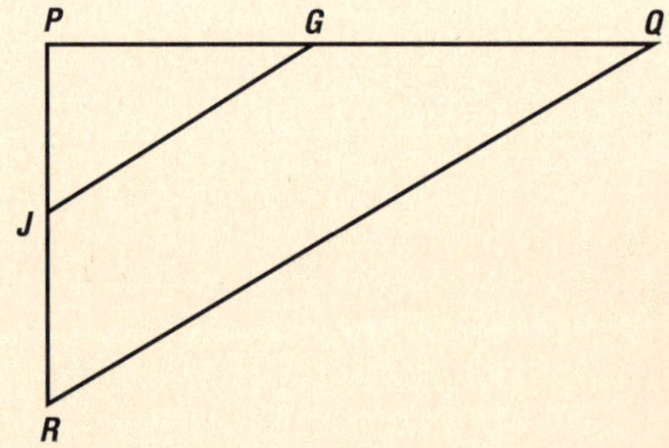

4. $RQ = 10$
$JG = 8$
$PJ = 8x - 5$
$JR = x$
$PG = 3y + 2$
$QG = y$ **$\frac{5}{4}$, 2**

5. In the figure at the right, $\overleftrightarrow{YA} \parallel \overleftrightarrow{OE} \parallel \overleftrightarrow{BR}$. Find the values of x and y if $YO = 4$, $ER = 16$, and $AR = 24$. **8, 8**

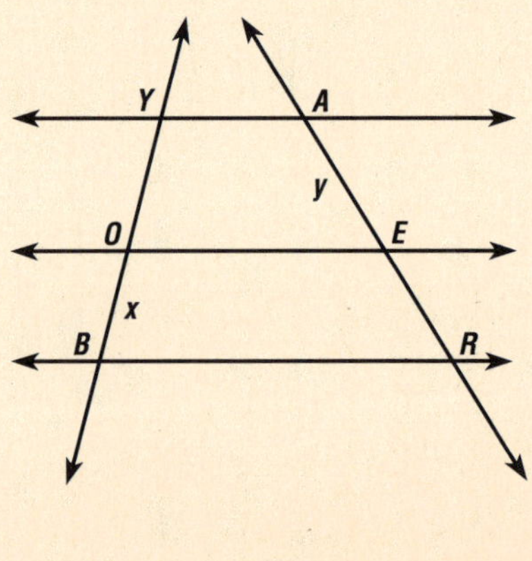

NAME ______________________ DATE ______________

7-6

Practice Worksheet

Parts of Similar Triangles

In the figure at the right, $\triangle ABC \sim \triangle DEF$, $\overline{BR} \cong \overline{RC}$, and $\overline{ES} \cong \overline{SF}$. Find the value of x.

1. $BC = 24$
$EF = 15$
$AR = x$
$DS = x - 6$

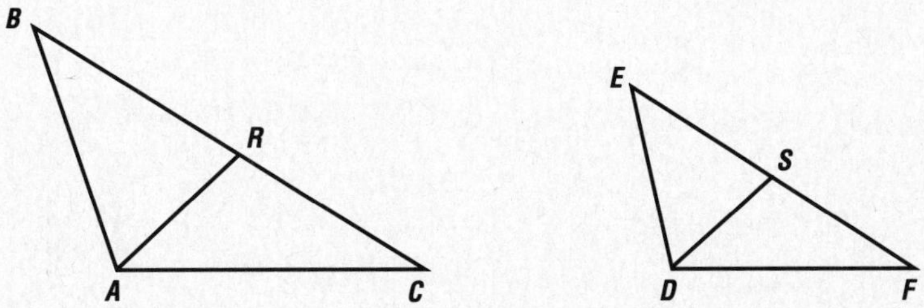

2. $AB = 2x + 5$
$DE = x + 7$
$AR = 24$
$DS = 18$

Find the value of x.

3.

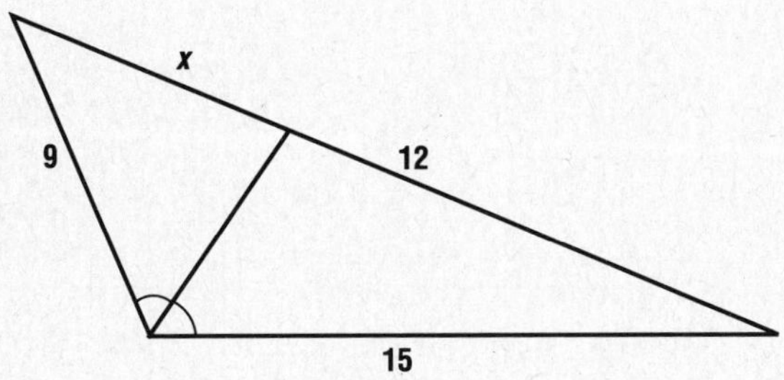

4.

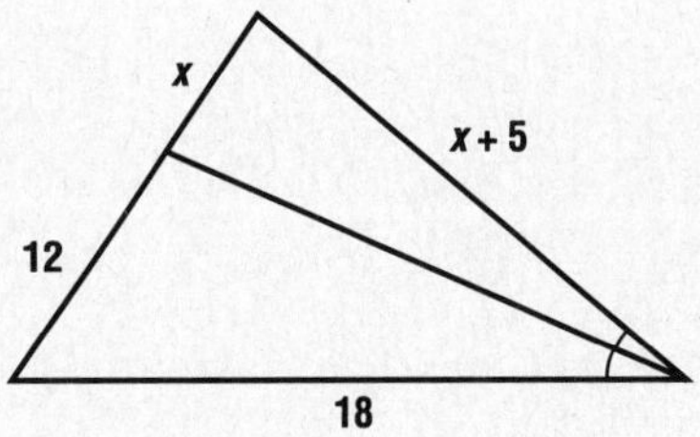

In the figure at the right, $\triangle ABC \sim \triangle DEF$, and $\overline{BX}$ and $\overline{EY}$ are altitudes. Find the value of x.

5. $AB = 25$
$DE = 16$
$BX = 18$
$EY = x$

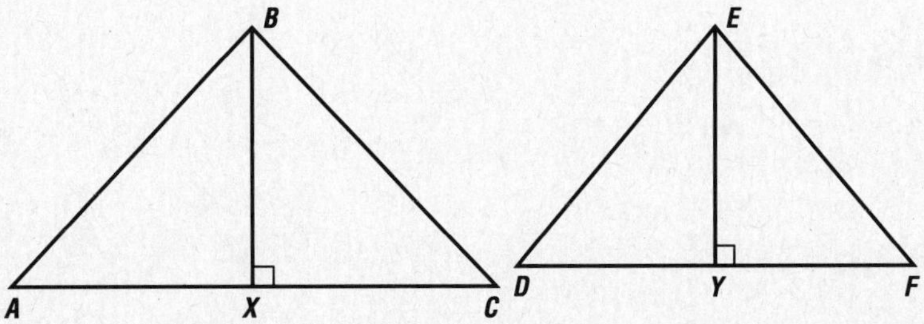

6. $AB = 30$
$DE = 25$
$BX = 2x + 5$
$EY = x + 10$

NAME ______________________ DATE ____________

7-6 Practice Worksheet

Parts of Similar Triangles

In the figure at the right, $\triangle ABC \sim \triangle DEF$, $\overline{BR} \cong \overline{RC}$, and $\overline{ES} \cong \overline{SF}$. Find the value of x.

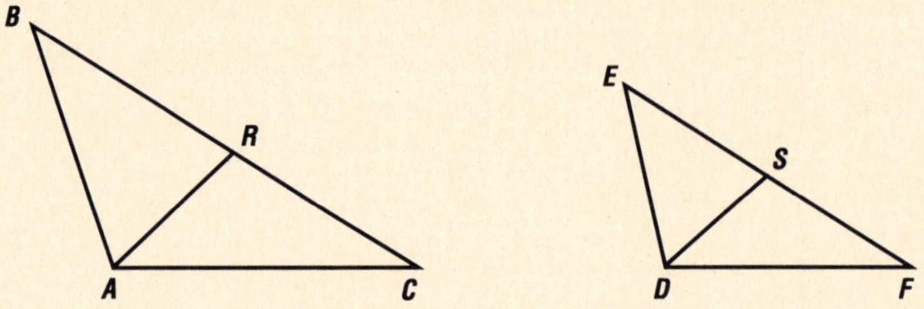

1. $BC = 24$
   $EF = 15$
   $AR = x$
   $DS = x - 6$ **16**

2. $AB = 2x + 5$
   $DE = x + 7$
   $AR = 24$
   $DS = 18$ **6.5**

Find the value of x.

3.

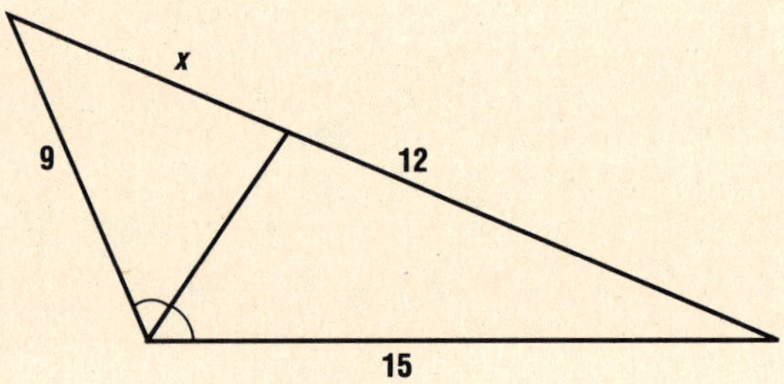

7.2

4.

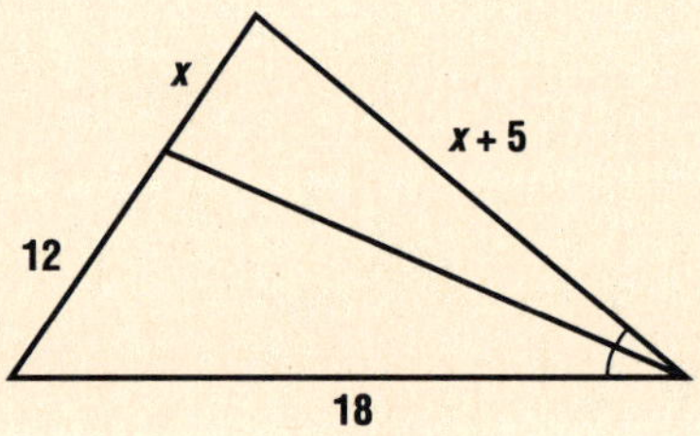

10

In the figure at the right, $\triangle ABC \sim \triangle DEF$, and $\overline{BX}$ and $\overline{EY}$ are altitudes. Find the value of x.

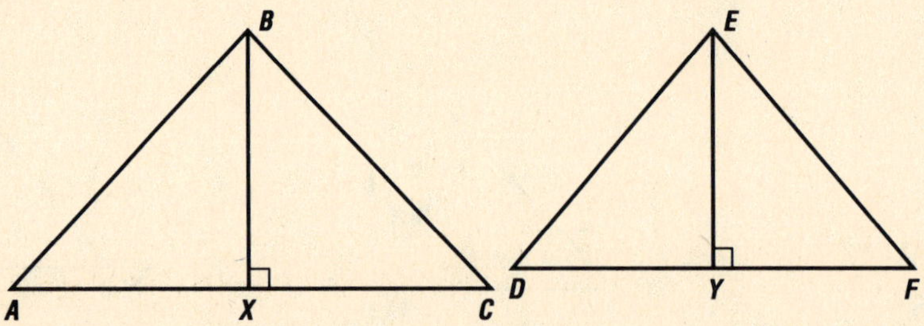

5. $AB = 25$
   $DE = 16$
   $BX = 18$
   $EY = x$ **11.52**

6. $AB = 30$
   $DE = 25$
   $BX = 2x + 5$
   $EY = x + 10$ **8.75**

7-7

NAME ______________________ DATE ______________

Practice Worksheet

Problem-Solving Strategy: Solve a Simpler Problem

Solve. Use any strategy.

1. Justin has 20 coins in his pocket. The change is worth $1.25. What combinations of coins might Justin have?

2. The hundreds digit of a four-digit whole number is the sum of the thousands digit and the tens digit. The tens digit is the sum of the thousands digit and the ones digit. The sum of the hundreds digit and the ones digit is 10. The sum of all four digits is 17.

3. Find the least whole number with exactly six divisors, itself included.

4. How many diagonals can be drawn for a polygon with 15 sides?

5. For a research project on repeating decimals, Howard calculated reciprocals of whole numbers from 1 to 100. He told Marion that for $\frac{1}{97}$, the repeating string of digits was 103 digits long. Marion said she was sure his decimal was wrong. Was she correct? Explain.

7-7

NAME ______________________________ DATE ______________

Practice Worksheet

Problem-Solving Strategy: Solve a Simpler Problem

Solve. Use any strategy.

1. Justin has 20 coins in his pocket. The change is worth \$1.25. What combinations of coins might Justin have? **half dollar, quarter, 8 nickels, 10 pennies or 9 dimes, 6 nickels, 5 pennies**

2. The hundreds digit of a four-digit whole number is the sum of the thousands digit and the tens digit. The tens digit is the sum of the thousands digit and the ones digit. The sum of the hundreds digit and the ones digit is 10. The sum of all four digits is 17. **2753**

3. Find the least whole number with exactly six divisors, itself included. **12**

4. How many diagonals can be drawn for a polygon with 15 sides? **90**

5. For a research project on repeating decimals, Howard calculated reciprocals of whole numbers from 1 to 100. He told Marion that for $\frac{1}{97}$, the repeating string of digits was 103 digits long. Marion said she was sure his decimal was wrong. Was she correct? Explain. **Yes, she was correct. There are only 96 nonzero remainders possible in the division process for $\frac{1}{97}$.**

NAME ______________________ DATE ____________

8-1 Practice Worksheet

The Geometric Mean

Find the geometric mean between each pair of numbers.

1. 5 and 10

2. 3 and 27

3. 6 and $\frac{1}{2}$

4. 16 and $\frac{1}{9}$

Find the values of x and y.

5.

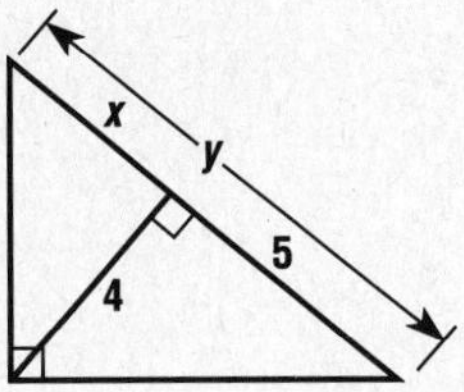

6.

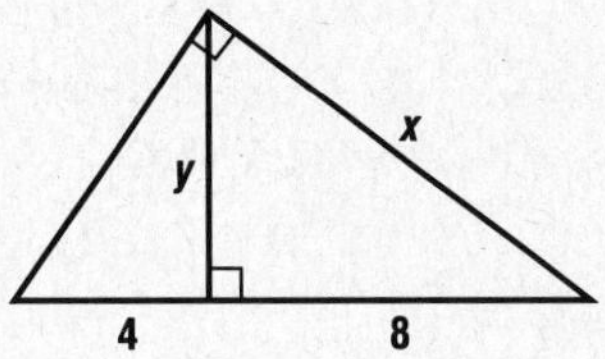

7.

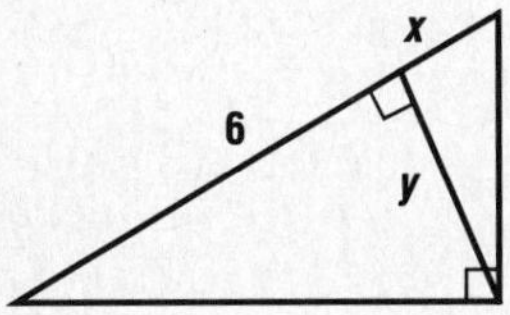

8.

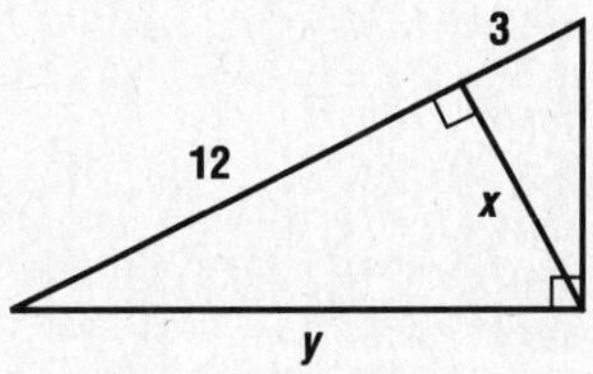

9.

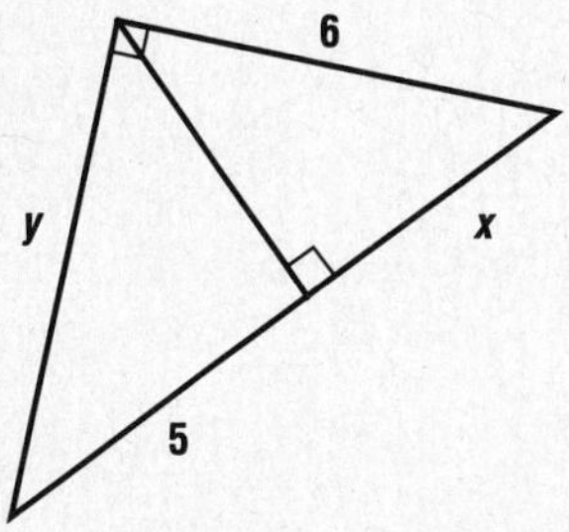

10.

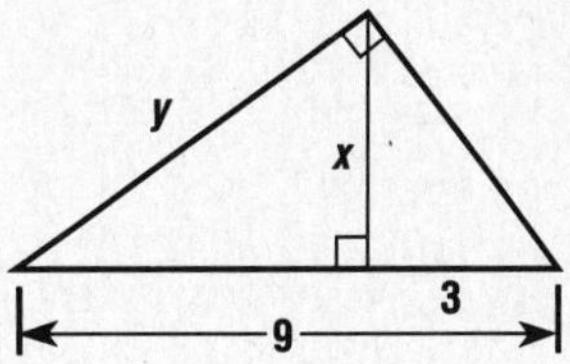

11.

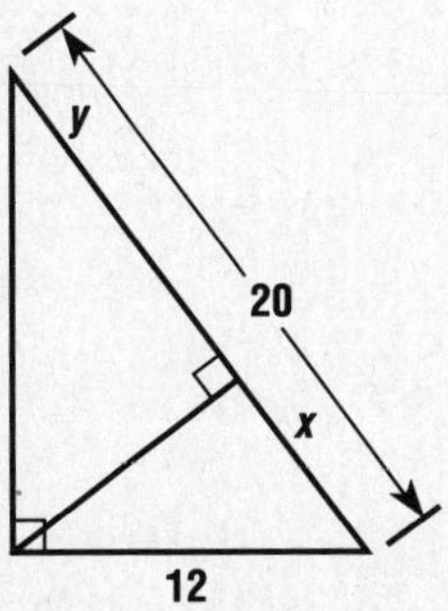

12.

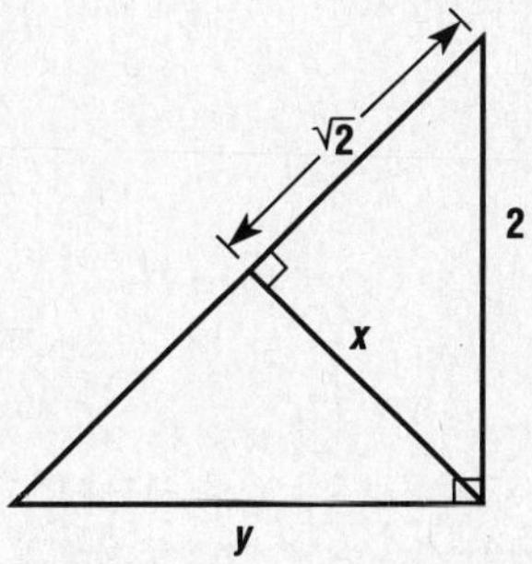

13. Find AC and AB if $AD = 9$ and $BC = 6$.

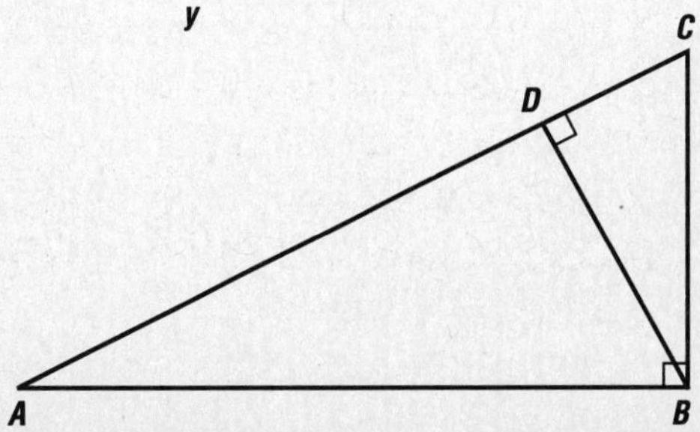

Glencoe Division, Macmillan/McGraw-Hill

NAME ______________________ DATE ____________

8-1

Practice Worksheet

The Geometric Mean

Find the geometric mean between each pair of numbers.

1. 5 and 10 $5\sqrt{2} \approx 7.1$

2. 3 and 27 9

3. 6 and $\frac{1}{2}$ $\sqrt{3} \approx 1.7$

4. 16 and $\frac{1}{9}$ $\frac{4}{3}$

Find the values of x and y.

5.

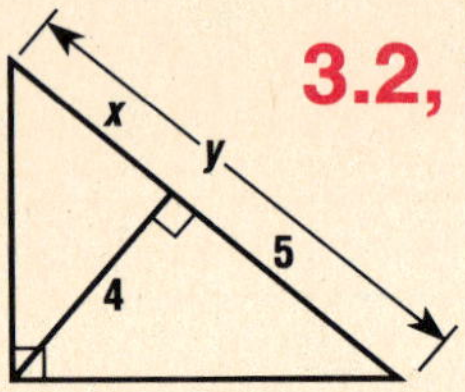

3.2, 8.2

6.

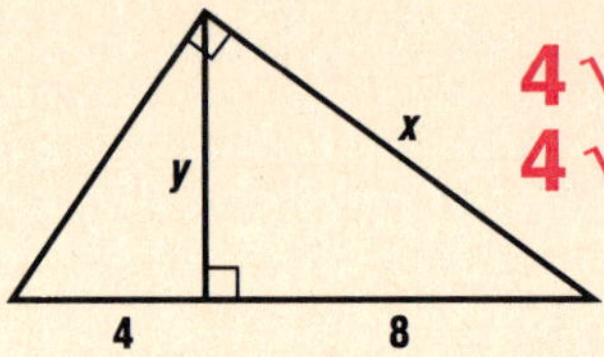

$4\sqrt{6} \approx 9.8$, $4\sqrt{2} \approx 5.7$

7.

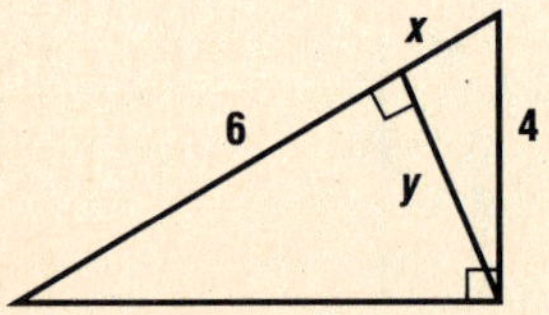

2, $2\sqrt{3} \approx 3.5$

8.

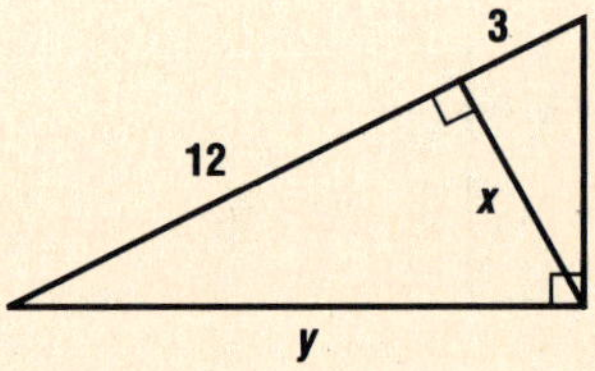

6, $6\sqrt{5} \approx 13.4$

9.

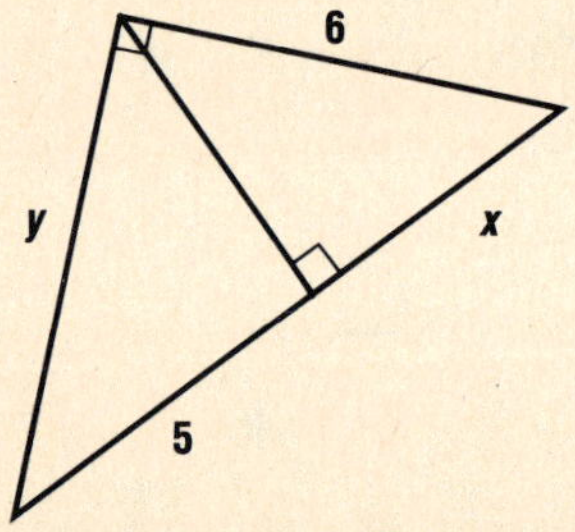

4, $3\sqrt{5} \approx 6.7$

10.

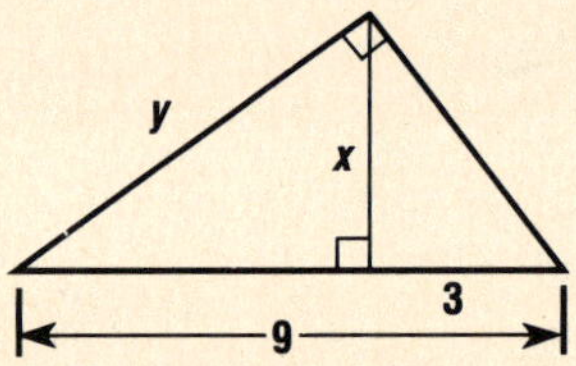

$3\sqrt{2} \approx 4.2$, $3\sqrt{6} \approx 7.3$

11.

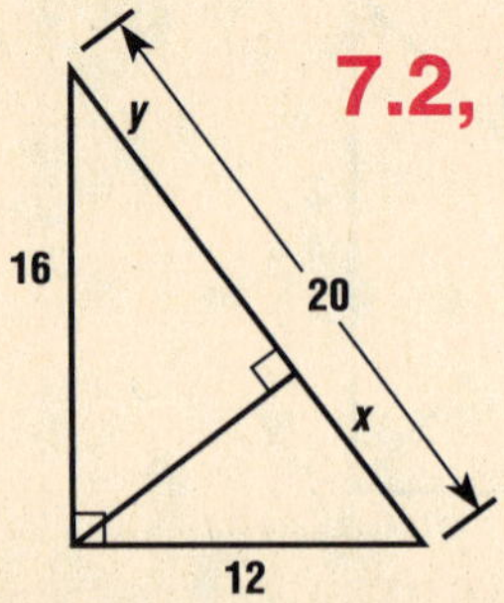

7.2, 12.8

12.

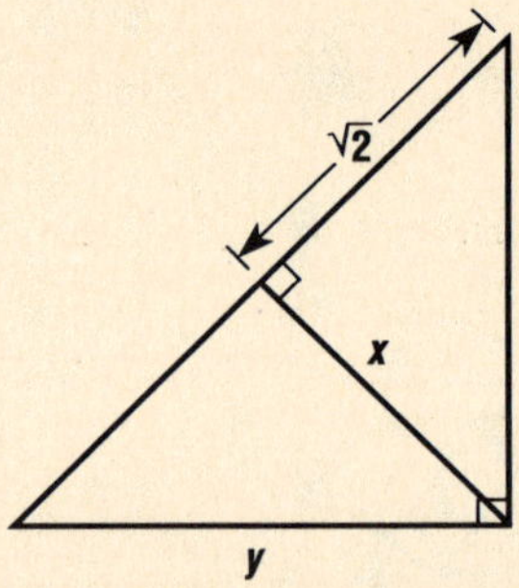

$\sqrt{2} \approx 1.4$, 2

13. Find *AC* and *AB* if *AD* = 9 and *BC* = 6.
12, $6\sqrt{3} \approx 10.4$

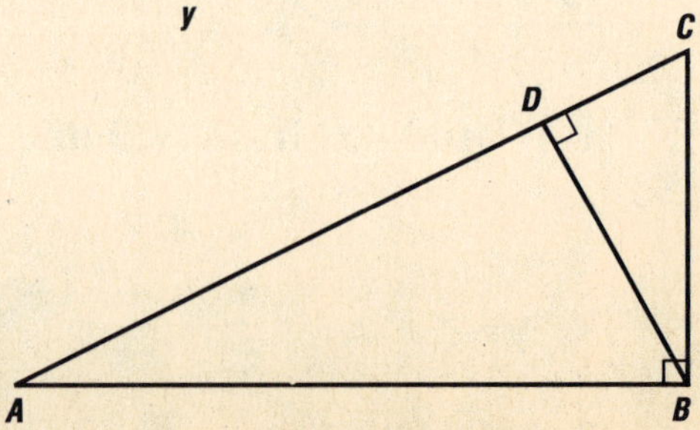

8-2

NAME ______________________ DATE ______________

Practice Worksheet

The Pythagorean Theorem

Determine whether a triangle with sides having the given measures is a right triangle.

1. 14, 48, 50

2. 50, 75, 85

3. 15, 36, 39

4. 45, 60, 80

Find the value of x. Round your answer to the nearest tenth.

5.

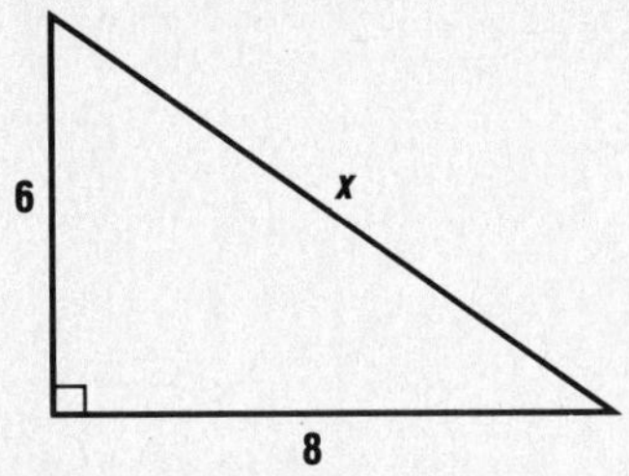

6.

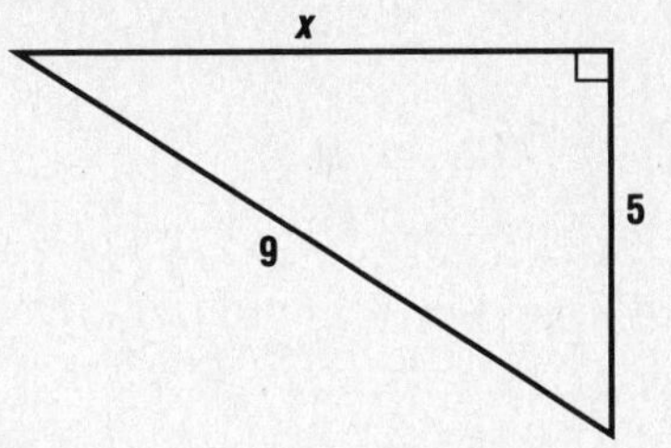

7.

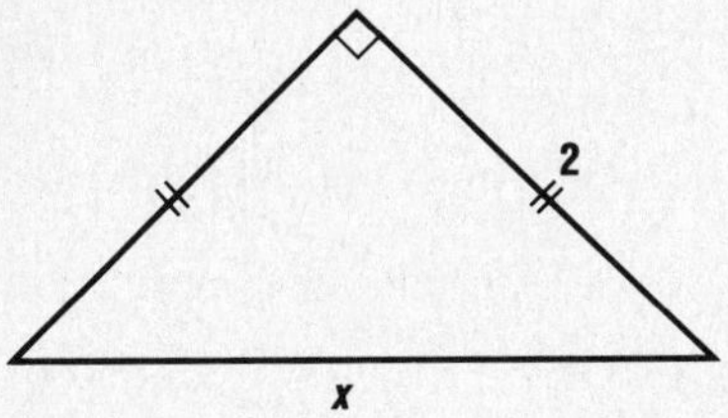

8.

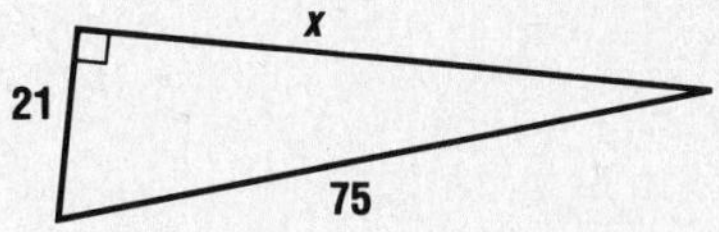

9.

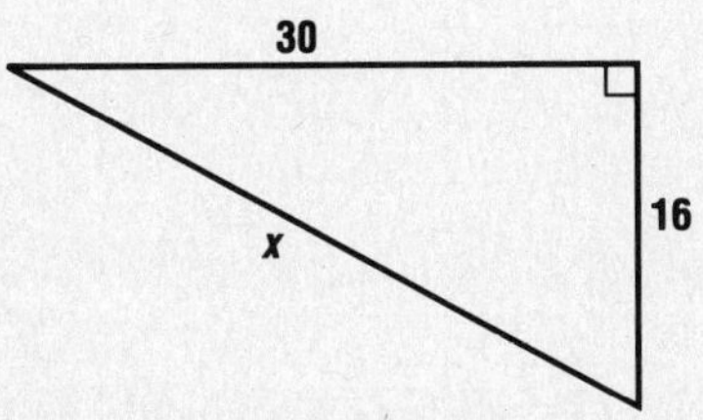

10.

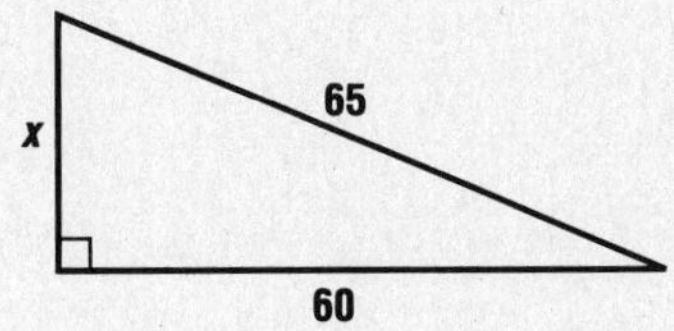

Glencoe Division, Macmillan/McGraw-Hill

NAME ______________________ DATE ____________

8-2

Practice Worksheet

The Pythagorean Theorem

Determine whether a triangle with sides having the given measures is a right triangle.

1. 14, 48, 50 **yes**

2. 50, 75, 85 **no**

3. 15, 36, 39 **yes**

4. 45, 60, 80 **no**

Find the value of x. Round your answer to the nearest tenth.

5.

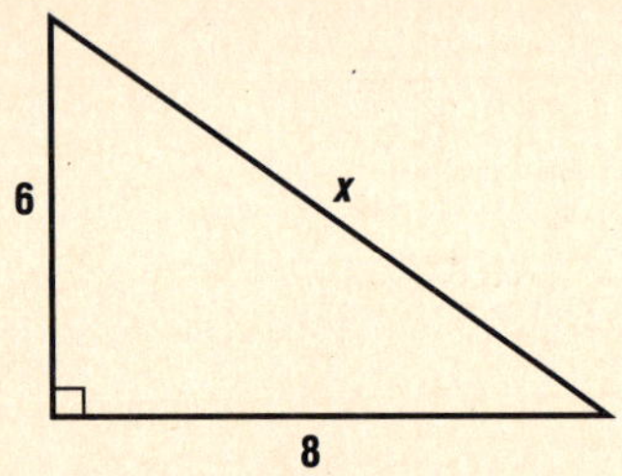

10

6.

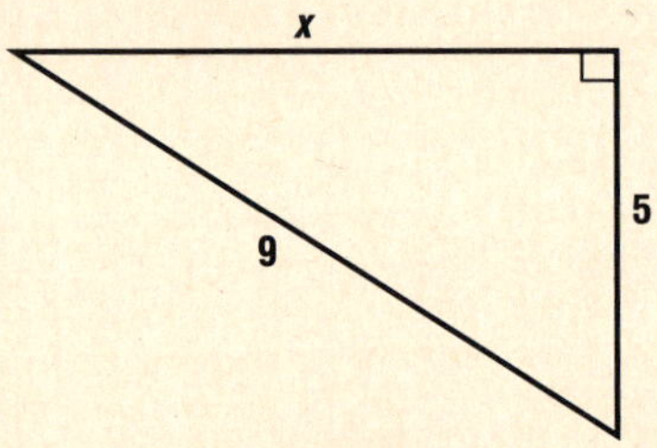

7.5

7.

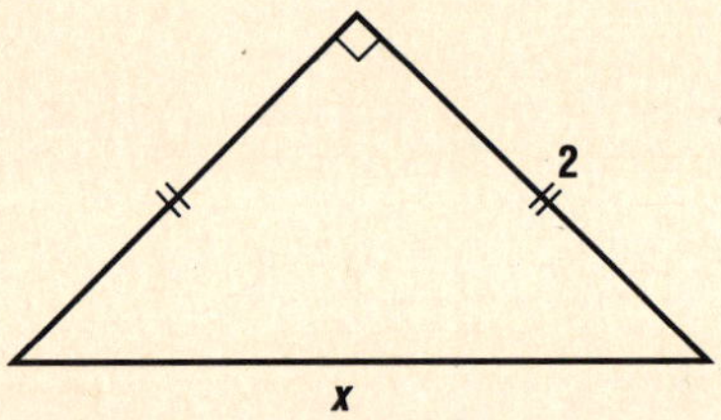

2.8

8.

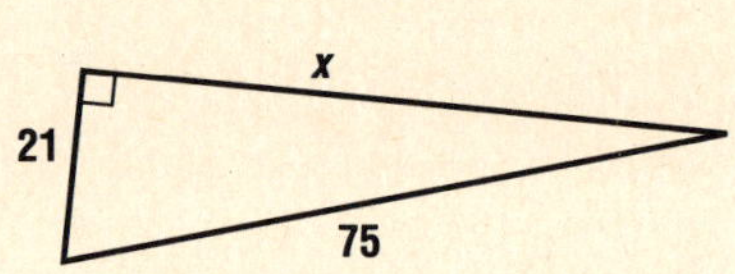

72

9.

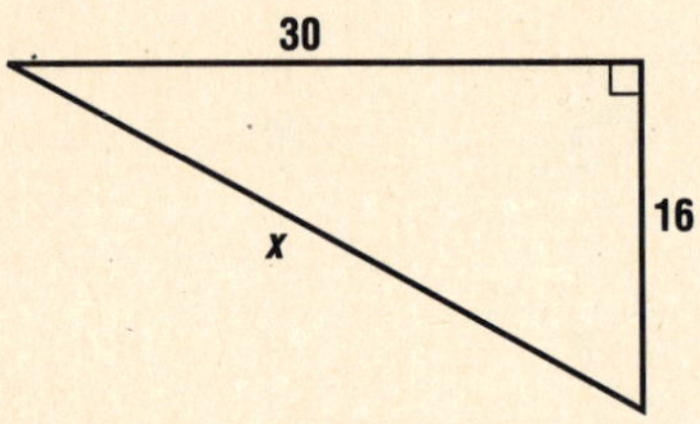

34

10.

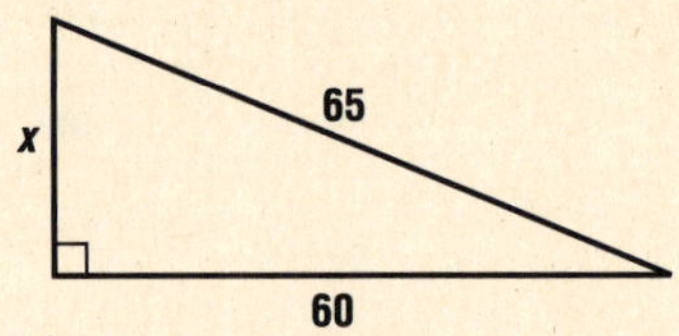

25

NAME ______________________ DATE __________

8-3

Practice Worksheet

Special Right Triangles

Find the values of x and y.

1.

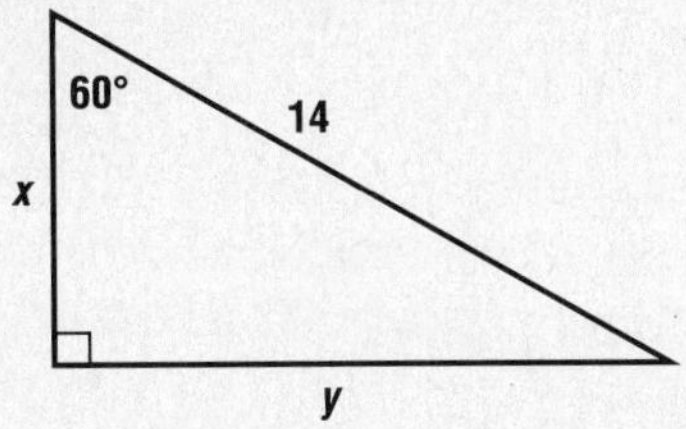

2.

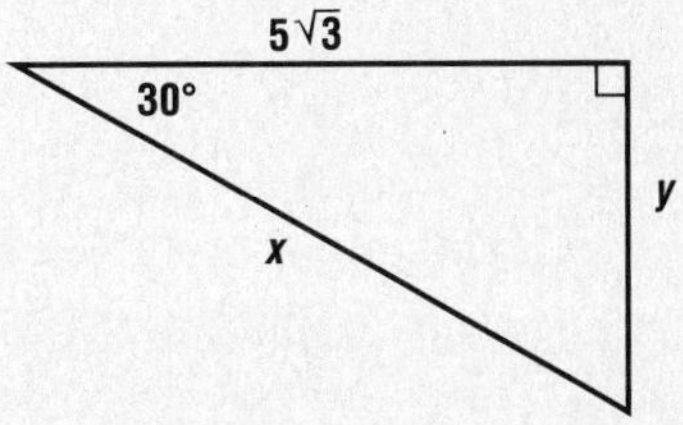

3.

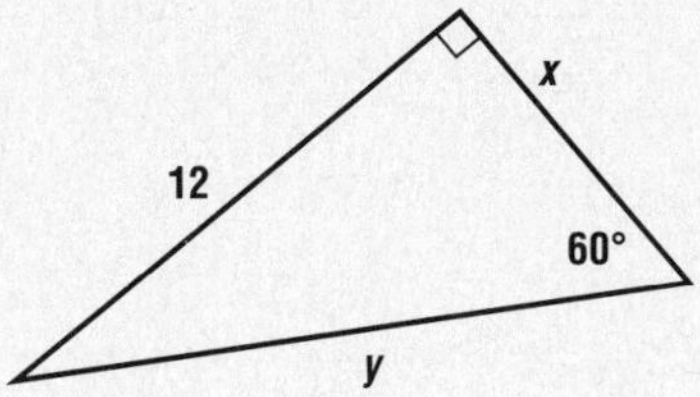

4.

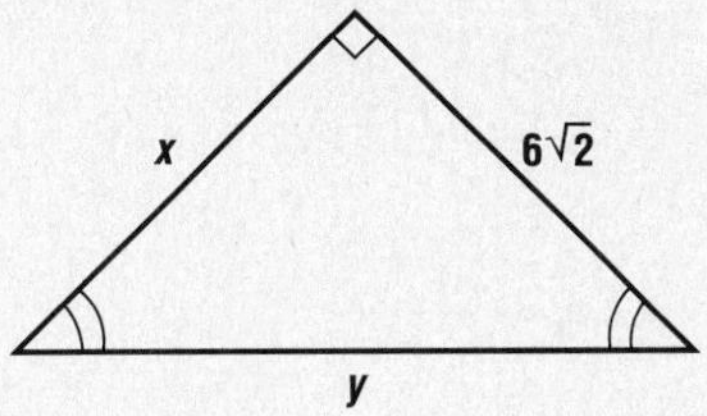

5.

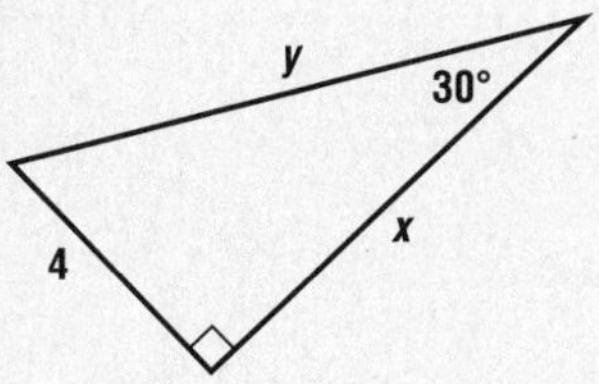

6.

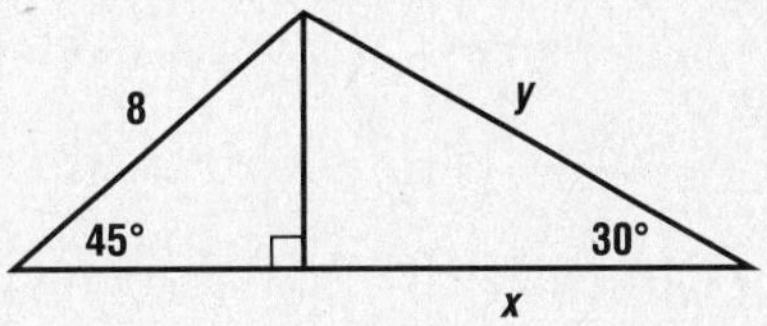

7. Find the length of a diagonal of a square with sides 10 in. long.

8. Find the length of a side of a square whose diagonal is 4 cm.

9. One side of an equilateral triangle measures 6 cm. Find the measure of an altitude of the triangle.

NAME ________________________ DATE ____________

8-3 Practice Worksheet

Special Right Triangles

Find the values of x and y.

1.

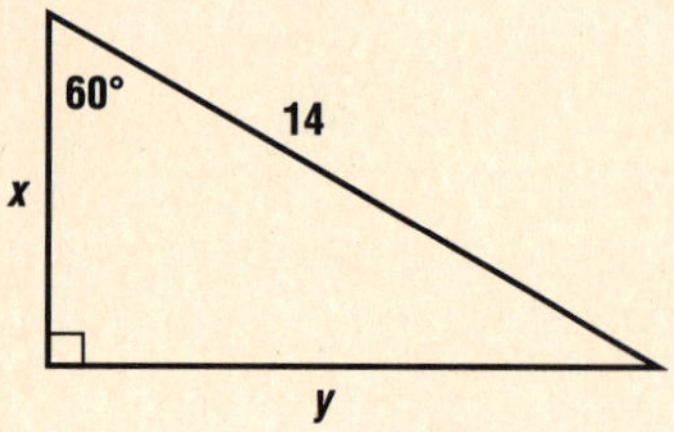

7, $7\sqrt{3} \approx 12.1$

2.

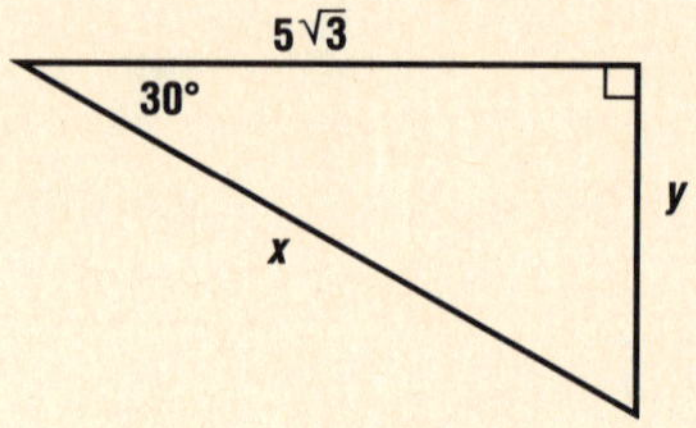

10, 5

3.

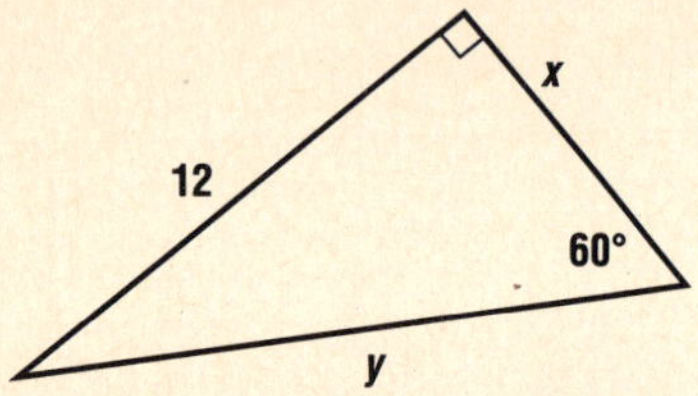

$4\sqrt{3} \approx 6.9$,
$8\sqrt{3} \approx 13.8$

4.

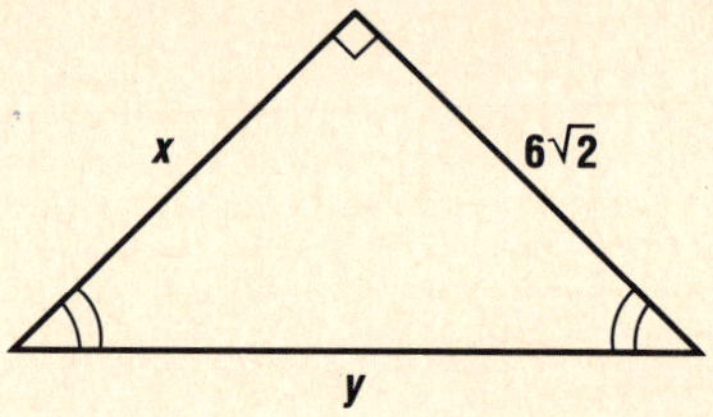

$6\sqrt{2} \approx 8.5$, 12

5.

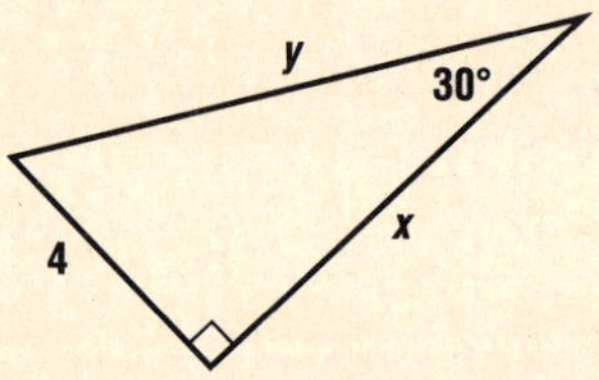

$4\sqrt{3} \approx 6.9$, 8

6.

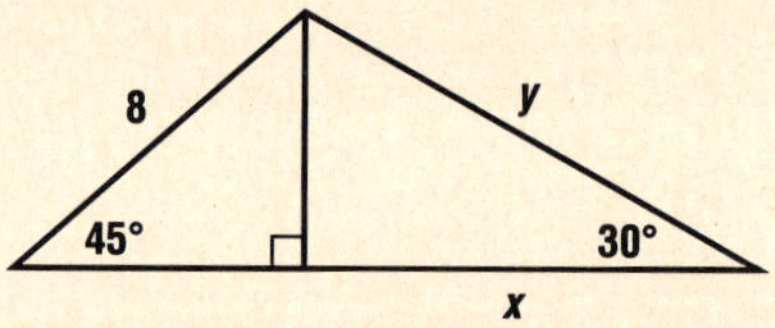

$4\sqrt{6} \approx 9.8$,
$8\sqrt{2} \approx 11.3$

7. Find the length of a diagonal of a square with sides 10 in. long. $10\sqrt{2} \approx 14.1$ in.

8. Find the length of a side of a square whose diagonal is 4 cm. $2\sqrt{2} \approx 2.8$ cm

9. One side of an equilateral triangle measures 6 cm. Find the measure of an altitude of the triangle. $3\sqrt{3} \approx 5.2$ cm

NAME ______________________________ DATE ______________

8-4 Practice Worksheet

Trigonometry

Find the values of x and y. Round measures of segments to the nearest tenth and angle measures to the nearest degree.

1.

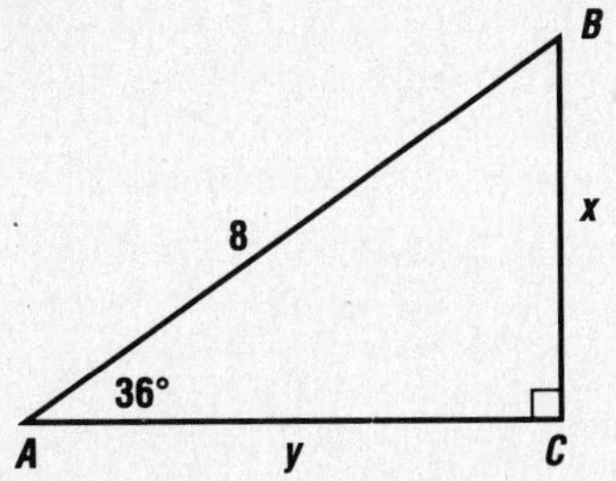

2.

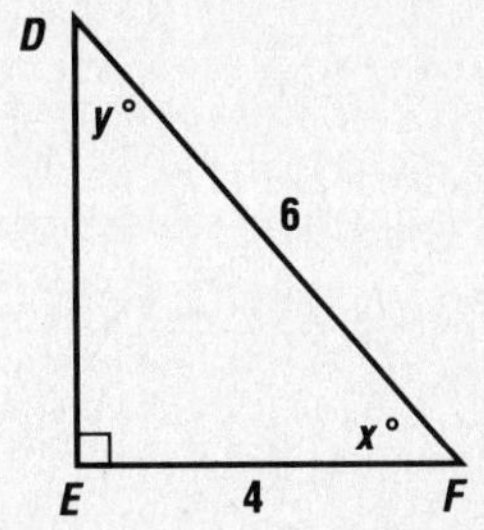

3.

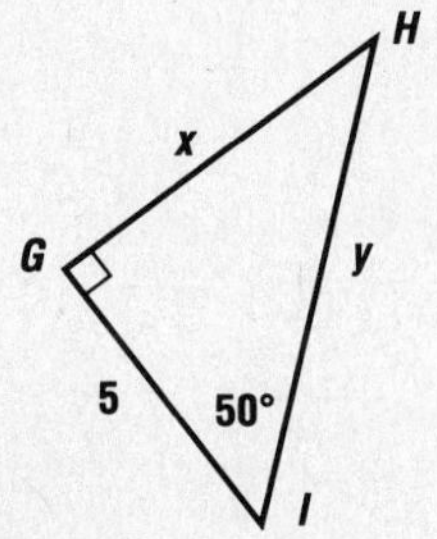

4.

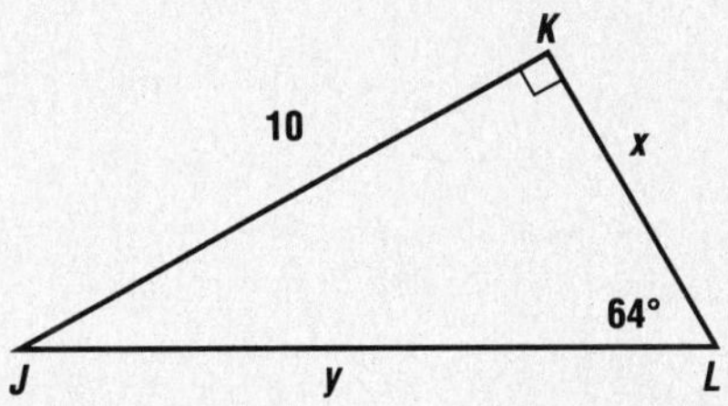

5.

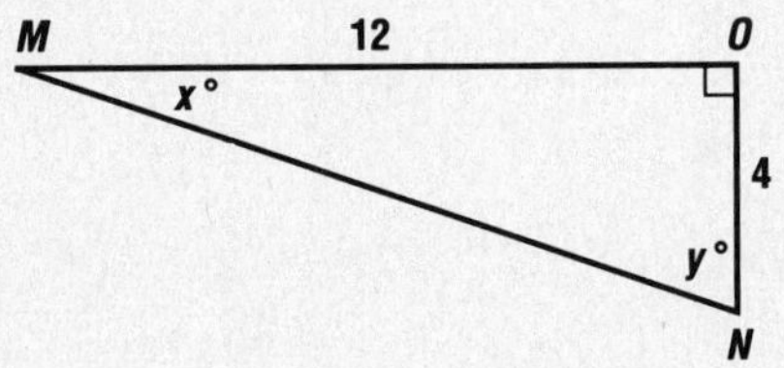

6.

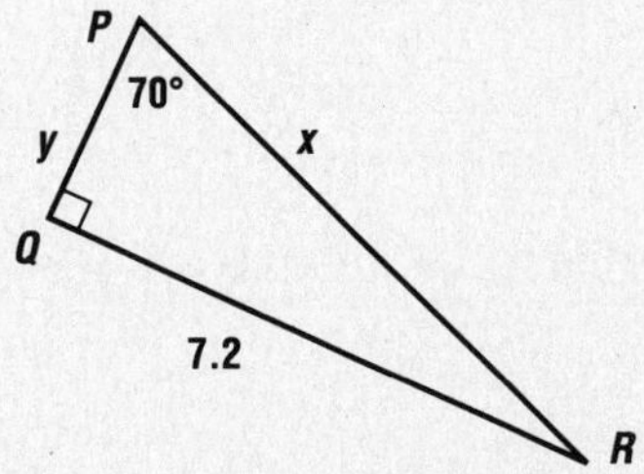

NAME ______________________ DATE ____________

8-4

Practice Worksheet

Trigonometry

Find the values of x and y. Round measures of segments to the nearest tenth and angle measures to the nearest degree.

1.

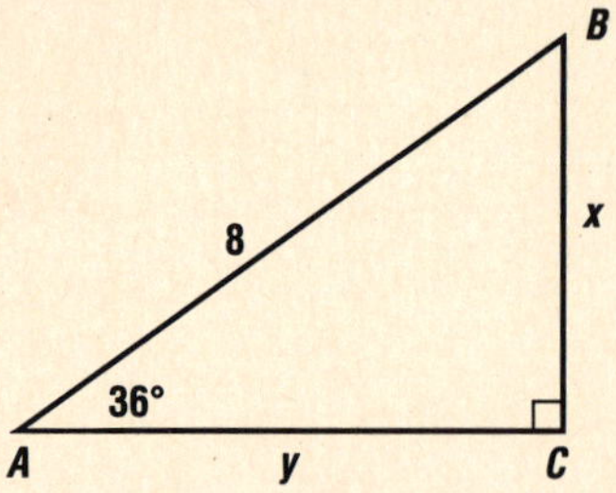

$x \approx 4.7$, $y \approx 6.5$

2.

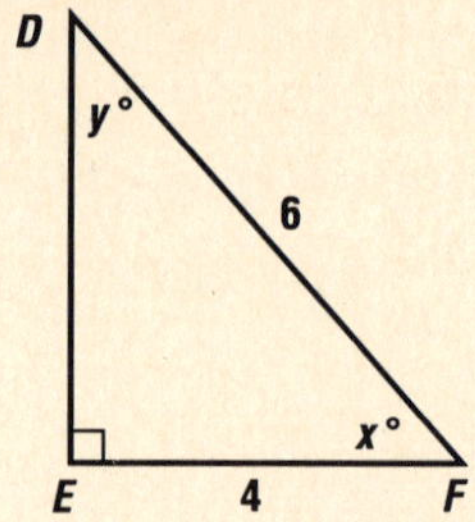

$x \approx 48$, $y \approx 42$

3.

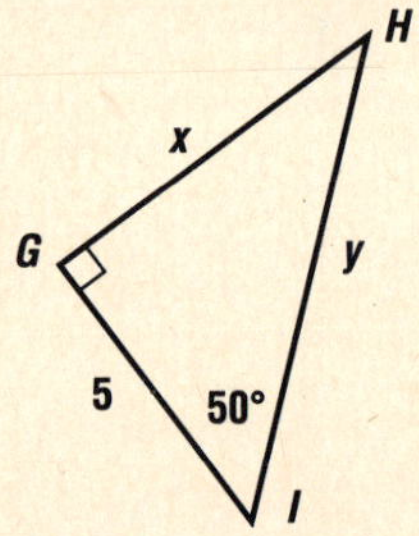

$x \approx 6.0$, $y \approx 7.8$

4.

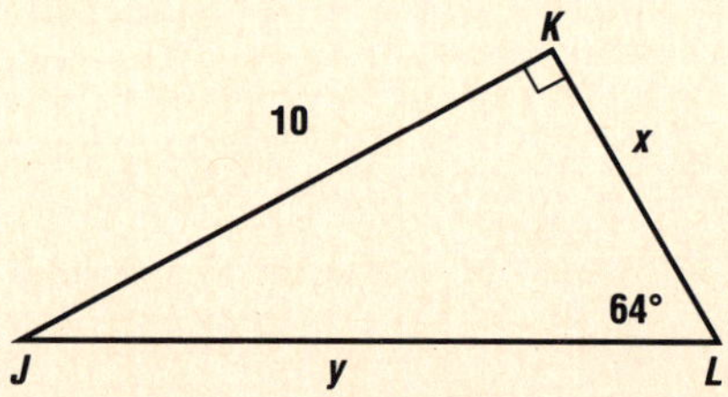

$x \approx 4.9$, $y \approx 11.1$

5.

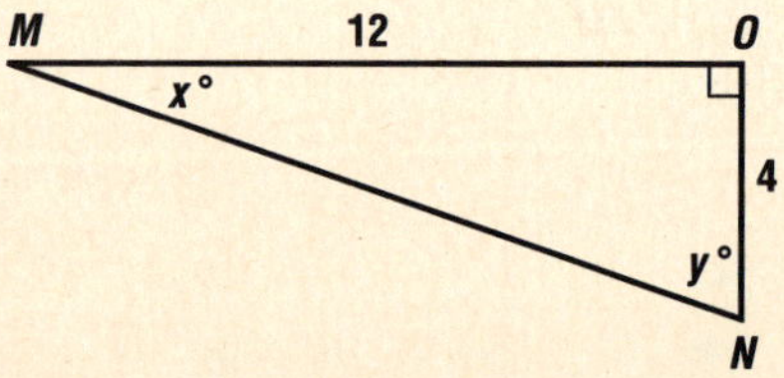

$x \approx 18$, $y \approx 72$

6.

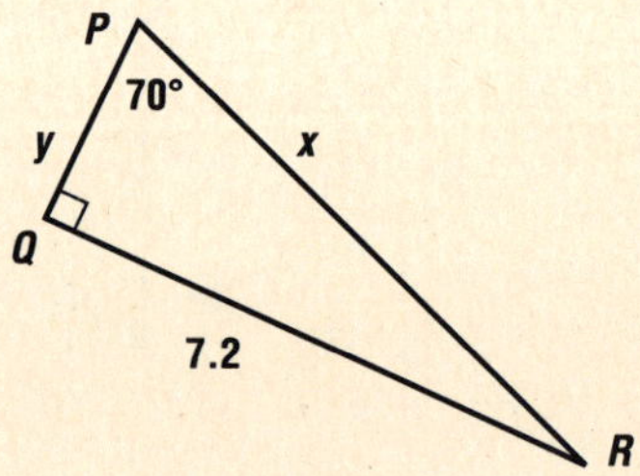

$x \approx 7.7$, $y \approx 2.6$

NAME ______________________ DATE ______________

8-5

Practice Worksheet

Application: Using Trigonometry

Solve each problem. Round measures of segments to the nearest hundredth and measures of angles to the nearest degree.

1. A 20-foot ladder leans against a wall so that the base of the ladder is 8 feet from the base of the building. What angle does the ladder make with the ground?

2. A 50-meter vertical tower is braced with a cable secured at the top of the tower and tied 30 meters from the base. What angle does the cable form with the vertical tower?

3. At a point on the ground 50 feet from the foot of a tree, the angle of elevation to the top of the tree is 53°. Find the height of the tree.

4. From the top of a lighthouse 210 feet high, the angle of depression of a boat is 27°. Find the distance from the boat to the foot of the lighthouse. The lighthouse was built at sea level.

5. Richard is flying a kite. The kite string makes an angle of 57° with the ground. If Richard is standing 100 feet from the point on the ground directly below the kite, find the length of the kite string.

6. An airplane rises vertically 1000 feet over a horizontal distance of 1 mile. What is the angle of elevation of the airplane's path?

Glencoe Division, Macmillan/McGraw-Hill

NAME ______________________________ DATE ____________

8-5

Practice Worksheet

Application: Using Trigonometry

Solve each problem. Round measures of segments to the nearest hundredth and measures of angles to the nearest degree.

1. A 20-foot ladder leans against a wall so that the base of the ladder is 8 feet from the base of the building. What angle does the ladder make with the ground? **66°**

2. A 50-meter vertical tower is braced with a cable secured at the top of the tower and tied 30 meters from the base. What angle does the cable form with the vertical tower? **31°**

3. At a point on the ground 50 feet from the foot of a tree, the angle of elevation to the top of the tree is 53°. Find the height of the tree. **66.35 ft**

4. From the top of a lighthouse 210 feet high, the angle of depression of a boat is 27°. Find the distance from the boat to the foot of the lighthouse. The lighthouse was built at sea level. **412.15 ft**

5. Richard is flying a kite. The kite string makes an angle of 57° with the ground. If Richard is standing 100 feet from the point on the ground directly below the kite, find the length of the kite string. **183.61 ft**

6. An airplane rises vertically 1000 feet over a horizontal distance of 1 mile. What is the angle of elevation of the airplane's path? **11°**

NAME ______________________ DATE ____________

8-6 Practice Worksheet

Law of Sines

Solve each triangle ABC. Round lengths of sides to the nearest tenth and measures of angles to the nearest degree.

1. $a = 12$, $m\angle B = 70$, $m\angle C = 15$

2. $a = 12$, $b = 5$, $m\angle A = 110$

3. $a = 8$, $m\angle A = 60$, $m\angle C = 40$

4. $a = 5$, $c = 4$, $m\angle A = 65$

5. $b = 6$, $m\angle A = 44$, $m\angle B = 68$

6. $a = 7$, $m\angle A = 37$, $m\angle B = 76$

7. $a = 9$, $b = 9$, $m\angle C = 20°$

8. A ship is sighted from two radar stations 43 km apart. The angle between the line segment joining the two stations and the radar beam of the first station is 37°. The angle between the line segment joining the two stations and the beam from the second station is 113°. How far is the ship from the second station?

Glencoe Division, Macmillan/McGraw-Hill

8-6

NAME ______________________ DATE ____________

Practice Worksheet

Law of Sines

Solve each triangle ABC. Round lengths of sides to the nearest tenth and measures of angles to the nearest degree.

1. $a = 12$, $m\angle B = 70$, $m\angle C = 15$ **$m\angle A = 95$, $b \approx 11.3$, $c \approx 3.1$**

2. $a = 12$, $b = 5$, $m\angle A = 110$ **$m\angle B \approx 23$, $m\angle C \approx 47$, $c \approx 9.3$**

3. $a = 8$, $m\angle A = 60$, $m\angle C = 40$ **$m\angle B = 80$, $b \approx 9.1$, $c \approx 5.9$**

4. $a = 5$, $c = 4$, $m\angle A = 65$ **$m\angle C \approx 46$, $m\angle B \approx 69$, $b \approx 5.2$**

5. $b = 6$, $m\angle A = 44$, $m\angle B = 68$ **$m\angle C = 68$, $a \approx 4.5$, $c = 6$**

6. $a = 7$, $m\angle A = 37$, $m\angle B = 76$ **$m\angle C = 67$, $b \approx 11.3$, $c \approx 10.7$**

7. $a = 9$, $b = 9$, $m\angle C = 20°$ **$m\angle A = 80$, $m\angle B = 80$, $c \approx 3.1$**

8. A ship is sighted from two radar stations 43 km apart. The angle between the line segment joining the two stations and the radar beam of the first station is 37°. The angle between the line segment joining the two stations and the beam from the second station is 113°. How far is the ship from the second station? **51.8 km**

8-7

NAME ______________________ DATE ______________

Practice Worksheet

Law of Cosines

Solve each triangle *ABC* described below. Round lengths of sides to the nearest tenth and measures of angles to the nearest degree.

1. $a = 16$, $b = 20$, $m\angle B = 40$

2. $a = 10$, $b = 15$, $c = 12$

3. $a = 42$, $c = 60$, $m\angle B = 58$

4. $m\angle A = 60$, $m\angle B = 72$, $c = 9$

5. $a = 7$, $b = 12$, $c = 15$

6. $m\angle A = 43$, $b = 23$, $c = 26$

7. $a = 16$, $m\angle A = 23$, $m\angle B = 87$

8. $c = 15.6$, $a = 12.9$, $b = 18.4$

8-7

NAME ______________________ DATE ____________

Practice Worksheet

Law of Cosines

Solve each triangle *ABC* described below. Round lengths of sides to the nearest tenth and measures of angles to the nearest degree.

1. $a = 16$, $b = 20$, $m\angle B = 40$

$m\angle A \approx 31$
$m\angle C \approx 109$
$c \approx 29.4$

2. $a = 10$, $b = 15$, $c = 12$

$m\angle A \approx 42$
$m\angle B \approx 85$
$m\angle C \approx 53$

3. $a = 42$, $c = 60$, $m\angle B = 58$

$m\angle A \approx 43$
$m\angle C \approx 79$
$b \approx 51.9$

4. $m\angle A = 60$, $m\angle B = 72$, $c = 9$

$m\angle C \approx 48$
$a \approx 10.5$
$b \approx 11.5$

5. $a = 7$, $b = 12$, $c = 15$

$m\angle A \approx 27$
$m\angle B \approx 52$
$m\angle C \approx 101$

6. $m\angle A = 43$, $b = 23$, $c = 26$

$m\angle B \approx 60$
$m\angle C \approx 77$
$a \approx 18.2$

7. $a = 16$, $m\angle A = 23$, $m\angle B = 87$

$m\angle C \approx 70$
$b \approx 40.9$
$c \approx 38.5$

8. $c = 15.6$, $a = 12.9$, $b = 18.4$

$m\angle A \approx 44$
$m\angle B \approx 80$
$m\angle C \approx 57$

Glencoe Division, Macmillan/McGraw-Hill

8-8

NAME ______________________ DATE ____________

Practice Worksheet

Problem-Solving Strategy: Decision-Making

Solve. Use any strategy.

1. Find the next number in the sequence 0, 3, 8, 15, 24 . . . and describe the pattern.

2. Toni cashed her paycheck. One-fourth of what she earned, she put in her savings account. One-third of what she had left went to pay her rent. She spent \$30 on food, and had \$120 left. How much was Toni's paycheck?

3. Complete the addition problem at the right. If a letter is used more than once, it represents the same digit each time.

$$\begin{array}{r} \text{SOME} \\ +\underline{\text{MORE}} \\ \text{SENSE} \end{array}$$

4. A square results when the length of a rectangle is reduced by 10% and the width is increased by 10%. How does the area of the square compare to the area of the rectangle? Express your answer as a percentage.

5. Suppose that n is a whole number. Which is always greater, $n\sqrt{n + 1}$ or $(n + 1)\sqrt{n}$?

NAME ______________________ DATE ____________

8-8

Practice Worksheet

Problem-Solving Strategy: Decision-Making

Solve. Use any strategy.

1. Find the next number in the sequence 0, 3, 8, 15, 24 . . . and describe the pattern. **35; n^{th} term is $n^2 - 1$**

2. Toni cashed her paycheck. One-fourth of what she earned, she put in her savings account. One-third of what she had left went to pay her rent. She spent \$30 on food, and had \$120 left. How much was Toni's paycheck? **\$300**

3. Complete the addition problem at the right. If a letter is used more than once, it represents the same digit each time.

SOME	**1780**
+MORE	**+8730**
SENSE	**10510**

4. A square results when the length of a rectangle is reduced by 10% and the width is increased by 10%. How does the area of the square compare to the area of the rectangle? Express your answer as a percentage. **The area of the square is 99% of the area of the rectangle.**

5. Suppose that n is a whole number. Which is always greater, $n\sqrt{n+1}$ or $(n+1)\sqrt{n}$? **$(n+1)\sqrt{n}$**

NAME ______________________ DATE ____________

9-1

Practice Worksheet

Parts of Circles

Refer to the figure at the right.

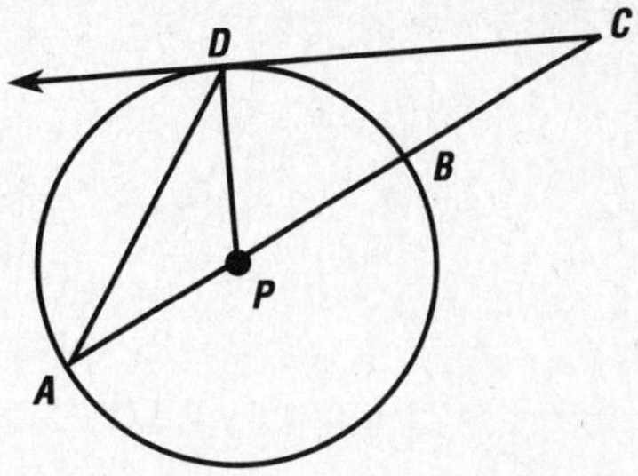

1. Name the center of $\odot P$.

2. Name three radii of the circle.

3. Name a diameter.

4. Name two chords.

5. Name a line tangent to $\odot P$.

6. Name a line that is a secant of $\odot P$ and contains a diameter.

7. Name a point in the interior of the circle.

8. Name a point in the exterior of the circle.

9. Name three points that lie on the circle.

10. If $AP = 4$ in., find AB.

The coordinates of the center and the radius of a circle are given. Write an equation of the circle.

11. (2, 3), 5

12. (1, –4), 9

Given $\odot M$ with equation $(x - 4)^2 + (y + 3)^2 = 14$, determine if each point is on the circle, in its interior, or in its exterior.

13. (2, 3)

14. (1, –3)

15. (1, –5)

16. (8, –1)

NAME ______________________ DATE ____________

9-1

Practice Worksheet

Parts of Circles

Refer to the figure at the right.

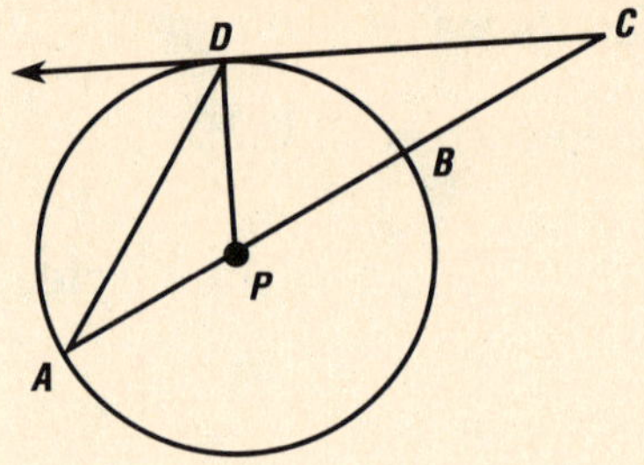

1. Name the center of ⊙P. **P**

2. Name three radii of the circle.
$\overline{PA}$, $\overline{PD}$, $\overline{PB}$

3. Name a diameter. **$\overline{AB}$**

4. Name two chords. **$\overline{AD}$, $\overline{AB}$**

5. Name a line tangent to ⊙P. **$\overleftrightarrow{CD}$**

6. Name a line that is a secant of ⊙P and contains a diameter. **$\overleftrightarrow{AC}$**

7. Name a point in the interior of the circle. **P**

8. Name a point in the exterior of the circle. **C**

9. Name three points that lie on the circle. **A, D, B**

10. If $AP = 4$ in., find AB. **8 in.**

The coordinates of the center and the radius of a circle are given. Write an equation of the circle.

11. (2, 3), 5
$(x - 2)^2 + (y - 3)^2 = 25$

12. (1, –4), 9
$(x - 1)^2 + (y + 4)^2 = 81$

Given ⊙M with equation $(x - 4)^2 + (y + 3)^2 = 14$, determine if each point is on the circle, in its interior, or in its exterior.

13. (2, 3)
exterior

14. (1, –3)
interior

15. (1, –5)
interior

16. (8, –1)
exterior

NAME ______________________________ DATE ______________

9-2 Practice Worksheet

Angles and Arcs

In $\odot P$, $m\angle 1 = 140$ and $\overline{AC}$ is a diameter. Find each measure.

1. $m\angle 2$

2. $m\overset{\frown}{BC}$

3. $m\overset{\frown}{AB}$

4. $m\overset{\frown}{ABC}$

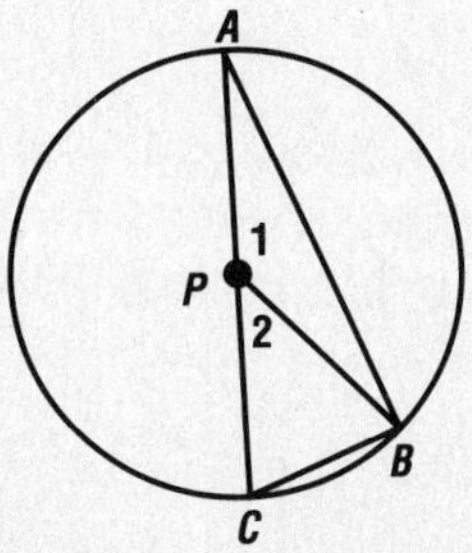

In $\odot P$, $m\angle 2 = m\angle 1$, $m\angle 2 = 4x + 35$, $m\angle 1 = 9x + 5$, and $\overline{BD}$ and $\overline{AC}$ are diameters. Find each value or measure.

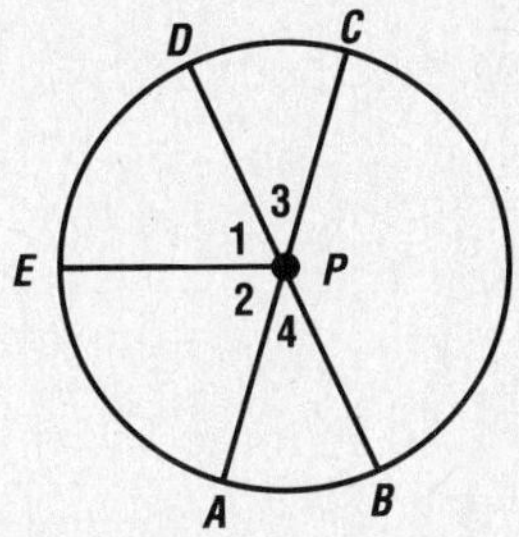

5. x

6. $m\overset{\frown}{AE}$

7. $m\overset{\frown}{ED}$

8. $m\angle 3$

9. $m\overset{\frown}{AB}$

10. $m\overset{\frown}{EC}$

11. $m\overset{\frown}{EB}$

12. $m\angle CPB$

13. $m\overset{\frown}{CB}$

14. $m\overset{\frown}{CEB}$

15. $m\overset{\frown}{DC}$

16. $m\overset{\frown}{CEA}$

17. Sketch three concentric circles and a fourth circle that intersects each of the other three, with the total number of intersection points being four.

Glencoe Division, Macmillan/McGraw-Hill

NAME ______________________________ DATE ____________

9-2 Practice Worksheet

Angles and Arcs

In ⊙P, m∠1 = 140 and $\overline{AC}$ is a diameter. Find each measure.

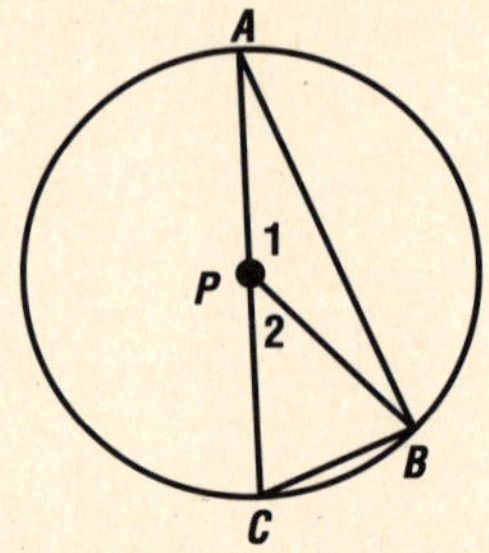

1. m∠2 40
2. m$\widehat{BC}$ 40
3. m$\widehat{AB}$ 140
4. m$\widehat{ABC}$ 180

In ⊙P, m∠2 = m∠1, m∠2 = 4x + 35, m∠1 = 9x + 5, and $\overline{BD}$ and $\overline{AC}$ are diameters. Find each value or measure.

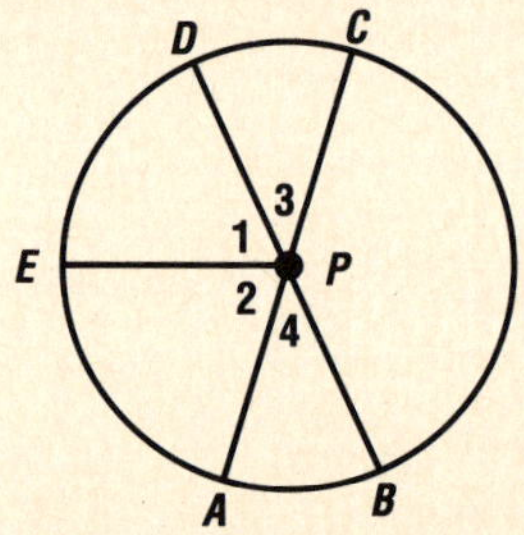

5. x 6
6. m$\widehat{AE}$ 59
7. m$\widehat{ED}$ 59
8. m∠3 62
9. m$\widehat{AB}$ 62
10. m$\widehat{EC}$ 121
11. m$\widehat{EB}$ 121
12. m∠CPB 118
13. m$\widehat{CB}$ 118
14. m$\widehat{CEB}$ 242
15. m$\widehat{DC}$ 62
16. m$\widehat{CEA}$ 180

17. Sketch three concentric circles and a fourth circle that intersects each of the other three, with the total number of intersection points being four.

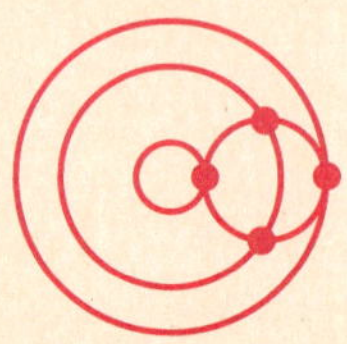

9-3

NAME ________________________ DATE ____________

Practice Worksheet

Arcs and Chords

In each figure, O is the center of the circle. Find each missing measure.

1. $m\widehat{BC}$

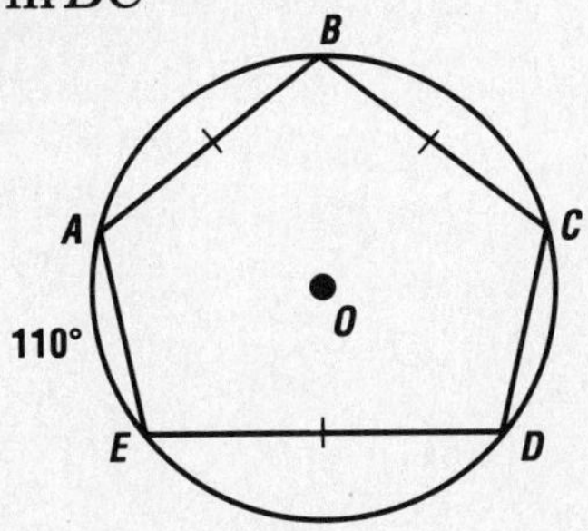

2. YQ

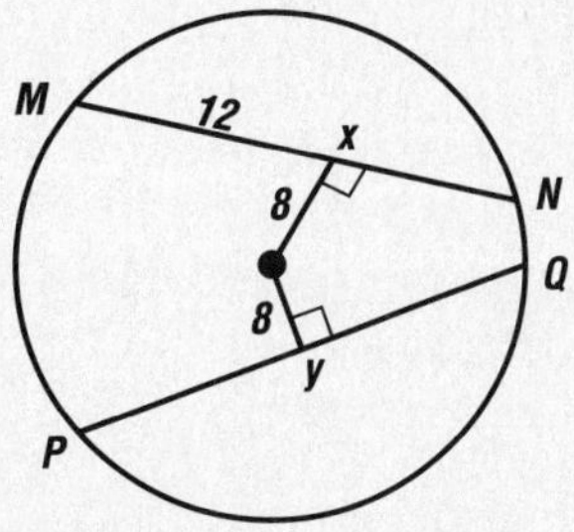

3. Suppose a chord of a circle is 16 inches long and is 6 inches from the center of the circle. Find the length of a radius.

4. Find the length of a chord that is 5 inches from the center of a circle with a radius of 13 inches.

5. Suppose a radius of a circle is 17 units and a chord is 30 units long. Find the distance from the center of the circle to the chord.

6. Find AB.

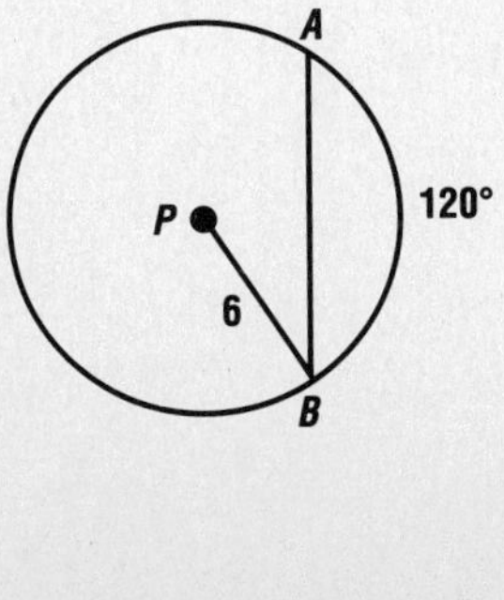

7. Find AB.

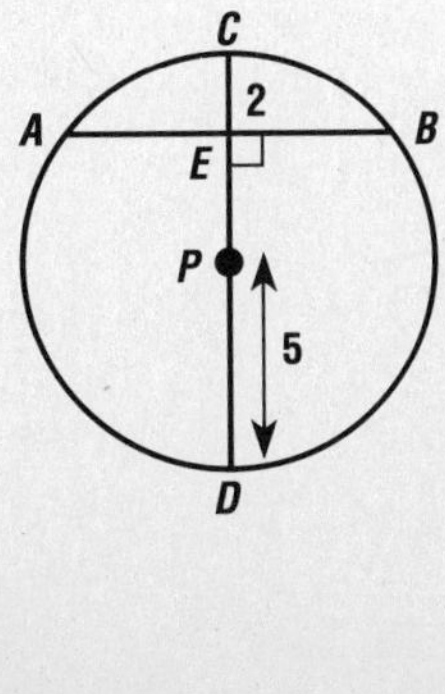

9-3

NAME ______________________ DATE ____________

Practice Worksheet

Arcs and Chords

In each figure, 0 is the center of the circle. Find each missing measure.

1. m$\overarc{BC}$

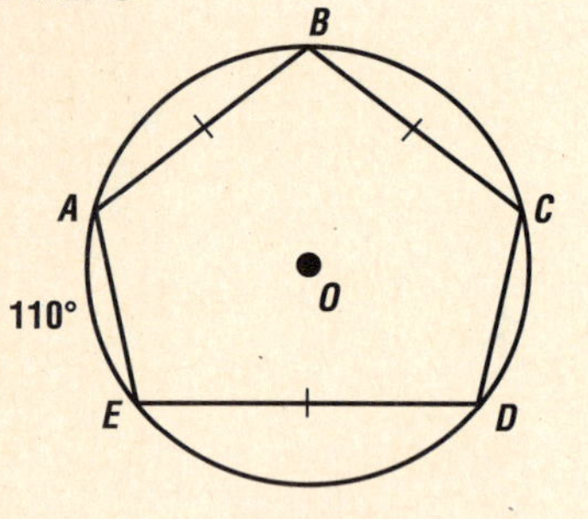

62.5

2. YQ

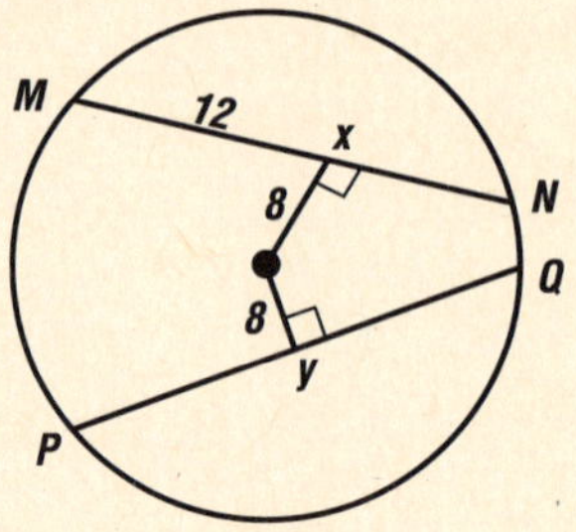

12

3. Suppose a chord of a circle is 16 inches long and is 6 inches from the center of the circle. Find the length of a radius. **10 in.**

4. Find the length of a chord that is 5 inches from the center of a circle with a radius of 13 inches. **24 in.**

5. Suppose a radius of a circle is 17 units and a chord is 30 units long. Find the distance from the center of the circle to the chord. **8 units**

6. Find AB.

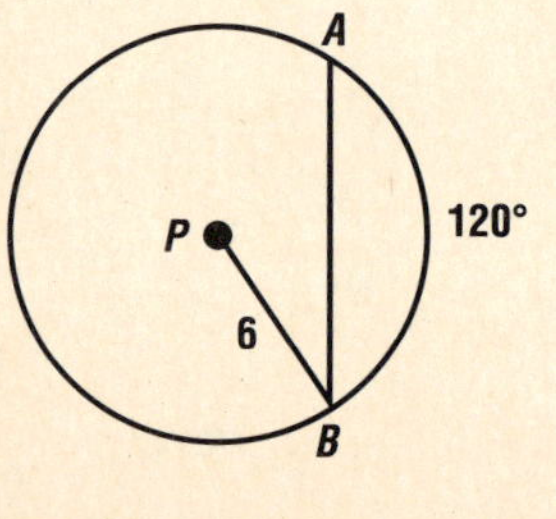

$6\sqrt{3}$ or about 10.4

7. Find AB.

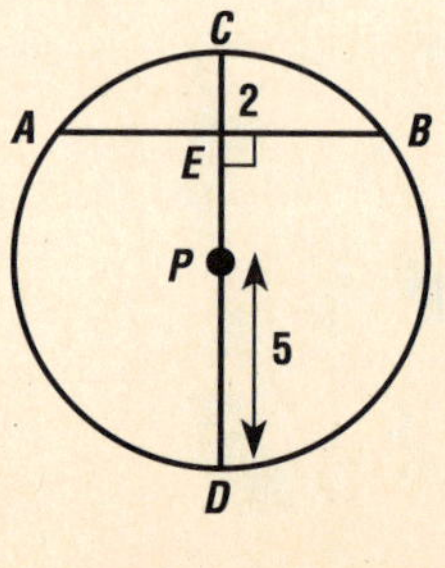

8

NAME ______________________ DATE ____________

9-4

Practice Worksheet

Inscribed Angles

In $\odot P$, $m\overset{\frown}{AB} = x$ and $m\overset{\frown}{BC} = 3x$. Find each measure.

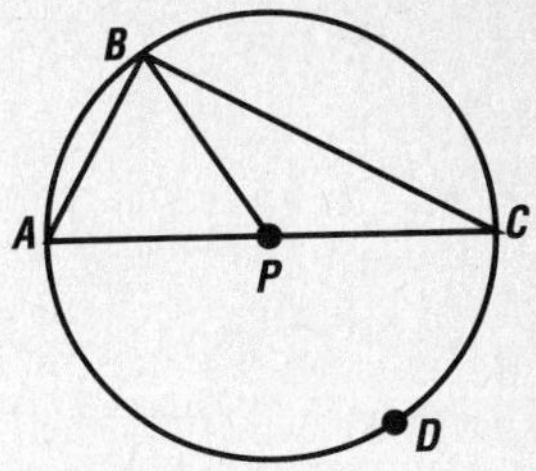

1. $m\overset{\frown}{ADC}$

2. $m\overset{\frown}{AB}$

3. $m\overset{\frown}{BC}$

4. $m\angle ABC$

5. $m\angle A$

6. $m\angle C$

In $\odot Q$, $m\angle ABC = 72$ and $m\overset{\frown}{CD} = 46$. Find each measure.

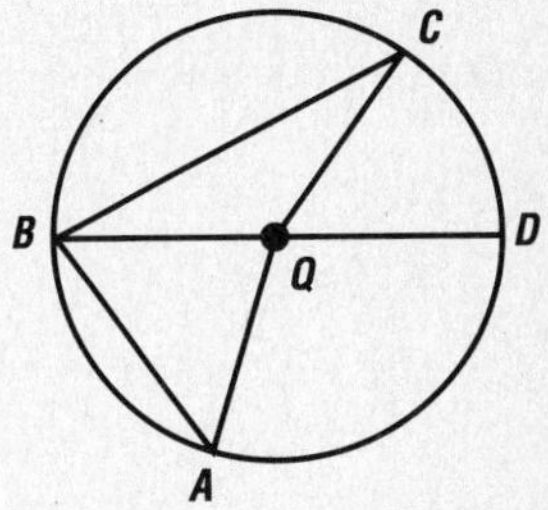

7. $m\overset{\frown}{CA}$

8. $m\overset{\frown}{AD}$

9. $m\angle ABD$

10. $m\overset{\frown}{BC}$

11. $m\angle C$

12. $m\angle A$

13. Suppose $ABCD$ is a trapezoid that has its vertices on $\odot P$, with $AB \parallel CD$. Write a paragraph proof to show that $ABCD$ is an isosceles trapezoid.

Glencoe Division, Macmillan/McGraw-Hill

NAME ______________________ DATE ____________

9-4 Practice Worksheet

Inscribed Angles

In ⊙P, m$\overset{\frown}{AB}$ = x and m$\overset{\frown}{BC}$ = 3x. Find each measure.

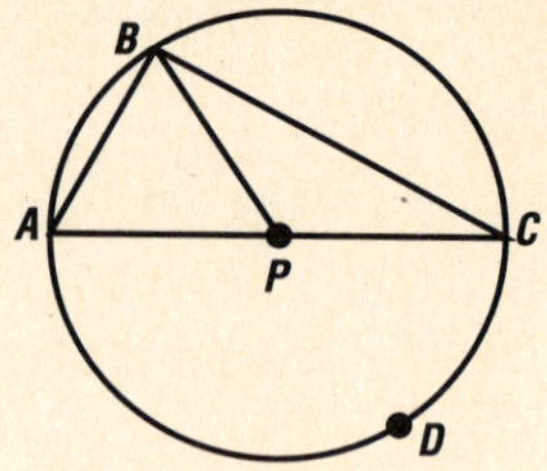

1. m$\overset{\frown}{ADC}$ **180**

2. m$\overset{\frown}{AB}$ **45**

3. m$\overset{\frown}{BC}$ **135**

4. m∠*ABC* **90**

5. m∠*A* **67.5**

6. m∠*C* **22.5**

In ⊙Q, m∠ABC = 72 and m$\overset{\frown}{CD}$ = 46. Find each measure.

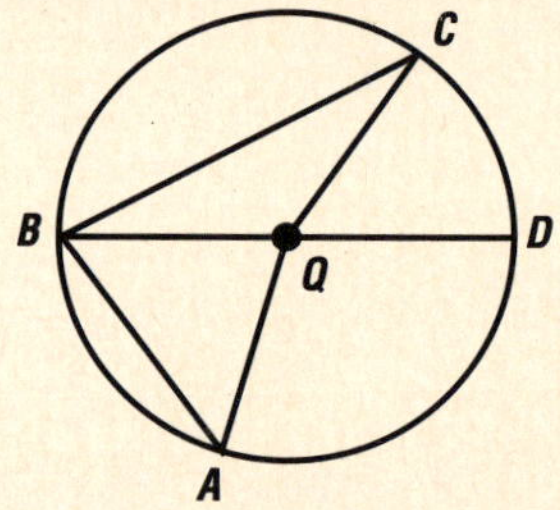

7. m$\overset{\frown}{CA}$ **144**

8. m$\overset{\frown}{AD}$ **98**

9. m∠*ABD* **49**

10. m$\overset{\frown}{BC}$ **134**

11. m∠*C* **23**

12. m∠*A* **49**

13. Suppose *ABCD* is a trapezoid that has its vertices on ⊙*P*, with *AB* ∥ *CD*. Write a paragraph proof to show that *ABCD* is an isosceles trapezoid.

∠*A* and ∠*D* are supplementary, since they are consecutive interior angles formed by two parallel lines and a transversal. ∠*B* and ∠*D* are supplementary, since they are opposite angles of a quadrilateral inscribed in a circle. It follows that $\overset{\frown}{ADC} \cong \overset{\frown}{DCB}$. By the arc addition postulate and the subtraction property of equality, m$\overset{\frown}{AD}$ = m$\overset{\frown}{BC}$. Thus $\overset{\frown}{AD} \cong \overset{\frown}{BC}$ and, since if two arcs in a circle are congruent then their chords are congruent, $\overline{AD} \cong \overline{BC}$. Trapezoid ABCD is isosceles by definition.

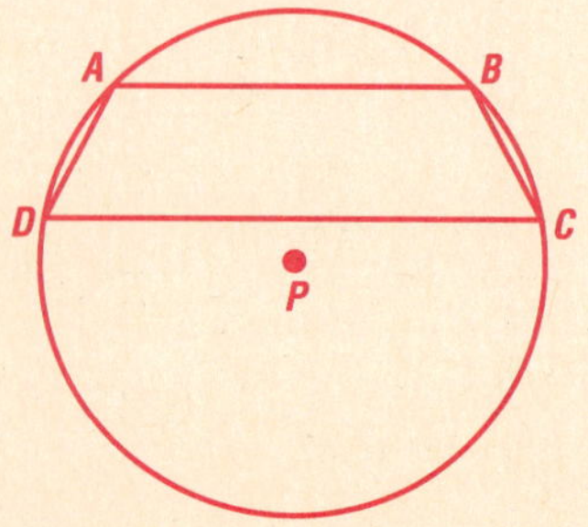

NAME ______________________ DATE ______________

9-5

Practice Worksheet

Tangents

Find the value of x. Assume P is the center of the circle.

1.

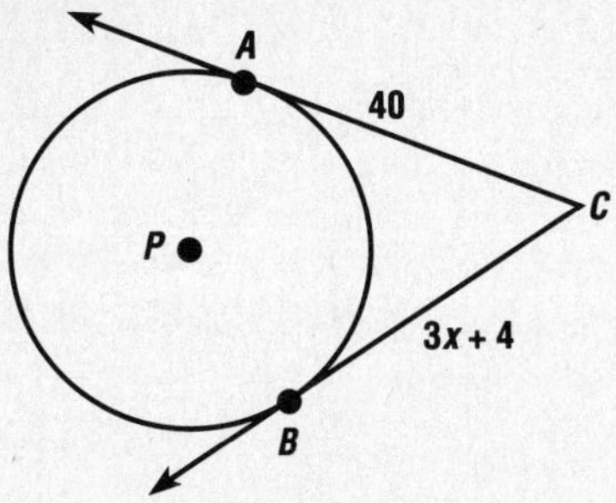

2.

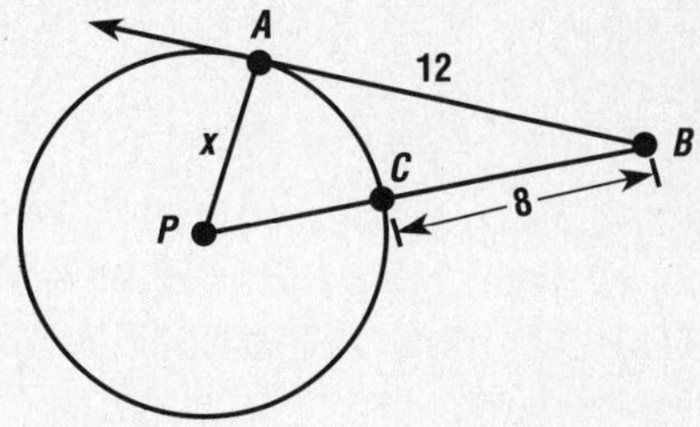

3.

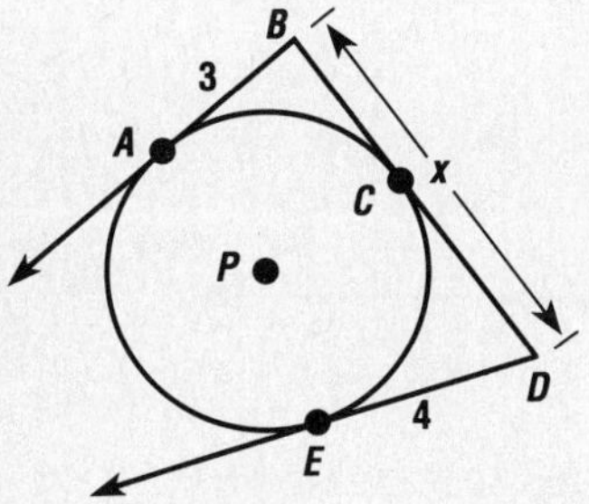

4.

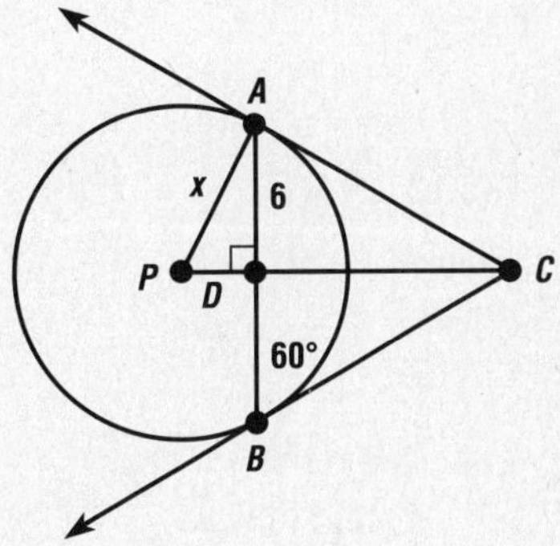

5.

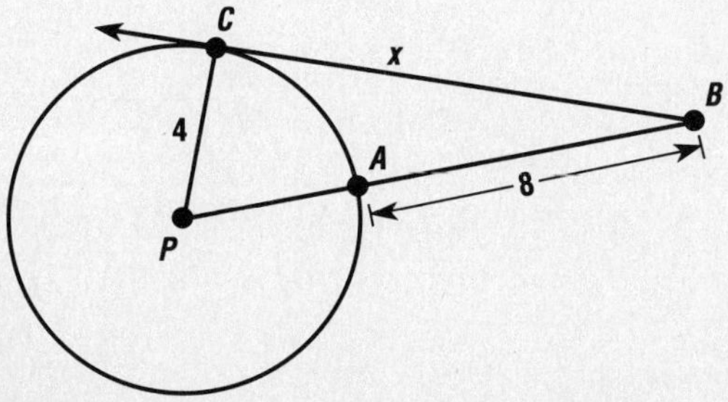

6.

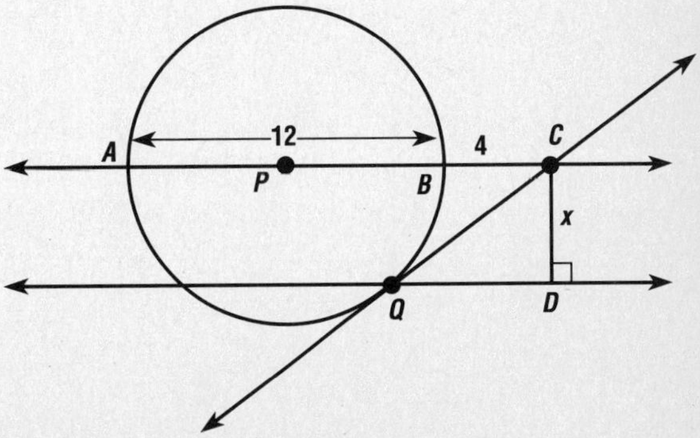

NAME ______________________ DATE ____________

9-5 Practice Worksheet

Tangents

Find the value of x. Assume P is the center of the circle.

1.

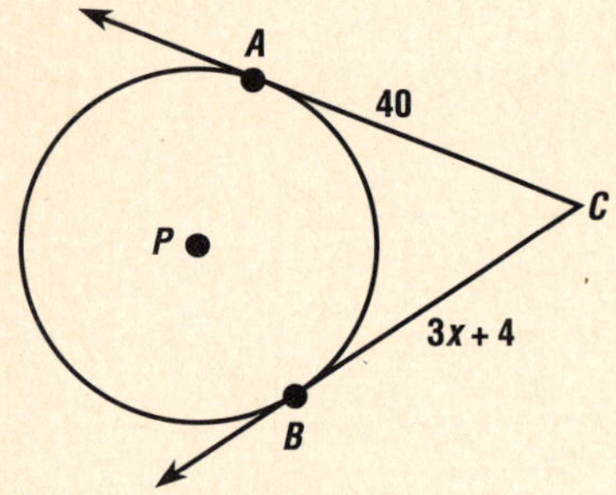

12

2.

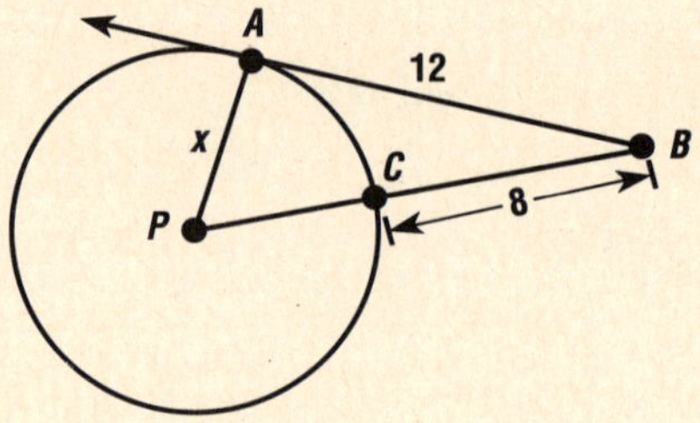

5

3.

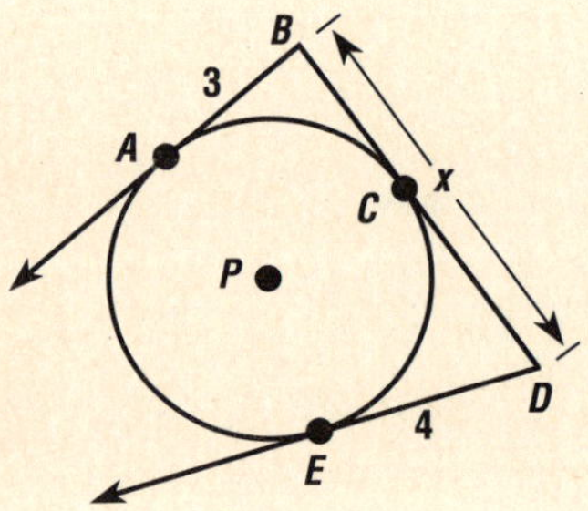

7

4.

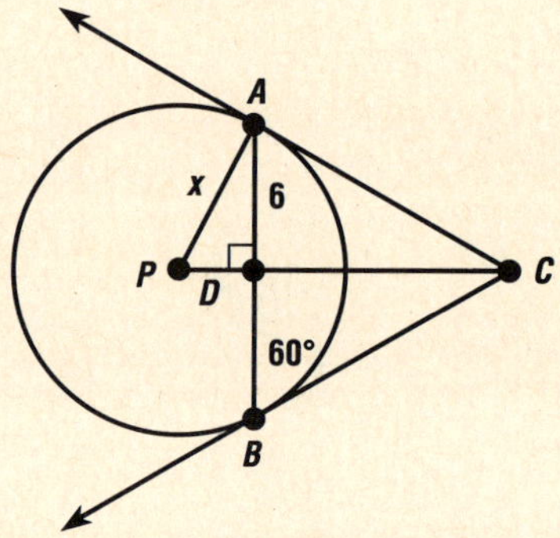

$4\sqrt{3} \approx 6.9$

5.

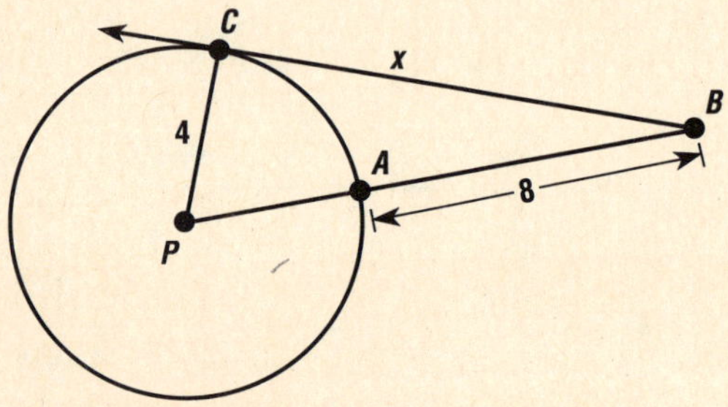

$8\sqrt{2} \approx 11.3$

6.

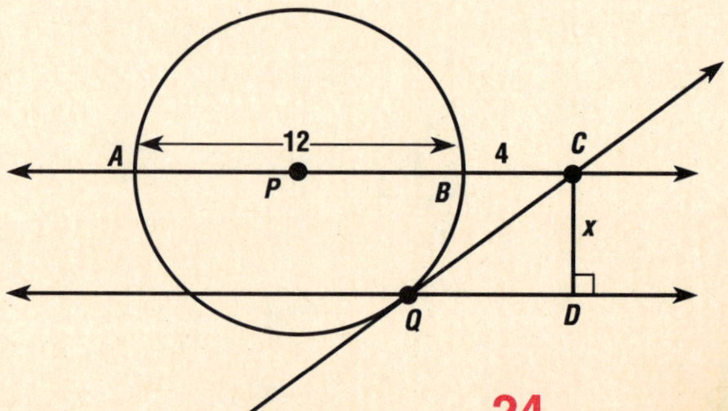

$\frac{24}{5}$

NAME ________________________ DATE ____________

9-6

Practice Worksheet

More Angle Measures

Assume that lines that appear to be tangents are tangents.

In ⊙Q, $m\angle CQD = 120$, $m\overset{\frown}{BC} = 30$, and $m\angle BEC = 25$. Find each measure.

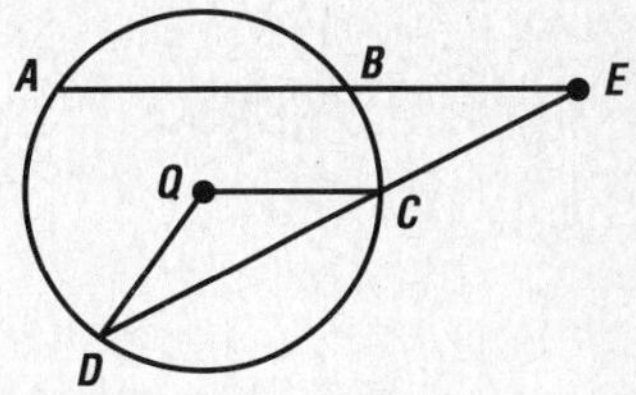

1. $m\overset{\frown}{DC}$

2. $m\overset{\frown}{AD}$

3. $m\overset{\frown}{AB}$

4. $m\angle QDC$

In ⊙Q, $m\overset{\frown}{AE} = 140$, $m\overset{\frown}{BD} = y$, $m\overset{\frown}{AB} = 2y$, and $m\overset{\frown}{DE} = 2y$. Find each measure.

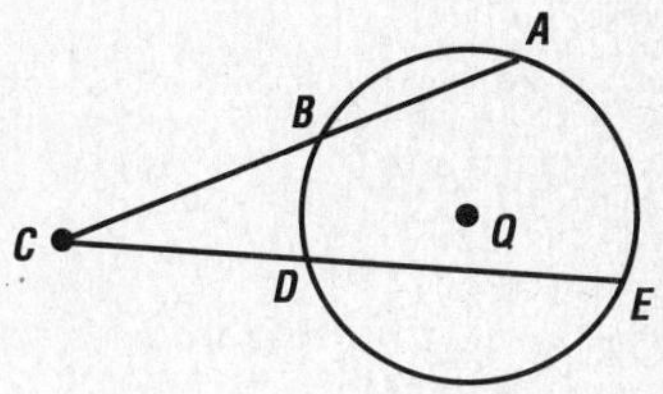

5. $m\overset{\frown}{BD}$

6. $m\overset{\frown}{AB}$

7. $m\overset{\frown}{DE}$

8. $m\angle BCD$

In ⊙P, $m\overset{\frown}{BC} = 4x - 50$, $m\overset{\frown}{DE} = x + 25$, $m\overset{\frown}{EF} = x - 15$, $m\overset{\frown}{CD} = x$, and $m\overset{\frown}{FB} = 50$. Find each measure.

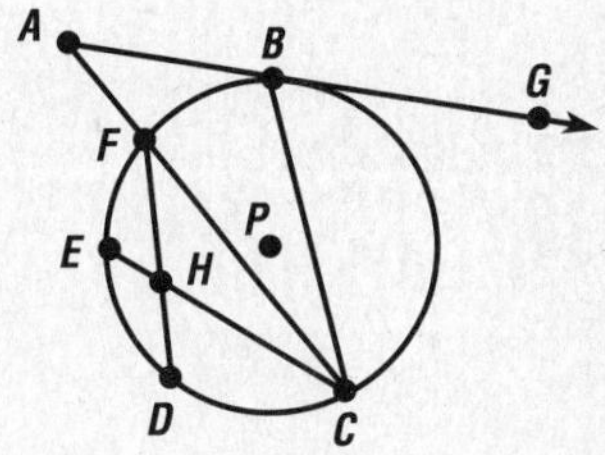

9. $m\angle A$

10. $m\angle BCA$

11. $m\angle ABC$

12. $m\angle GBC$

13. $m\angle FHE$

14. $m\angle CFD$

In ⊙P, $m\angle A = 62$ and $m\overset{\frown}{BD} = 120$. Find each measure.

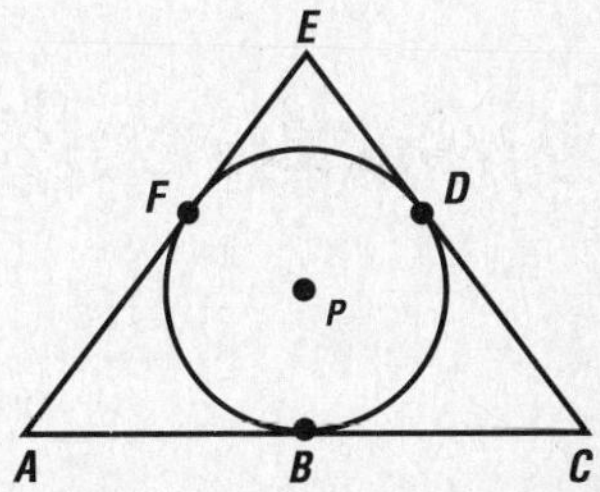

15. $m\angle C$

16. $m\overset{\frown}{DF}$

17. $m\angle E$

NAME ______________________ DATE ____________

Practice Worksheet

More Angle Measures

Assume that lines that appear to be tangents are tangents.

In ⊙Q, $m\angle CQD = 120$, $m\overset{\frown}{BC} = 30$, and $m\angle BEC = 25$. Find each measure.

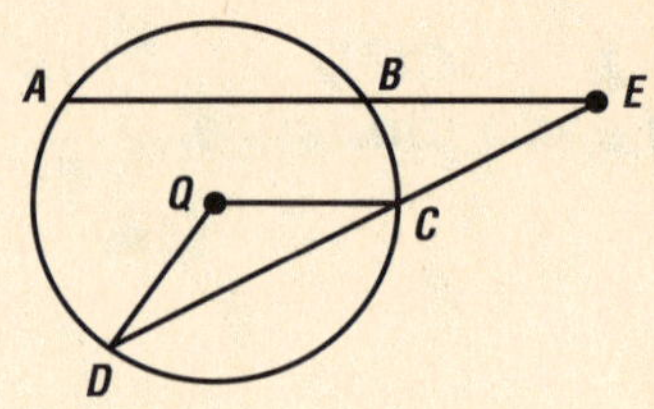

1. $m\overset{\frown}{DC}$ **120**

2. $m\overset{\frown}{AD}$ **80**

3. $m\overset{\frown}{AB}$ **130**

4. $m\angle QDC$ **30**

In ⊙Q, $m\overset{\frown}{AE} = 140$, $m\overset{\frown}{BD} = y$, $m\overset{\frown}{AB} = 2y$, and $m\overset{\frown}{DE} = 2y$. Find each measure.

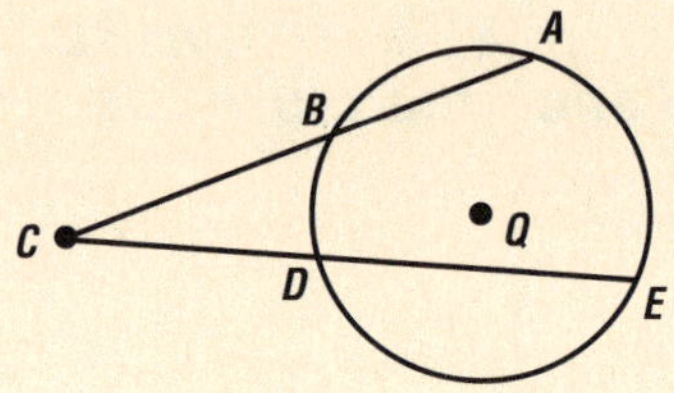

5. $m\overset{\frown}{BD}$ **44**

6. $m\overset{\frown}{AB}$ **88**

7. $m\overset{\frown}{DE}$ **88**

8. $m\angle BCD$ **48**

In ⊙P, $m\overset{\frown}{BC} = 4x - 50$, $m\overset{\frown}{DE} = x + 25$, $m\overset{\frown}{EF} = x - 15$, $m\overset{\frown}{CD} = x$, and $m\overset{\frown}{FB} = 50$. Find each measure.

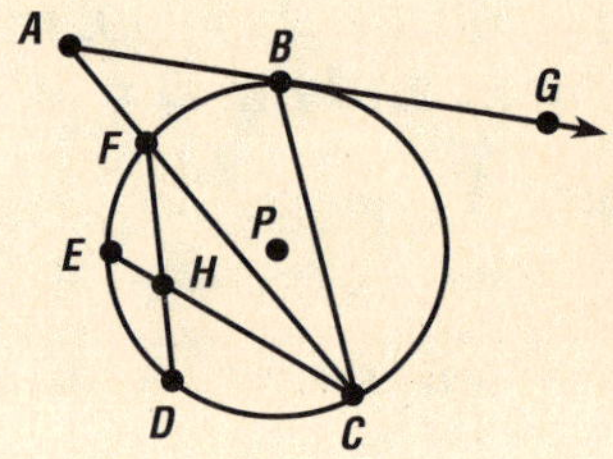

9. $m\angle A$ **50**

10. $m\angle BCA$ **25**

11. $m\angle ABC$ **105**

12. $m\angle GBC$ **75**

13. $m\angle FHE$ **42.5**

14. $m\angle CFD$ **25**

In ⊙P, $m\angle A = 62$ and $m\overset{\frown}{BD} = 120$. Find each measure.

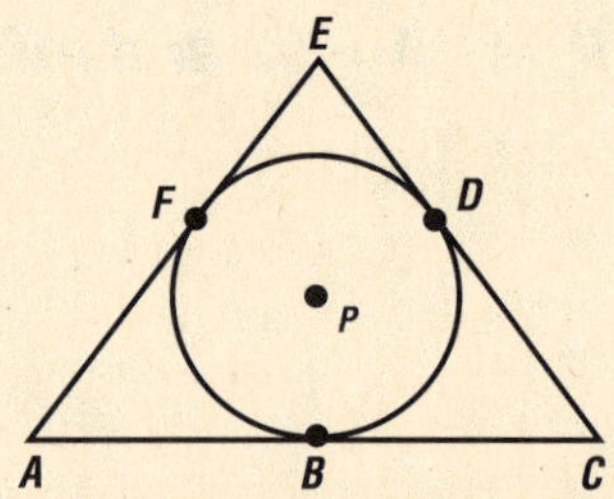

15. $m\angle C$ **60**

16. $m\overset{\frown}{DF}$ **122**

17. $m\angle E$ **58**

NAME ______________________________ DATE ______________

9-7 Practice Worksheet

Special Segments in a Circle

In ⊙P, find the value of x.

1.

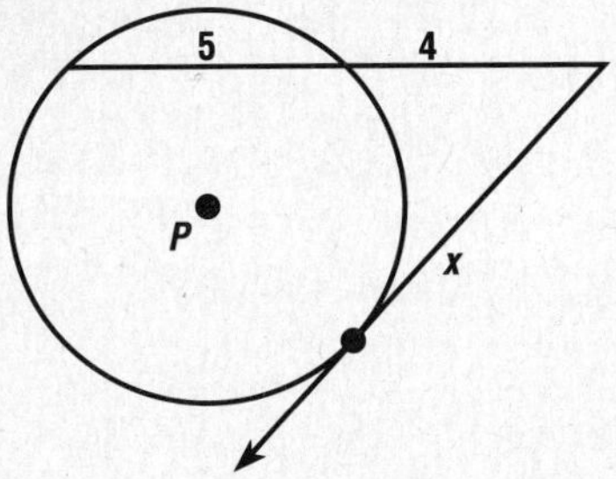

2.

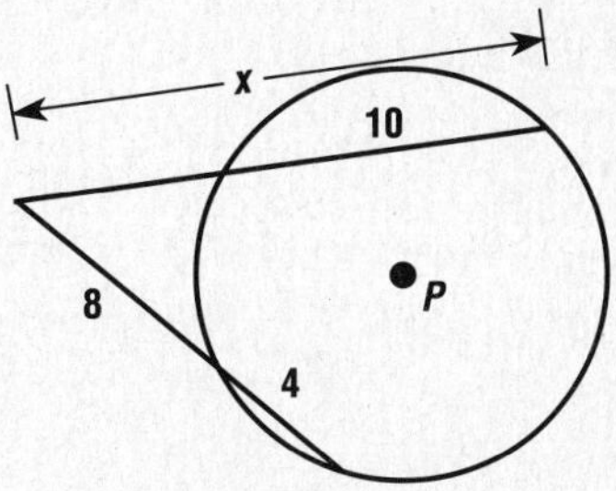

3.

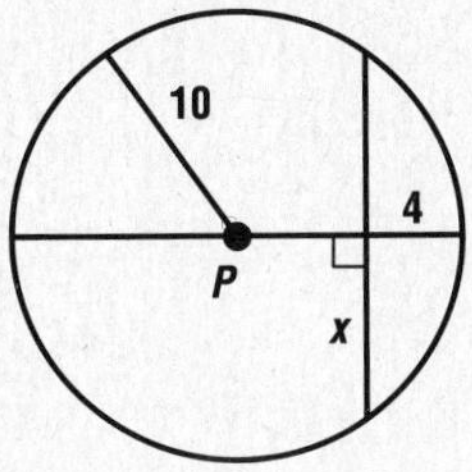

4.

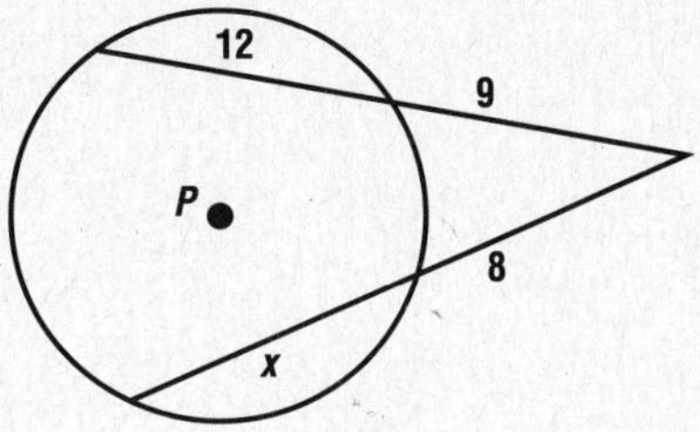

In ⊙P, CE = 6, CD = 16, and AB = 17. Find each of the following.

5. *EB*

6. *AE*

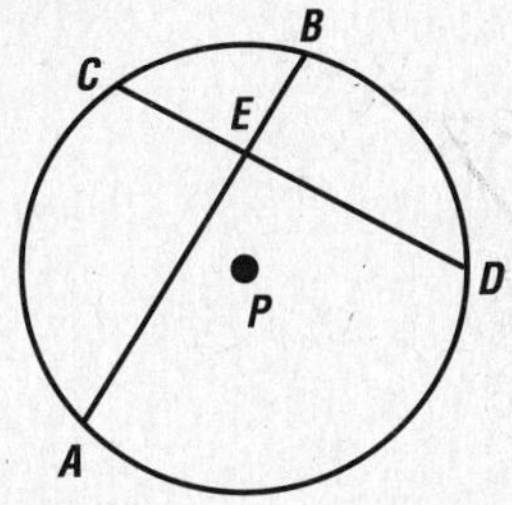

In ⊙P, AC = 3, BC = 5, and AD = 2. Find each of the following.

7. *PD*

8. *ED*

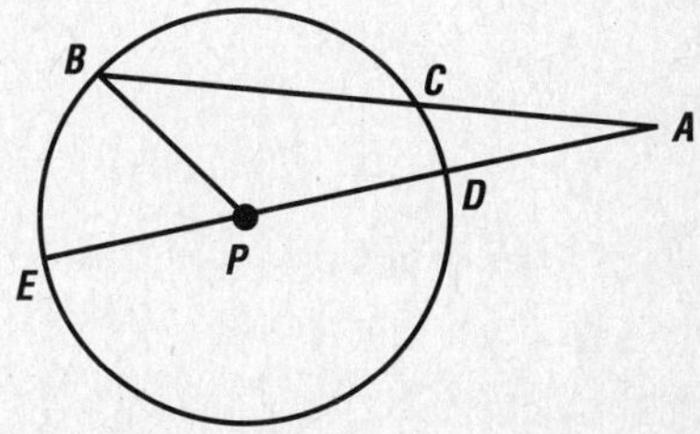

9. *PB*

NAME ______________________________ DATE ______________

9-7 Practice Worksheet

Special Segments in a Circle

In ⊙P, find the value of x.

1.

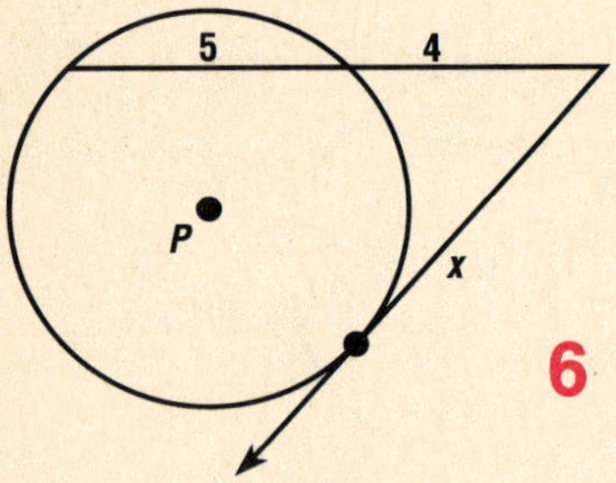

6

2.

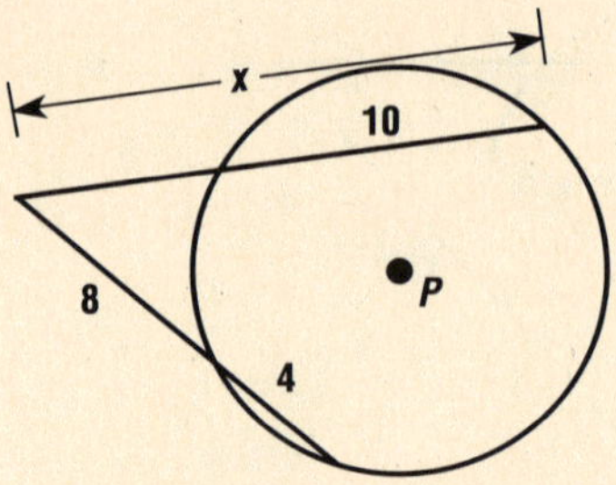

16

3.

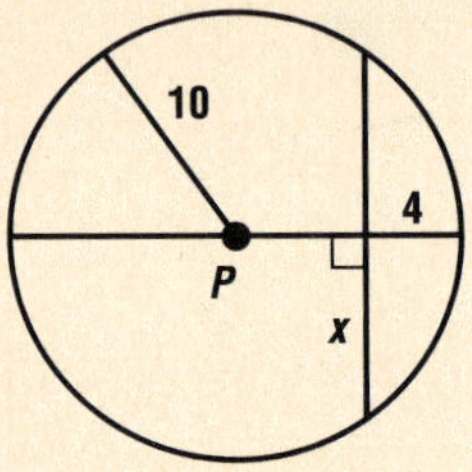

8

4.

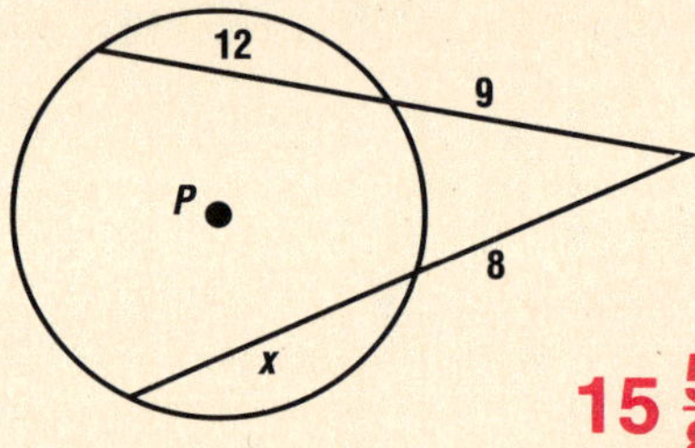

$15\frac{5}{8}$

In ⊙P, CE = 6, CD = 16, and AB = 17. Find each of the following.

5. *EB* 5

6. *AE* 12

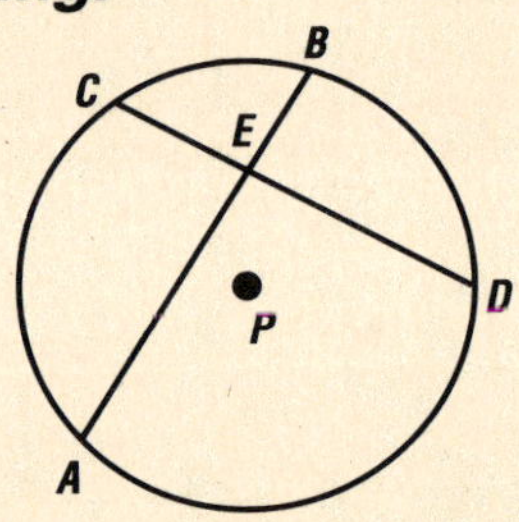

In ⊙P, AC = 3, BC = 5, and AD = 2. Find each of the following.

7. *PD* 5

8. *ED* 10

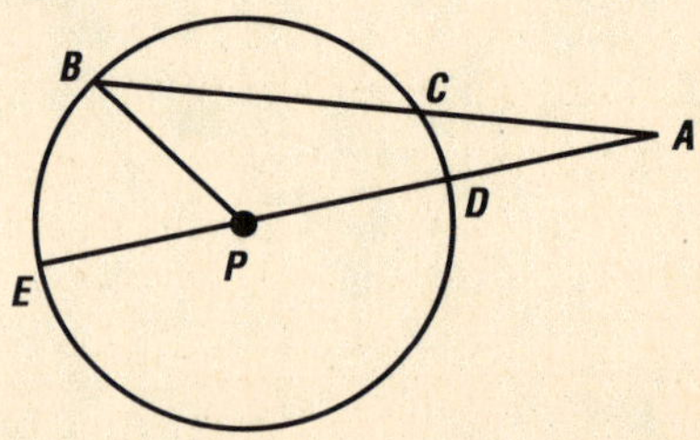

9. *PB* 5

NAME ______________________ DATE ____________

9-8 Practice Worksheet

Problem-Solving Strategy: Using Graphs

The table at the right shows how Federal funds were spent on education in 1990.

1990 Federal Funds Spent for Education

Elementary/Secondary	$ 7,945,177
Education/handicapped	4,204,099
Post secondary education	12,645,630
Public Library Services	145,367
Other	760,616
Total	$25,700,889

1. a. Use the information to make a circle graph.

b. Out of the $12,645,630 spent on post secondary education, $10,801,185 went to post secondary financial assistance. What percent is that of the $12,645,630?

2. Bill is 1,350,000,000 seconds old. How old is he in years?

3. Find the next number in the pattern 3, 6, 18, 72, 360, __?__ . Describe the pattern.

4. Suppose you mark three noncollinear points *inside* a circle. Mark a fourth point so that it is not collinear with any two of the first three. Continue: Mark a fifth point, being careful that it is not collinear with any of the other two already marked. For what number of points will it no longer be possible to do this?

5. How many sets of 100 points can you select from a set of 101 points if the order of selection is unimportant.

Glencoe Division, Macmillan/McGraw-Hill

9-8

NAME ______________________ DATE ______________

Practice Worksheet

Problem-Solving Strategy: Using Graphs

The table at the right shows how Federal funds were spent on education in 1990.

1990 Federal Funds Spent for Education

Elementary/Secondary	$ 7,945,177
Education/handicapped	4,204,099
Post secondary education	12,645,630
Public Library Services	145,367
Other	760,616
Total	$25,700,889

1. a. Use the information to make a circle graph.

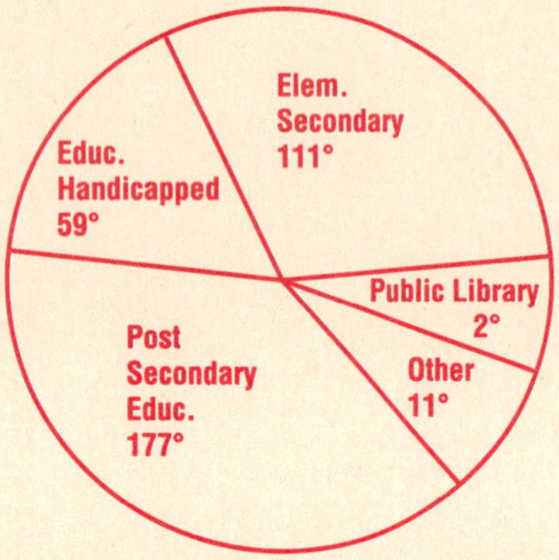

b. Out of the $12,645,630 spent on post secondary education, $10,801,185 went to post secondary financial assistance. What percent is that of the $12,645,630? **85.4%**

2. Bill is 1,350,000,000 seconds old. How old is he in years?
about 43 years old

3. Find the next number in the pattern 3, 6, 18, 72, 360, __?__ . Describe the pattern.
2160; Multiply by 2, 3, 4, 5, 6, . . .

4. Suppose you mark three noncollinear points *inside* a circle. Mark a fourth point so that it is not collinear with any two of the first three. Continue: Mark a fifth point, being careful that it is not collinear with any of the other two already marked. For what number of points will it no longer be possible to do this?
It will always be possible, since otherwise it would be possible to fill the circle and its interior with a finite number of segments.

5. How many sets of 100 points can you select from a set of 101 points if the order of selection is unimportant. **101**

NAME ______________________ DATE ______________

10-1 Practice Worksheet

Polygons and Polyhedra

State the number of sides for each convex polygon.

1. quadrilateral

2. octagon

3. 83-gon

4. heptagon

5. decagon

6. hexagon

Use polygon ABCDEFG to answer each question.

7. Name the vertices of the polygon.

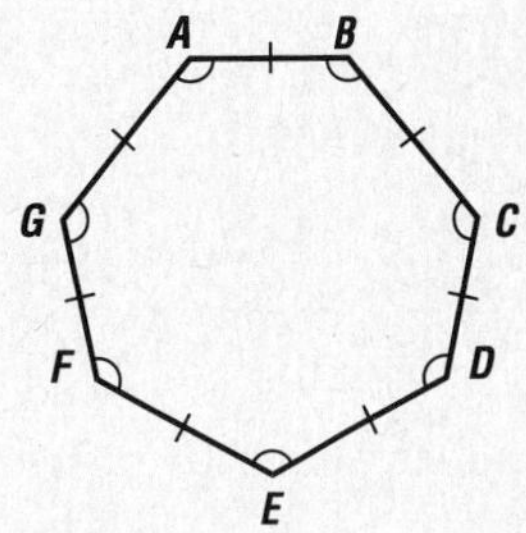

8. Name the angles of the polygon.

9. Name the sides of the polygon.

10. Is the polygon convex or concave?

11. Name the polygon according to the number of sides it has.

12. Is the polygon regular? Explain.

If possible, draw a polygon that fits each description.

13. a regular triangle

14. a quadrilateral with all angles congruent, but not all sides congruent

15. a concave hexagon with all sides congruent

16. a triangle with all sides congruent, but not all angles congruent

Glencoe Division, Macmillan/McGraw-Hill

10-1

NAME ____________________ DATE __________

Practice Worksheet

Polygons and Polyhedra

State the number of sides for each convex polygon.

1. quadrilateral **4** **2.** octagon **8** **3.** 83-gon **83**

4. heptagon **7** **5.** decagon **10** **6.** hexagon **6**

Use polygon ABCDEFG to answer each question.

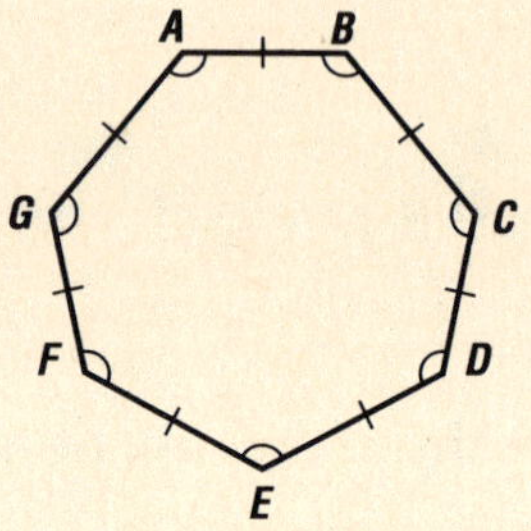

7. Name the vertices of the polygon.
A, B, C, D, E, F, G

8. Name the angles of the polygon.
∠A, ∠B, ∠C, ∠D, ∠E, ∠F, ∠G

9. Name the sides of the polygon.
$\overline{AB}, \overline{BC}, \overline{CD}, \overline{DE}, \overline{EF}, \overline{FG}, \overline{GA}$

10. Is the polygon convex or concave?
convex

11. Name the polygon according to the number of sides it has.
heptagon

12. Is the polygon regular? Explain.
yes; Its sides are all congruent, and its angles are all congruent, and it is convex.

If possible, draw a polygon that fits each description.

13. a regular triangle

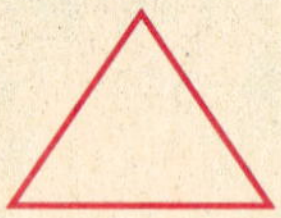

14. a quadrilateral with all angles congruent, but not all sides congruent

15. a concave hexagon with all sides congruent

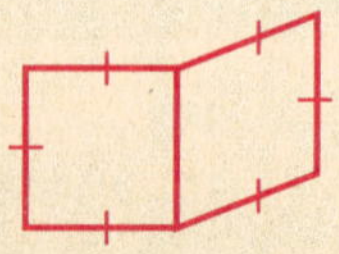

16. a triangle with all sides congruent, but not all angles congruent
impossible

NAME ______________________ DATE ____________

10-2 Practice Worksheet

Angles of Polygons

1. Find m$\angle E$ if m$\angle A$ = 160, m$\angle B$ = 50, and m$\angle C$ = m$\angle D$ = 140.

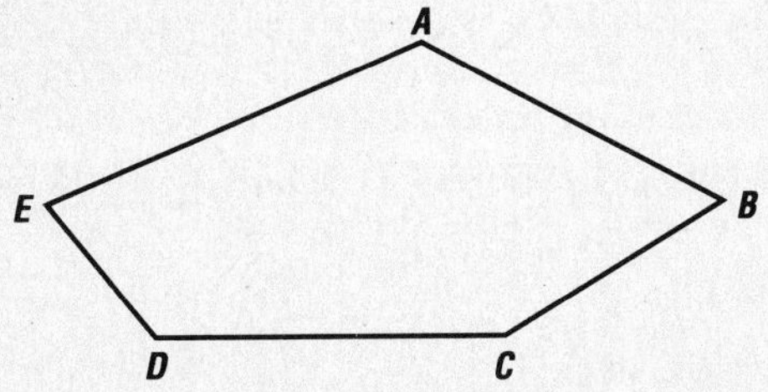

2. Find m$\angle 1$ if m$\angle G$ = 80, m$\angle F$ = 110, and m$\angle H$ = 74.

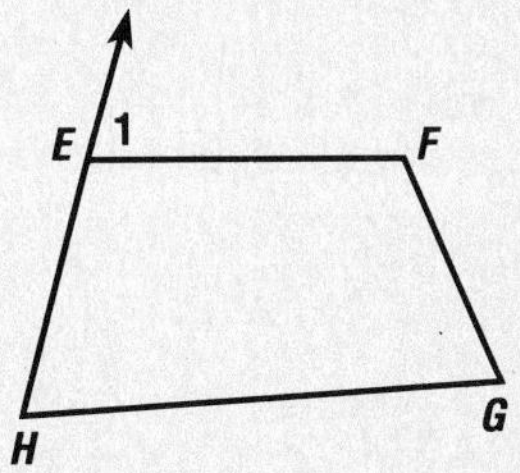

3. What is the fewest number of sides a polygon can have?

4. In what convex polygon is the sum of the measures of the exterior angles, one per vertex, equal to the sum of the measures of the interior angles?

Find the sum of the measures of the interior angles of each convex polygon.

5. heptagon
6. octagon
7. 13-gon

The number of sides of a regular polygon is given. Find the measure of an interior and an exterior angle of the polygon.

8. 5
9. 9
10. 10

The measure of one exterior angle of a regular polygon is given. Find the number of sides.

11. 30
12. 40
13. 90

Glencoe Division, Macmillan/McGraw-Hill

NAME ______________________ DATE ____________

10-2 **Practice Worksheet**

Angles of Polygons

1. Find $m\angle E$ if $m\angle A = 160$, $m\angle B = 50$, and $m\angle C = m\angle D = 140$. **50**

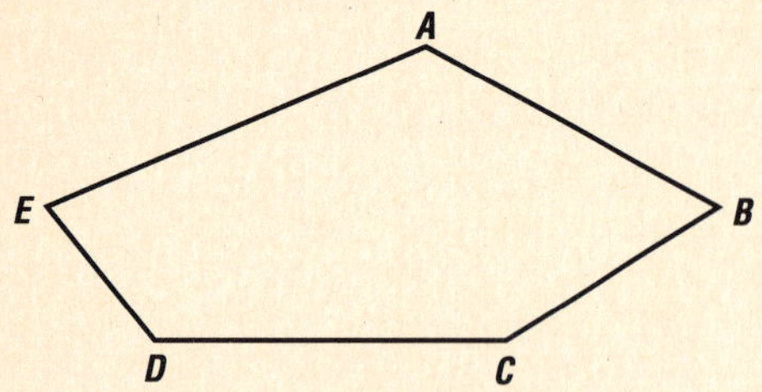

2. Find $m\angle 1$ if $m\angle G = 80$, $m\angle F = 110$, and $m\angle H = 74$. **84**

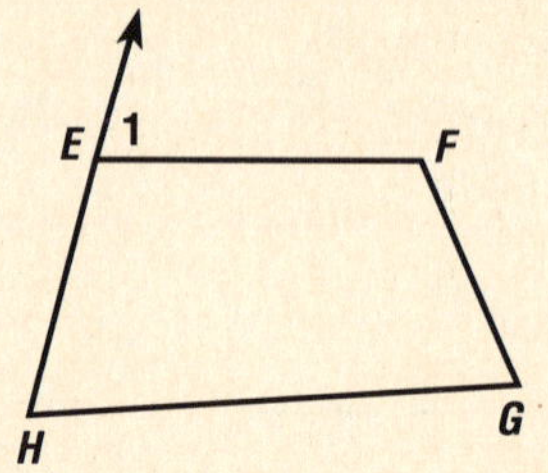

3. What is the fewest number of sides a polygon can have? **3**

4. In what convex polygon is the sum of the measures of the exterior angles, one per vertex, equal to the sum of the measures of the interior angles? **quadrilateral**

Find the sum of the measures of the interior angles of each convex polygon.

5. heptagon **900**

6. octagon **1080**

7. 13-gon **1980**

The number of sides of a regular polygon is given. Find the measure of an interior and an exterior angle of the polygon.

8. 5 **108, 72**

9. 9 **140,40**

10. 10 **144, 36**

The measure of one exterior angle of a regular polygon is given. Find the number of sides.

11. 30 **12**

12. 40 **9**

13. 90 **4**

NAME ________________________________ DATE ________________

10-3

Practice Worksheet

Problem Solving Strategy: Guess and Check

Solve. Use any strategy.

1. The house numbers on the Browns' house were nailed on a plaque upside down and backwards. The house number is four digits long. If the number that shows is 720 greater than their actual house number, what is the Browns' actual house number?

2. Which two consecutive whole numbers have cubes that differ by 2107?

3. Use the digits 1-9, once each, to fill in each circle so that the sum of each row, each column, and each diagonal is the same.

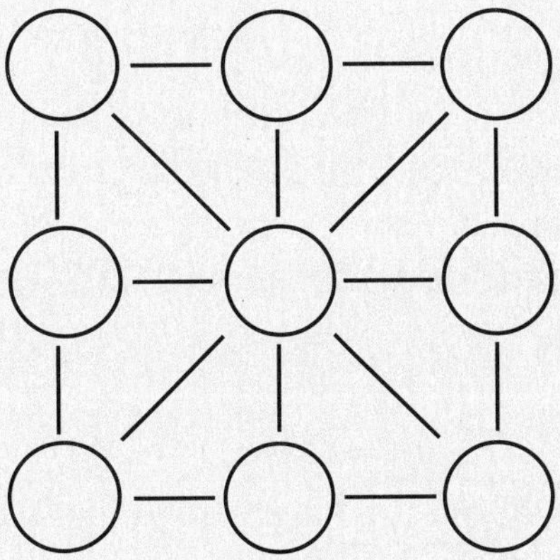

4. Insert operation symbols (+, −, ×, ÷) between the digits to form a true equation.

9 ____ 7 ____ 3 ____ 4 ____ 2 = 10

5. How many digits will there be in the product of 10000001 and 9999999? Try to answer *without* multiplying to get the actual product.

NAME ______________________________ DATE ______________

10-3 Practice Worksheet

Problem Solving Strategy: Guess and Check

Solve. Use any strategy.

1. The house numbers on the Browns' house were nailed on a plaque upside down and backwards. The house number is four digits long. If the number that shows is 720 greater than their actual house number, what is the Browns' actual house number? **6089**

2. Which two consecutive whole numbers have cubes that differ by 2107? **26, 27**

3. Use the digits 1-9, once each, to fill in each circle so that the sum of each row, each column, and each diagonal is the same.

Answers may vary. Sample answer:

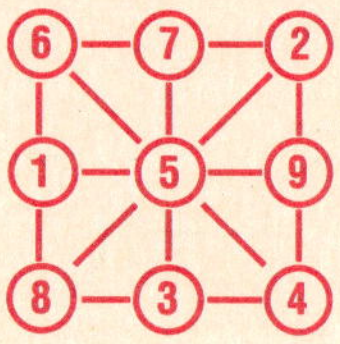

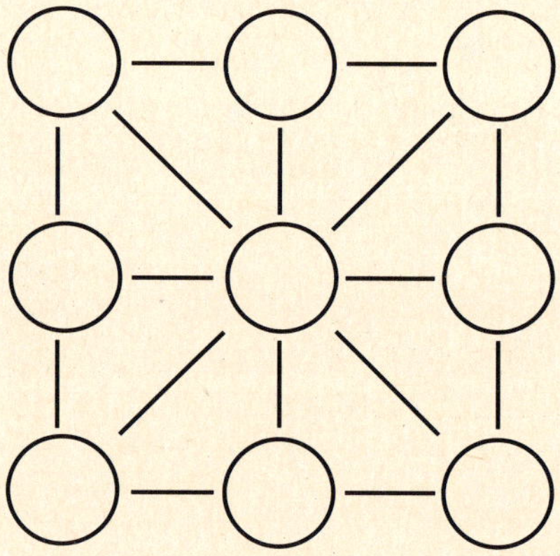

4. Insert operation symbols (+, −, ×, ÷) between the digits to form a true equation.

9 ____ 7 ____ 3 ____ 4 ____ 2 = 10

9 + 7 − 3 × 4 ÷ 2 = 10

5. How many digits will there be in the product of 10000001 and 9999999? Try to answer *without* multiplying to get the actual product. **14 digits (If one factor is $10^n + 1$ and the other is $10^n - 1$, the product will be $10^{2n} - 1$, which will have $2n$ digits, for any whole number n that is greater than 0).**

Glencoe Division, Macmillan/McGraw-Hill

NAME ______________________ DATE ______________

10-4 Practice Worksheet

Area of Parallelograms

Divide each figure into rectangles. Find the area of each figure.

1. 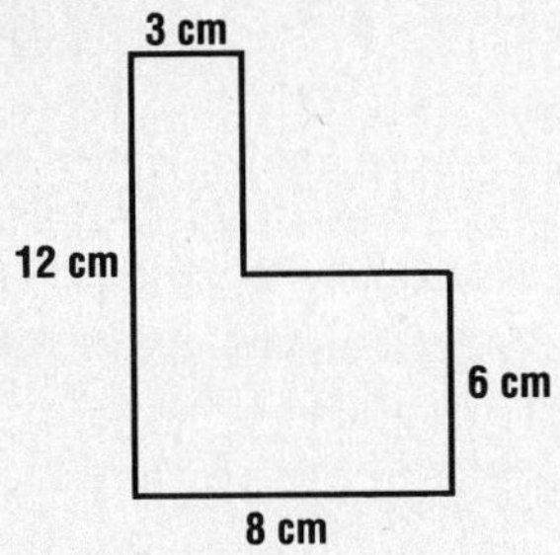

2. 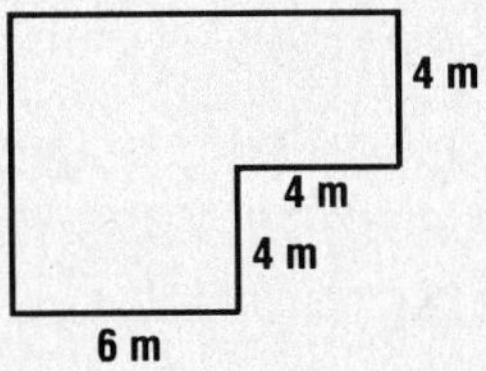

3. 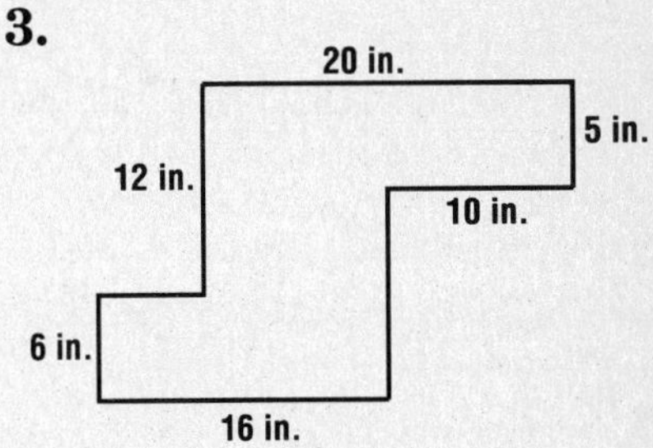

Find the area of each shaded region.

4. 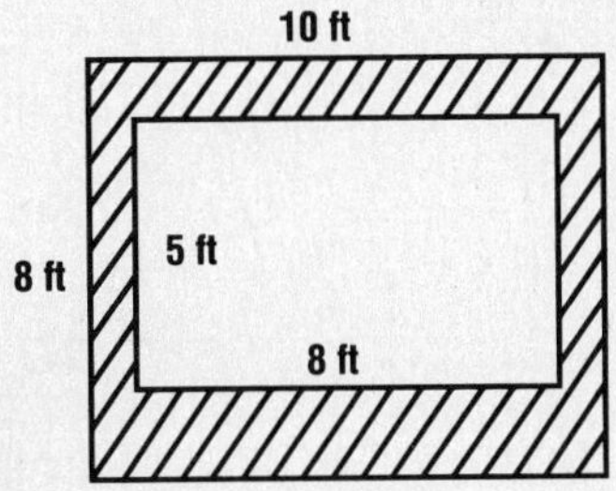

5. 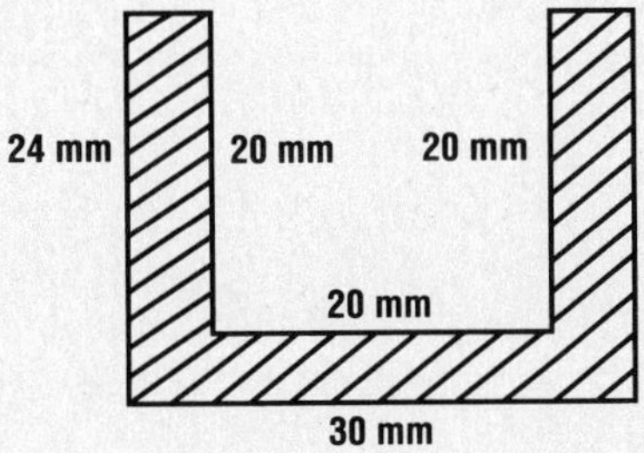

6. 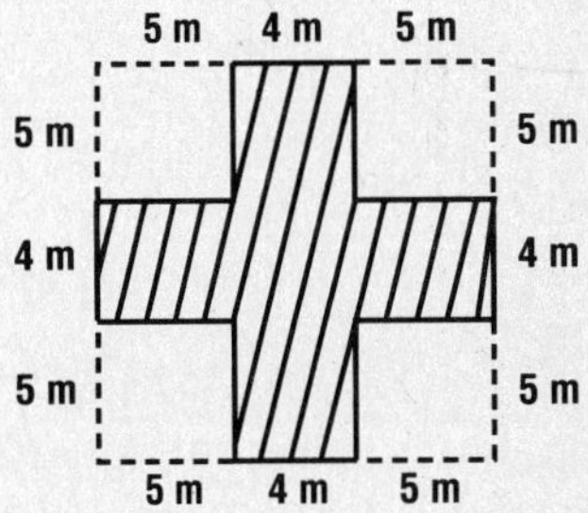

7. The sides of a parallelogram have lengths 8 inches and 16 inches and one of the angles of the parallelogram has a measure of 45°. Find the area of the parallelogram.

8. Find the area of the parallelogram that has vertices $A(0, 0)$, $B(2, 7)$, $C(10, 7)$, and $D(8, 0)$.

9. Find the area of the parallelogram that has vertices $W(-4, 15)$, $X(1, 15)$, $Y(4, 10)$, and $Z(-1, 10)$.

NAME ____________________ DATE __________

10-4 Practice Worksheet

Area of Parallelograms

Divide each figure into rectangles. Find the area of each figure.

1.

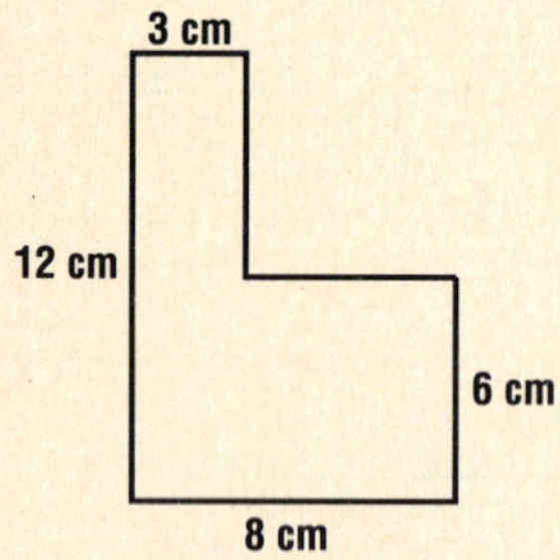

66 cm^2

2.

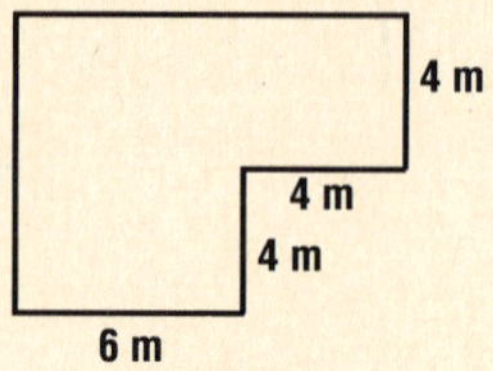

64 m^2

3.

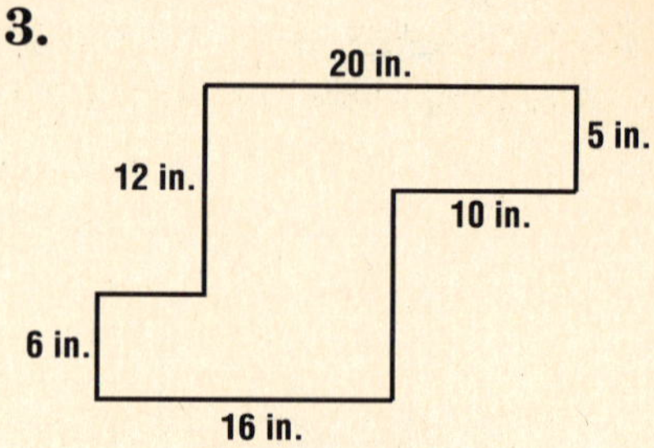

266 $in.^2$

Find the area of each shaded region.

4.

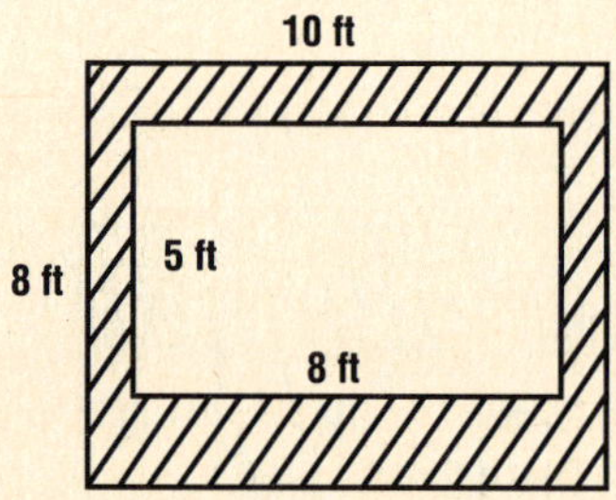

40 ft^2

5.

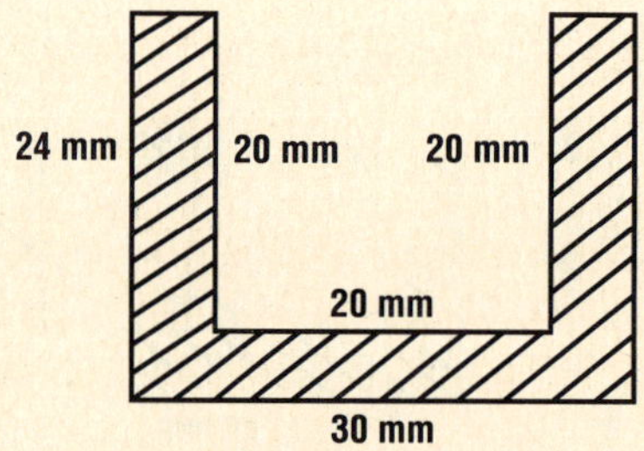

320 mm^2

6.

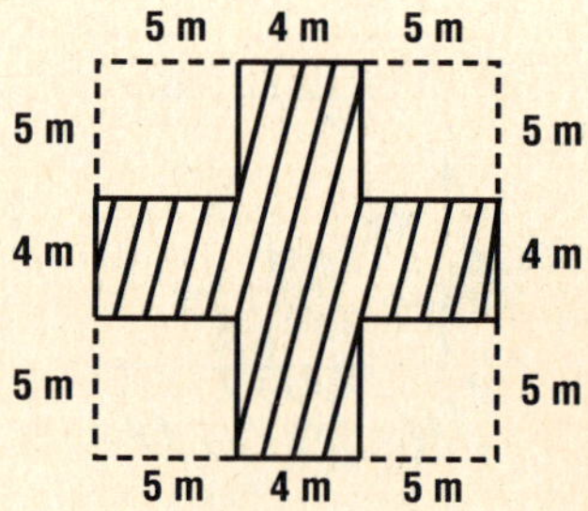

96 m^2

7. The sides of a parallelogram have lengths 8 inches and 16 inches and one of the angles of the parallelogram has a measure of 45°. Find the area of the parallelogram.
$64\sqrt{2}$ or about 90.5 in^2.

8. Find the area of the parallelogram that has vertices $A(0, 0)$, $B(2, 7)$, $C(10, 7)$, and $D(8, 0)$. 56 $units^2$

9. Find the area of the parallelogram that has vertices $W(-4, 15)$, $X(1, 15)$, $Y(4, 10)$, and $Z(-1, 10)$. 25 $units^2$

NAME ______________________________ DATE ______________

10-5 Practice Worksheet

Area of Triangles, Rhombi, and Trapezoids

Find each missing measure.

1. The area of a triangle is 216 square units. If the height is 18 units, what is the length of the base?

2. The diagonals of a rhombus are 21 and 16 centimeters long. Find the area of the rhombus.

3. The area of a trapezoid is 80 square units. If its height is 8 units, find the length of its median.

4. The height of a trapezoid is 9 cm. The bases are 8 cm and 12 cm long. Find the area.

5. A trapezoid has an area of 908.5 cm^2. If the altitude measures 23 cm and one base measures 36 cm, find the length of the other base.

6. The measures of the consecutive sides of an isosceles trapezoid are in the ratio 8:5:2:5. The perimeter of the trapezoid is 140 inches. If its height is 28 inches, find the area of the trapezoid.

NAME ______________________________ DATE ______________

10-5 Practice Worksheet

Area of Triangles, Rhombi, and Trapezoids

Find each missing measure.

1. The area of a triangle is 216 square units. If the height is 18 units, what is the length of the base?
   24 units

2. The diagonals of a rhombus are 21 and 16 centimeters long. Find the area of the rhombus.
   168 cm^2

3. The area of a trapezoid is 80 square units. If its height is 8 units, find the length of its median.
   10 units

4. The height of a trapezoid is 9 cm. The bases are 8 cm and 12 cm long. Find the area.
   90 cm^2

5. A trapezoid has an area of 908.5 cm^2. If the altitude measures 23 cm and one base measures 36 cm, find the length of the other base.
   43 cm

6. The measures of the consecutive sides of an isosceles trapezoid are in the ratio 8:5:2:5. The perimeter of the trapezoid is 140 inches. If its height is 28 inches, find the area of the trapezoid.
   980 in^2

NAME ________________________________ DATE ______________

10-6

Practice Worksheet

Area of Regular Polygons

Find the area of each regular polygon. Round your answers to the nearest tenth.

1. an octagon with an apothem 4.8 centimeters long and a side 4 centimeters long

2. a square with a side 24 inches long and an apothem 12 inches long

3. a hexagon with a side 23.1 meters long and an apothem 20.0 meters long

4. a pentagon with an apothem 316.6 millimeters long and a side 460 millimeters long

Find the perimeter, the length of the apothem, and the area of each regular polygon. Round your answers to the nearest tenth.

5.

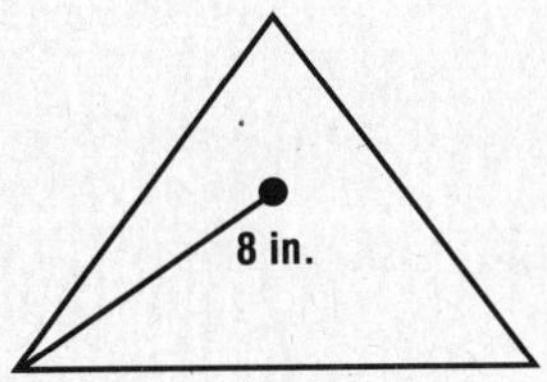

6.

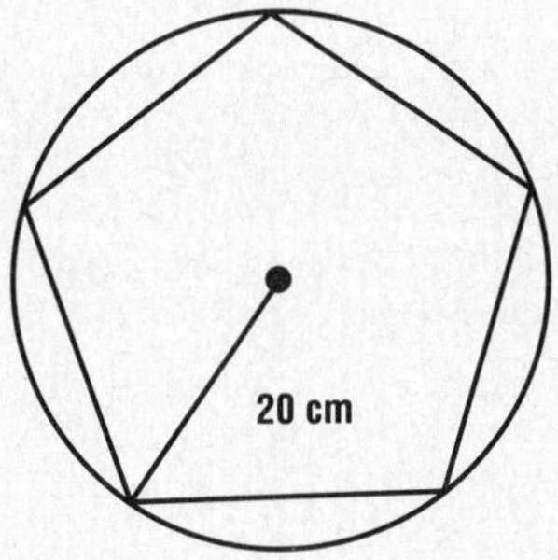

7.

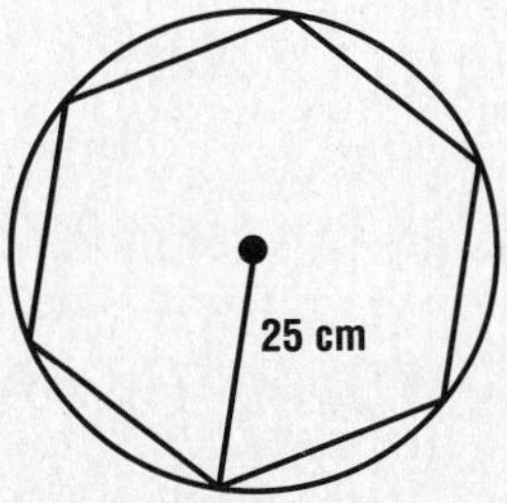

8.

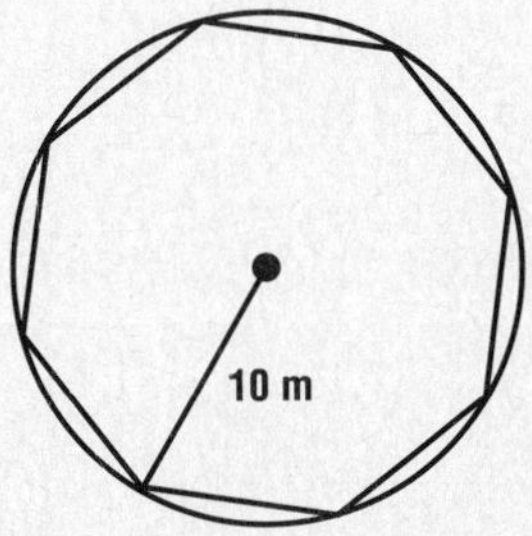

NAME ______________________ DATE __________

10-6 Practice Worksheet

Area of Regular Polygons

Find the area of each regular polygon. Round your answers to the nearest tenth.

1. an octagon with an apothem 4.8 centimeters long and a side 4 centimeters long
76.8 cm^2

2. a square with a side 24 inches long and an apothem 12 inches long
576 $in.^2$

3. a hexagon with a side 23.1 meters long and an apothem 20.0 meters long
1386 m^2

4. a pentagon with an apothem 316.6 millimeters long and a side 460 millimeters long
364,090 mm^2

Find the perimeter, the length of the apothem, and the area of each regular polygon. Round your answers to the nearest tenth.

5.

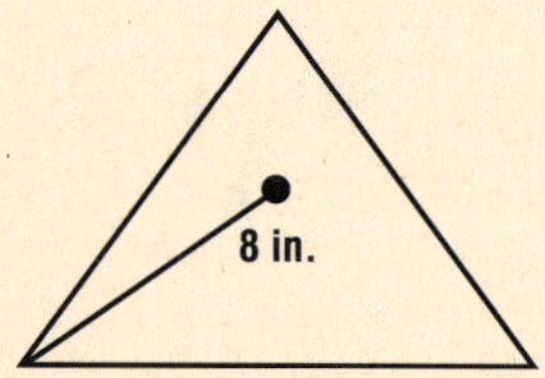

6.

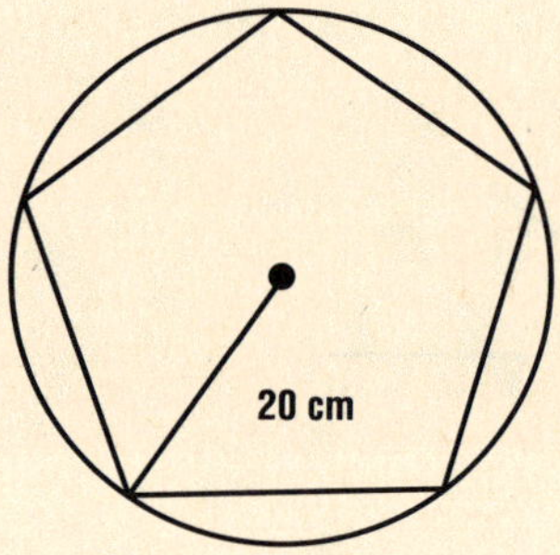

41.6 in., 4 in., 83.1 $in.^2$
117.6 cm, 16.2 cm, 951.1 cm^2

7.

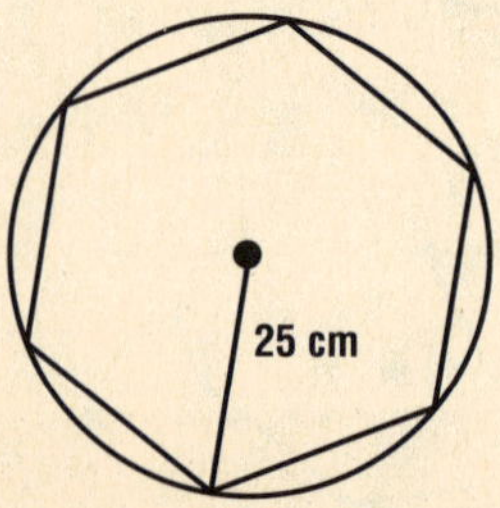

150 cm, 21.7 cm, 1623.8 cm^2

8.

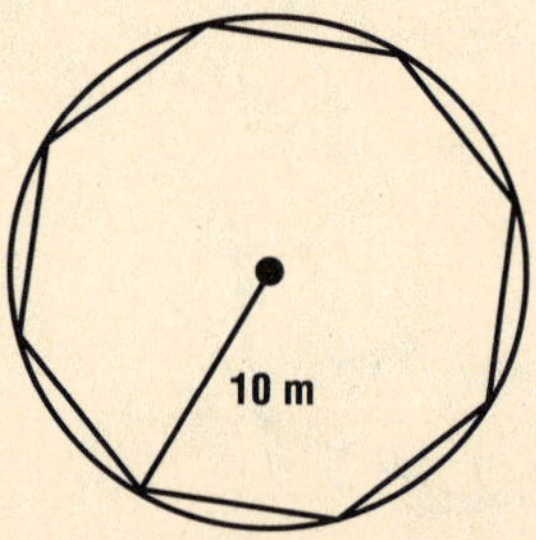

61.2 m, 9.2 m, 282.8 m^2

NAME ______________________ DATE ____________

10-7 Practice Worksheet

Area and Circumference of a Circle

Find the circumference of a circle with a radius of the given length. Round your answers to the nearest tenth.

1. 3 cm

2. 2 ft

3. 34 mm

4. 4.5 m

Find the area of a circle with a radius of the given length. Round your answers to the nearest tenth.

5. 15 cm

6. 2.6 m

7. 3 yd

8. 14 in.

State the measure of the radius of a circle with each equation. Then find the area of each circle. Give exact answers in terms of π.

9. $x^2 + y^2 = 49$

10. $x^2 + (y + 8)^2 = 81$

11. $(x - 3)^2 + (y + \frac{1}{2})^2 = \frac{9}{25}$

12. $(x - 2)^2 + (y - 3)^2 = 64$

Find the area of each shaded region. Express each answer as an exact number in terms of π and as a decimal rounded to the nearest hundredth.

13.

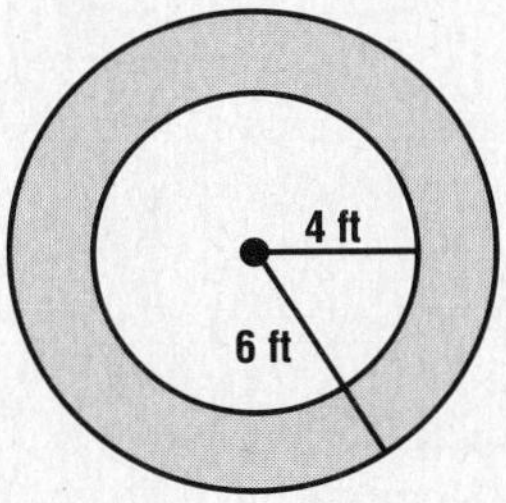

14.

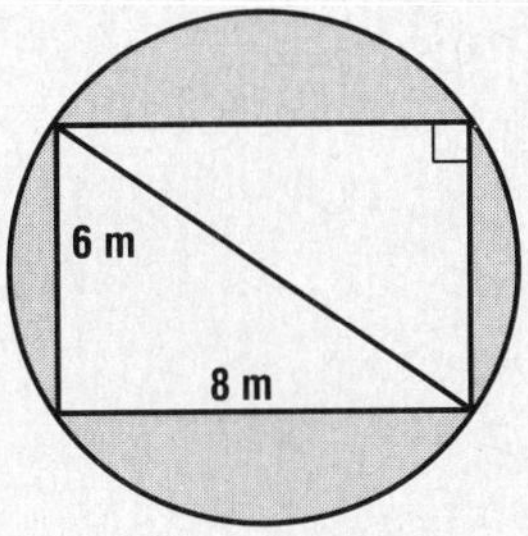

15.

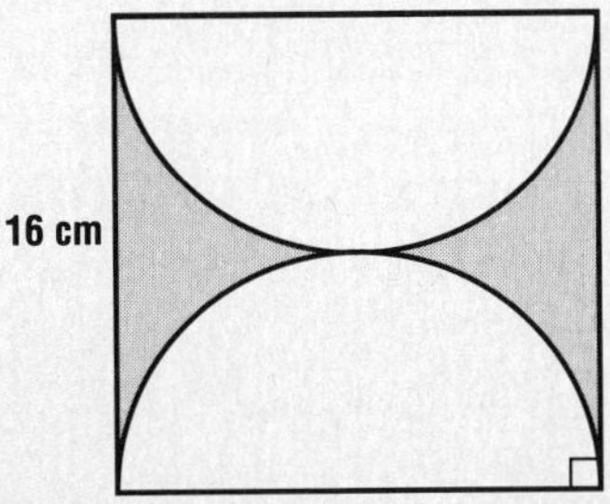

16.

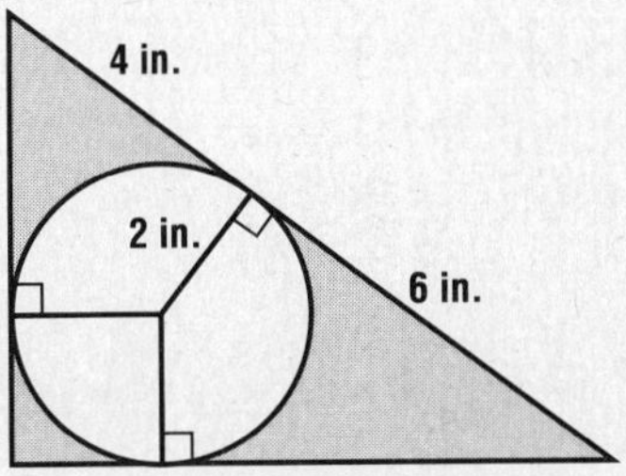

NAME ______________________ DATE ____________

10-7

Practice Worksheet

Area and Circumference of a Circle

Find the circumference of a circle with a radius of the given length. Round your answers to the nearest tenth.

1. 3 cm **18.8 cm**

2. 2 ft **12.6 ft**

3. 34 mm **213.6 mm**

4. 4.5 m **28.3 m**

Find the area of a circle with a radius of the given length. Round your answers to the nearest tenth.

5. 15 cm **706.9 cm^2**

6. 2.6 m **21.2 m^2**

7. 3 yd **28.3 yd^2**

8. 14 in. **615.8 in.^{2}**

State the measure of the radius of a circle with each equation. Then find the area of each circle. Give exact answers in terms of π.

9. $x^2 + y^2 = 49$ **7; 49π units^2**

10. $x^2 + (y + 8)^2 = 81$ **9; 81π units^2**

11. $(x - 3)^2 + (y + \frac{1}{2})^2 = \frac{9}{25}$ **$\frac{3}{5}$; $\frac{9}{25}\pi$ units^2**

12. $(x - 2)^2 + (y - 3)^2 = 64$ **8; 64π units^2**

Find the area of each shaded region. Express each answer as an exact number in terms of π and as a decimal rounded to the nearest hundredth.

13.

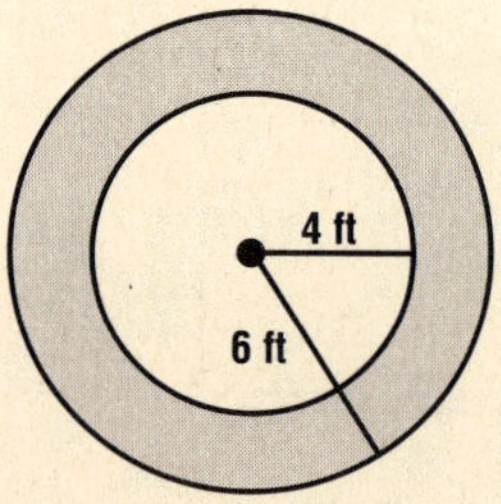

20π ft^2, 62.83 ft^2

14.

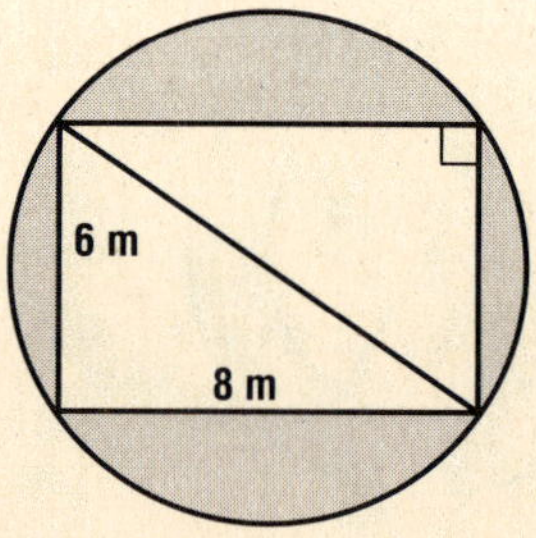

25π – 48 m^2, 30.54 m^2

15.

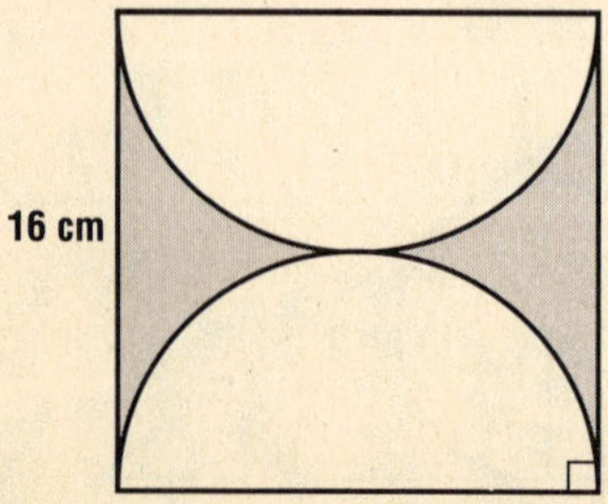

256 – 64π cm^2, 54.94 cm^2

16.

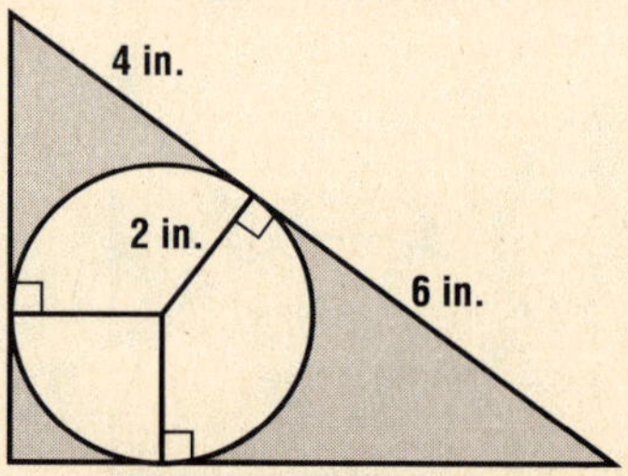

24 – 4π in.^{2}, 11.43 in.^{2}

NAME ______________________ DATE ____________

10-8 Practice Worksheet

Geometric Probability

A point is chosen at random on $\overline{AB}$. C is the midpoint of $\overline{AB}$, D is the midpoint of $\overline{AC}$, and E is the midpoint of $\overline{AD}$. Find each probability.

1. The point is on $\overline{AC}$.

2. The point is on $\overline{CB}$.

3. The point is on $\overline{AD}$.

4. The point is on $\overline{DB}$.

5. The point is on $\overline{ED}$.

6. The point is on $\overline{EC}$.

Find the probability that a point chosen at random in each figure is in the shaded region. Assume polygons that appear to be regular are regular. Round your answers to the nearest hundredth.

7.

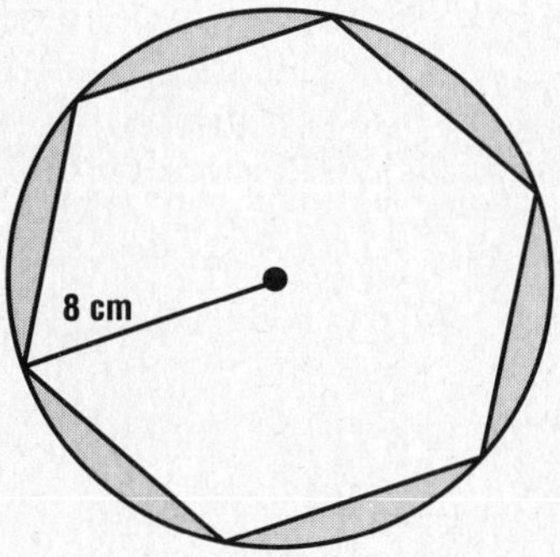

8.

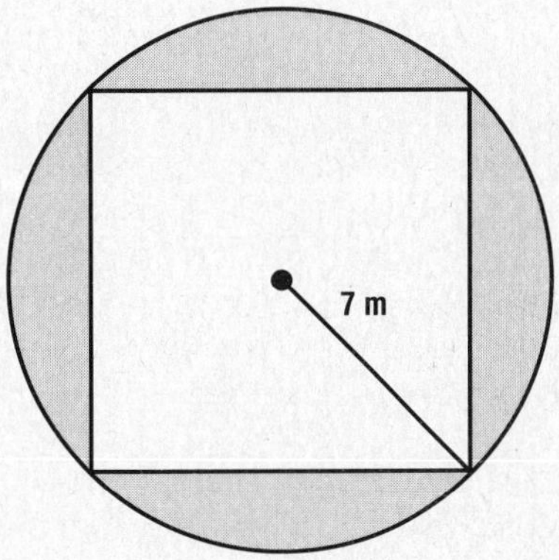

9.

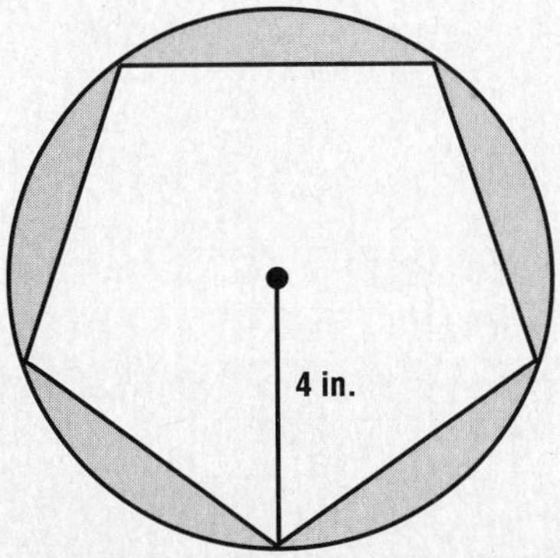

10.

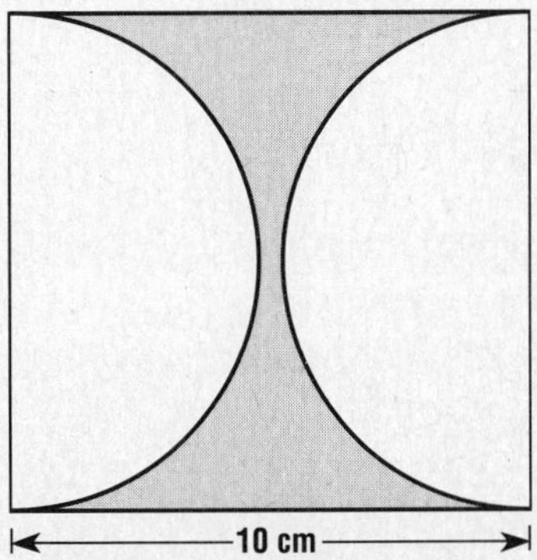

Glencoe Division, Macmillan/McGraw-Hill

NAME ______________________________ DATE ______________

10-8 Practice Worksheet

Geometric Probability

A point is chosen at random on $\overline{AB}$. C is the midpoint of $\overline{AB}$, D is the midpoint of $\overline{AC}$, and E is the midpoint of $\overline{AD}$. Find each probability.

1. The point is on $\overline{AC}$. $\frac{1}{2}$

2. The point is on $\overline{CB}$. $\frac{1}{2}$

3. The point is on $\overline{AD}$. $\frac{1}{4}$

4. The point is on $\overline{DB}$. $\frac{3}{4}$

5. The point is on $\overline{ED}$. $\frac{1}{8}$

6. The point is on $\overline{EC}$. $\frac{3}{8}$

Find the probability that a point chosen at random in each figure is in the shaded region. Assume polygons that appear to be regular are regular. Round your answers to the nearest hundredth.

7.

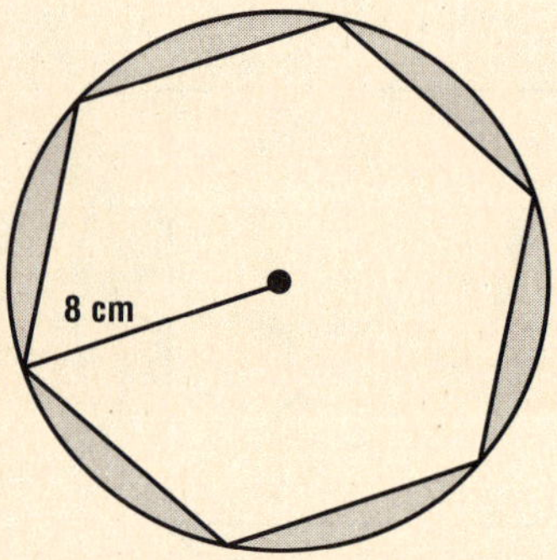

0.17

8.

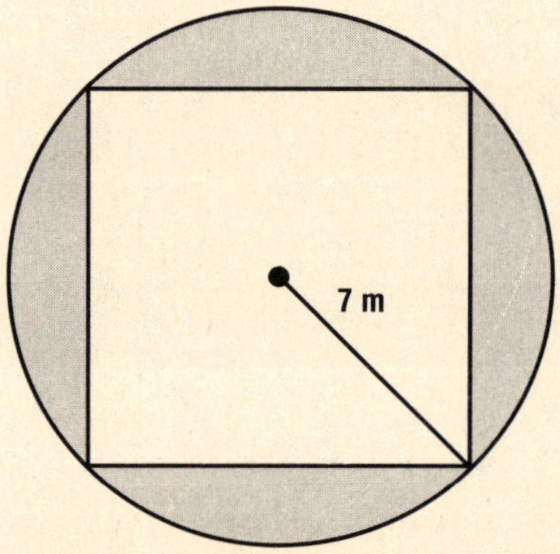

0.36

9.

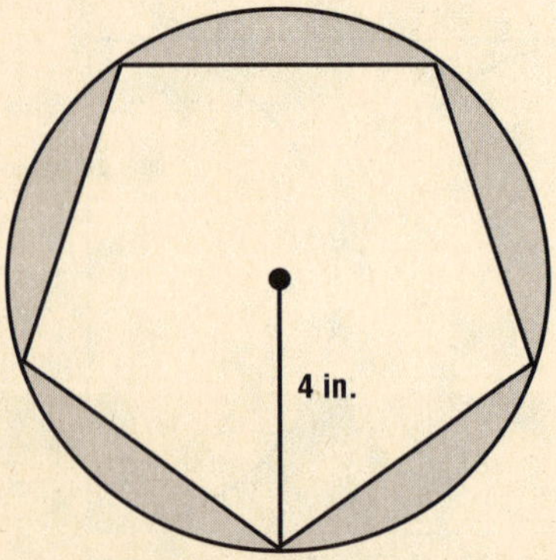

0.24

10.

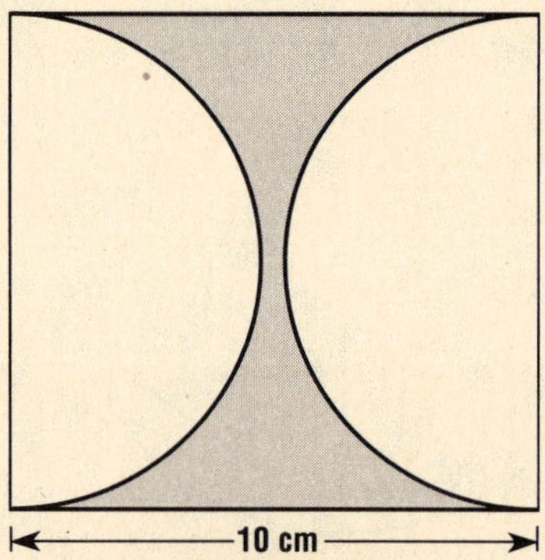

0.21

Glencoe Division, Macmillan/McGraw-Hill

NAME ______________________ DATE ____________

10-9

Practice Worksheet

Polygons As Networks

Find the degree of each node in the network.

1.

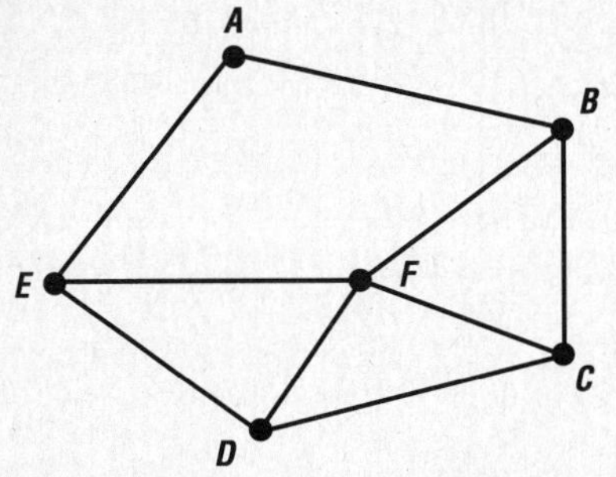

2.

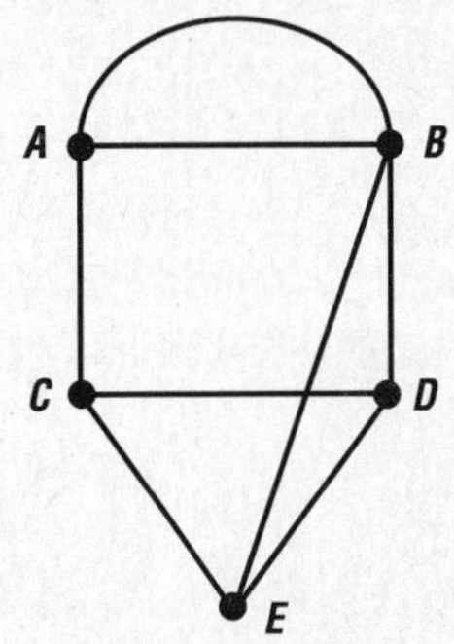

Name the additional edges that must be drawn for each network to be complete.

3.

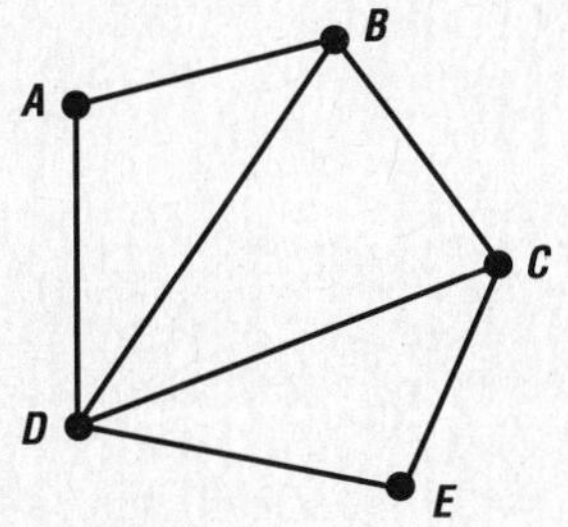

4.

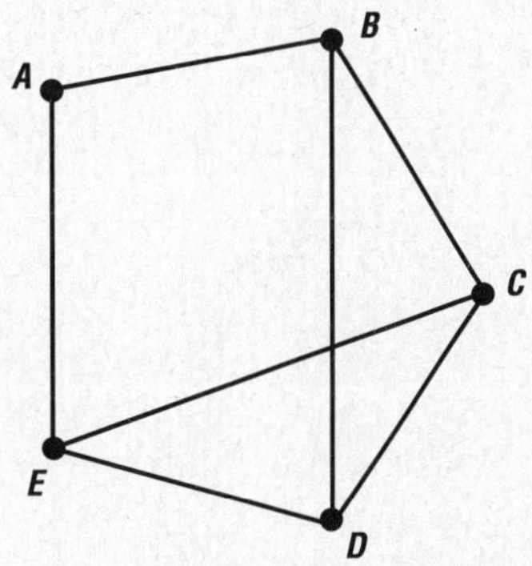

Determine whether each network is traceable. If a network is traceable, use arrows to trace the network.

5.

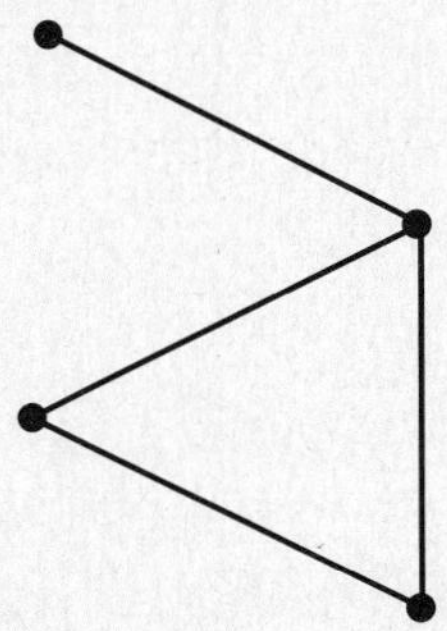

6.

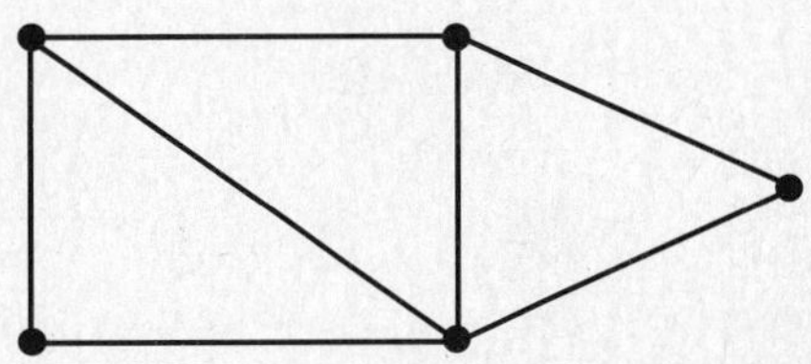

7. Draw two networks of your own that are traceable. Then draw two networks that are not traceable.

Glencoe Division, Macmillan/McGraw-Hill

NAME ________________ DATE ________

10-9 Practice Worksheet

Polygons As Networks

Find the degree of each node in the network.

1.

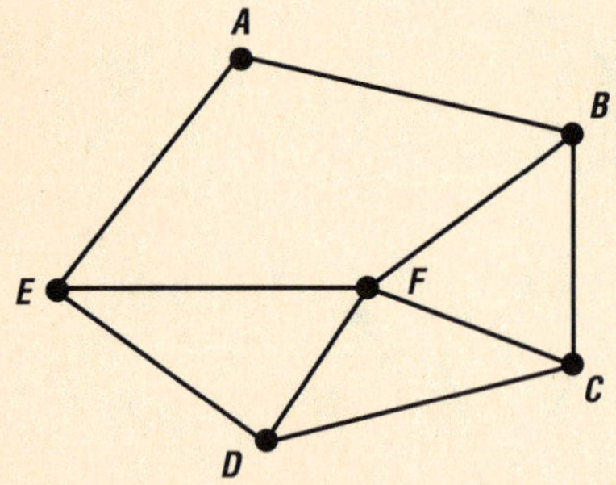

A - 2, *B* - 3, *C* - 3, *D* - 3, *F* - 4

2.

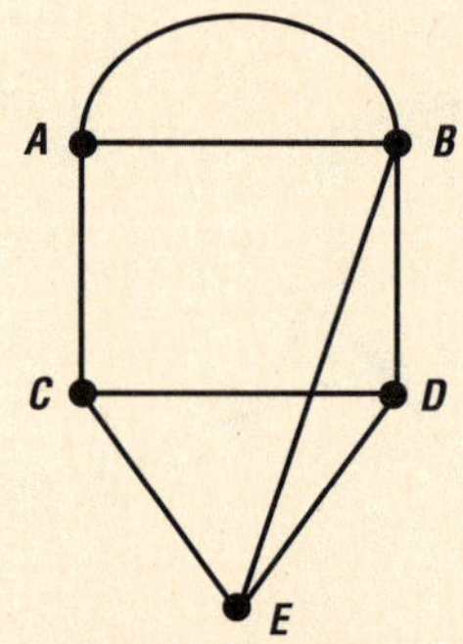

A - 3, *B* - 4, *C* - 3, *D* - 3, *E* - 3

Name the additional edges that must be drawn for each network to be complete.

3.

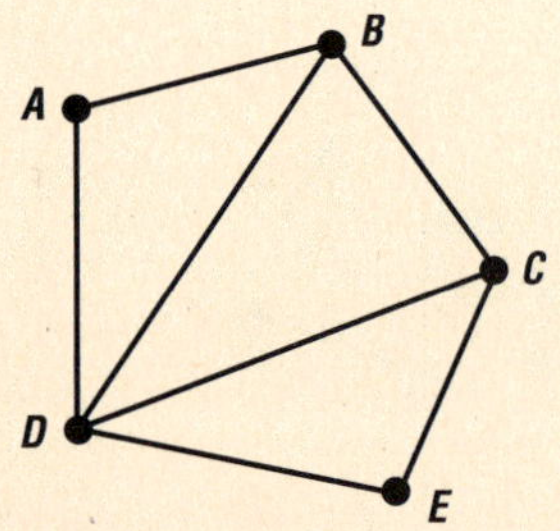

edges between: *A* and *C*, *A* and *E*, and *B* and *E*

4.

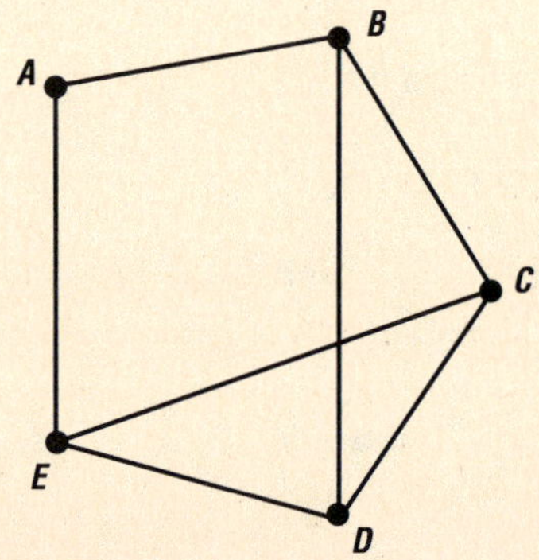

edges between: *A* and *C*, *A* and *D*, and *B* and *E*

Determine whether each network is traceable. If a network is traceable, use arrows to trace the network.

5.

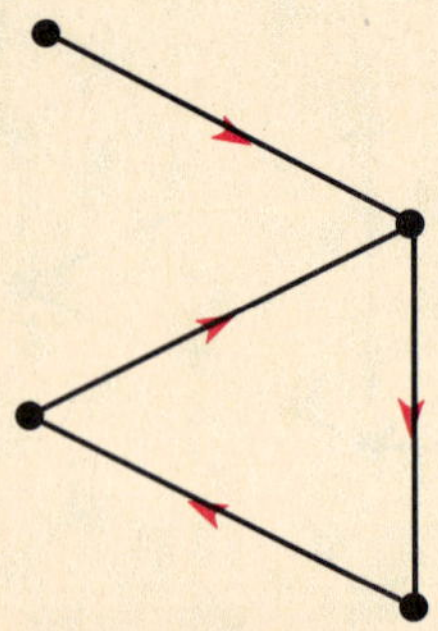

traceable

6.

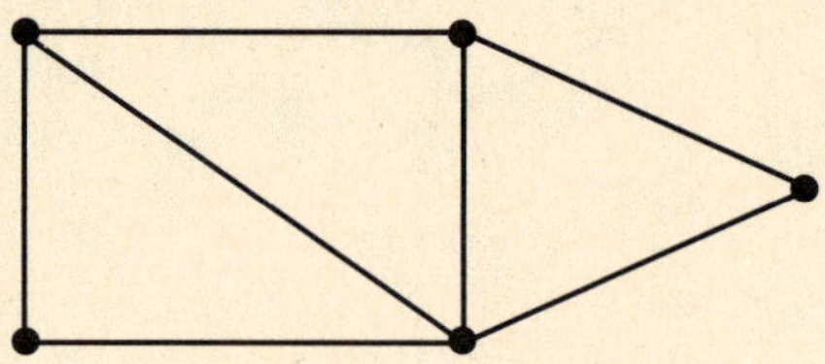

not traceable

7. Draw two networks of your own that are traceable. Then draw two networks that are not traceable. See students' work.

NAME ______________________ DATE ____________

11-1 **Practice Worksheet**

Problem-Solving Strategy: Make a Model

Solve. Use any strategy.

1. Insert one addition, one subtraction, one multiplication, and one division sign on the left side of the equals sign to make the statement true.

1 2 3 4 5 6 7 8 9 = 158

2. A certain bacteria doubles its population every 6 hours. At the end of the second full day, there were 800,000 bacteria. How many bacteria were there at the beginning of the first day?

3. Consider the perfect squares between 5000 and 10,001. What percent of these numbers are even numbers?

4. Two pennies, two nickels, three dimes, and a quarter are in a change purse. Three coins are removed at random. What are the possible values of the coins?

5. Suppose you draw five straight lines across a rectangle and cut along each line. What is the greatest number of sides possible for the polygonal pieces you obtain? What is maximum possible number of such pieces?

NAME ______________________________ DATE ______________

11-1

Practice Worksheet

Problem-Solving Strategy: Make a Model

Solve. Use any strategy.

1. Insert one addition, one subtraction, one multiplication, and one division sign on the left side of the equals sign to make the statement true.

1 2 3 4 5 6 7 8 9 = 158

$12 \div 3 \times 45 + 67 - 89 = 158$

2. A certain bacteria doubles its population every 6 hours. At the end of the second full day, there were 800,000 bacteria. How many bacteria were there at the beginning of the first day?

3125

3. Consider the perfect squares between 5000 and 10,001. What percent of these numbers are even numbers?

50%

4. Two pennies, two nickels, three dimes, and a quarter are in a change purse. Three coins are removed at random. What are the possible values of the coins?

7¢, 11¢, 12¢ (2 pennies, 1 dime), 16¢, 20¢, 21¢, 25¢, 27¢, 30¢, 31¢, 35¢, 36¢, 40¢, 45¢

5. Suppose you draw five straight lines across a rectangle and cut along each line. What is the greatest number of sides possible for the polygonal pieces you obtain? What is maximum possible number of such pieces?

9; 1

NAME ______________________ DATE ______________

11-2 **Practice Worksheet**

Exploring Surface Area

Given each polyhedron, label the remaining vertices of its net.

1.

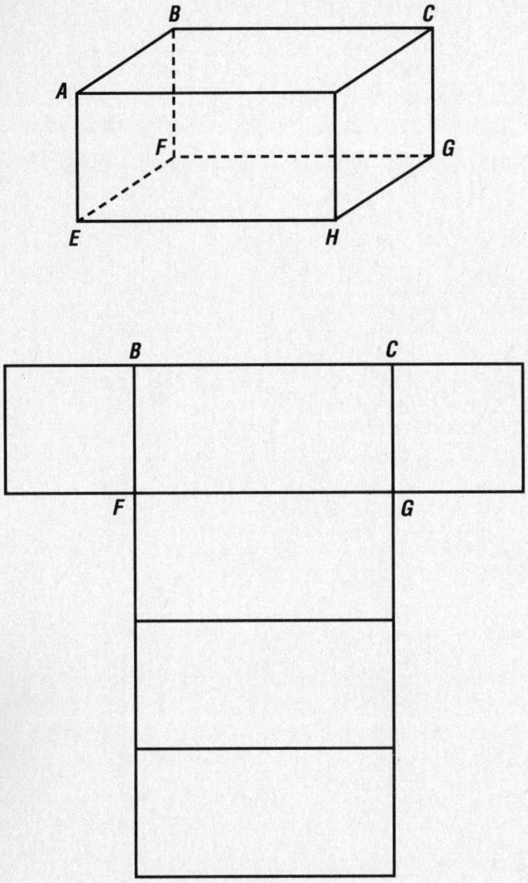

2.

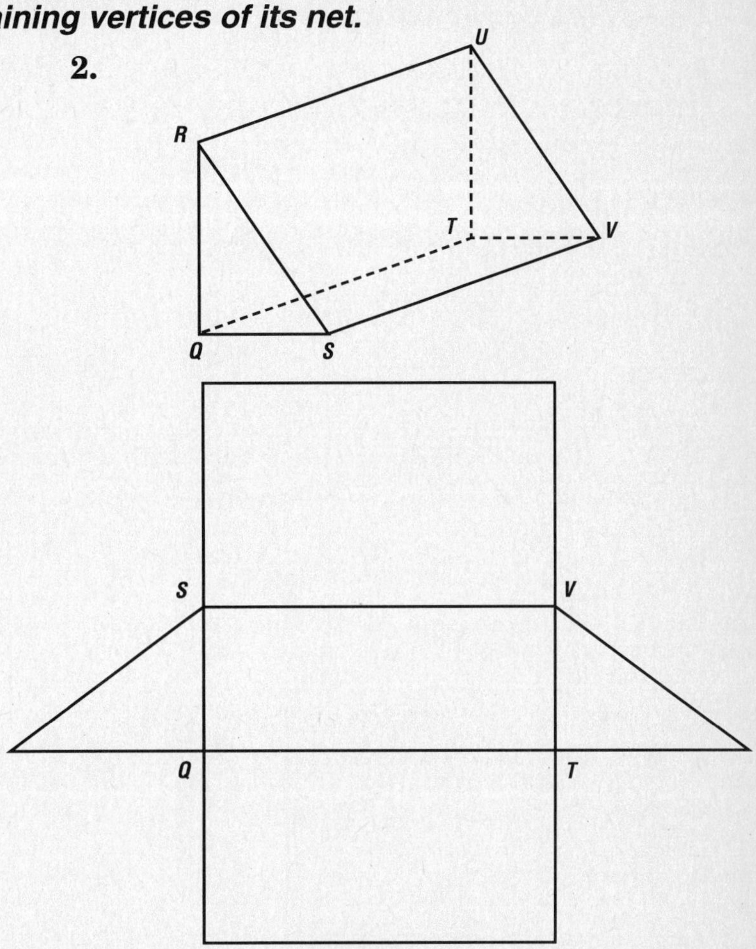

Identify the number and type of polygons that are faces in each polyhedron. Assume that edges that appear for all are parallel.

3.

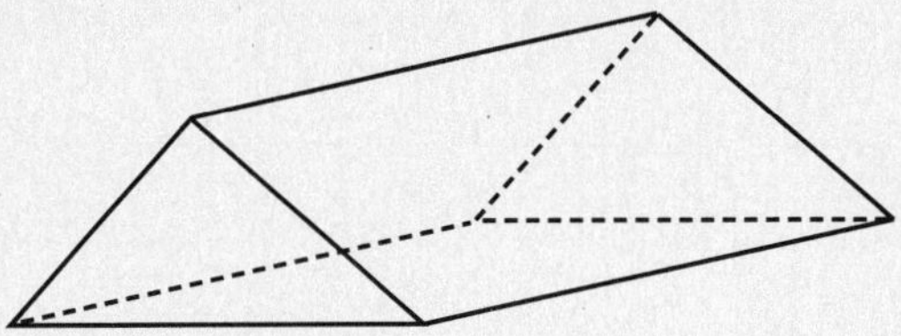

4.

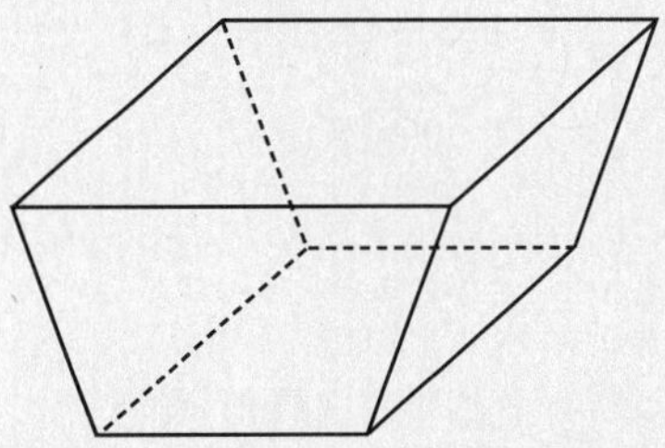

5.

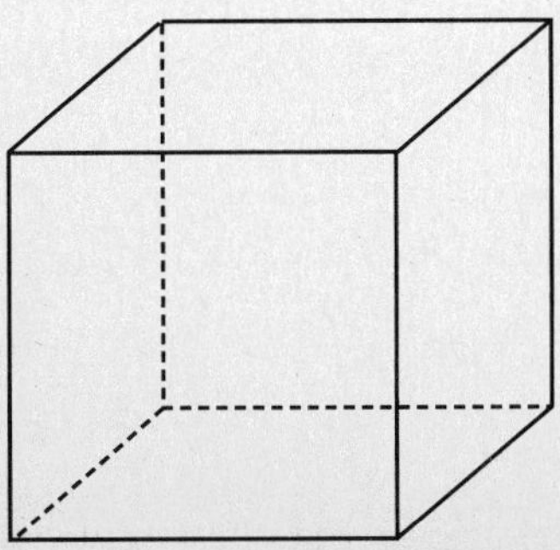

6.

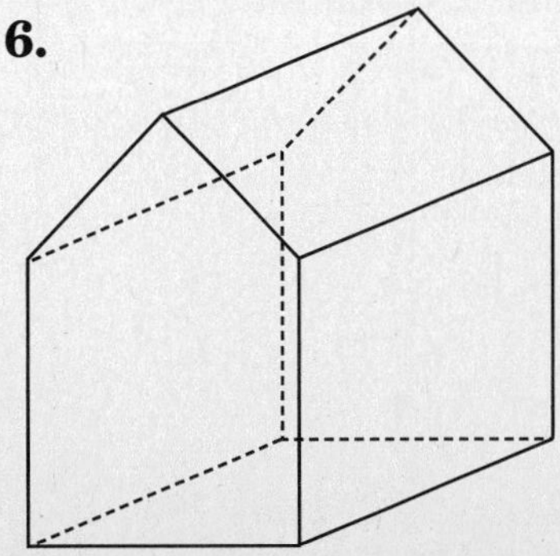

NAME ______________________ DATE ____________

11-2 Practice Worksheet

Exploring Surface Area

Given each polyhedron, label the remaining vertices of its net.

1.

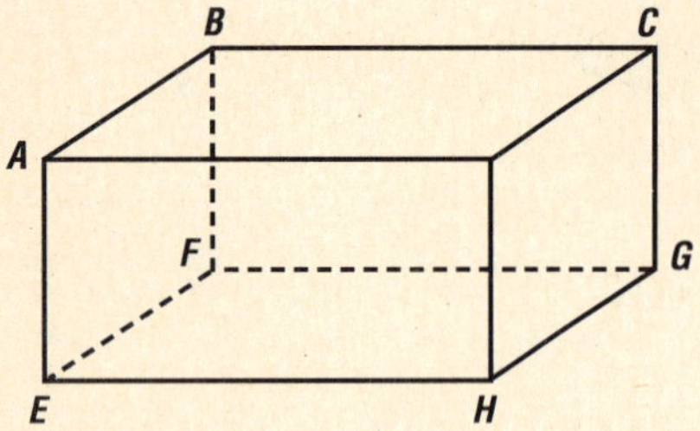

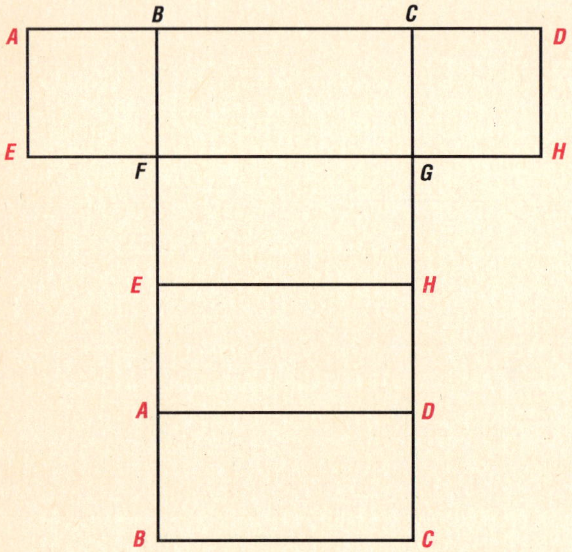

2.

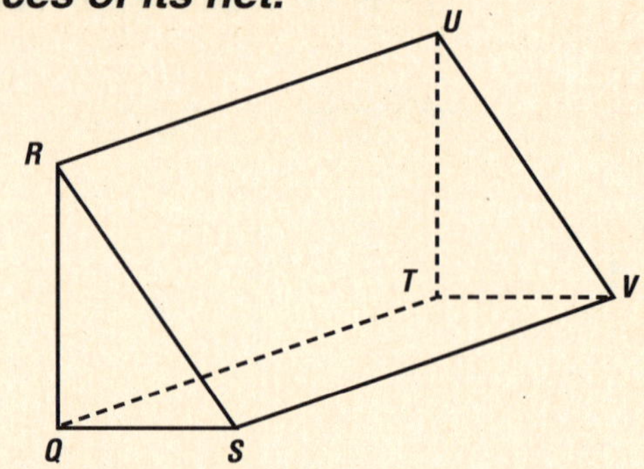

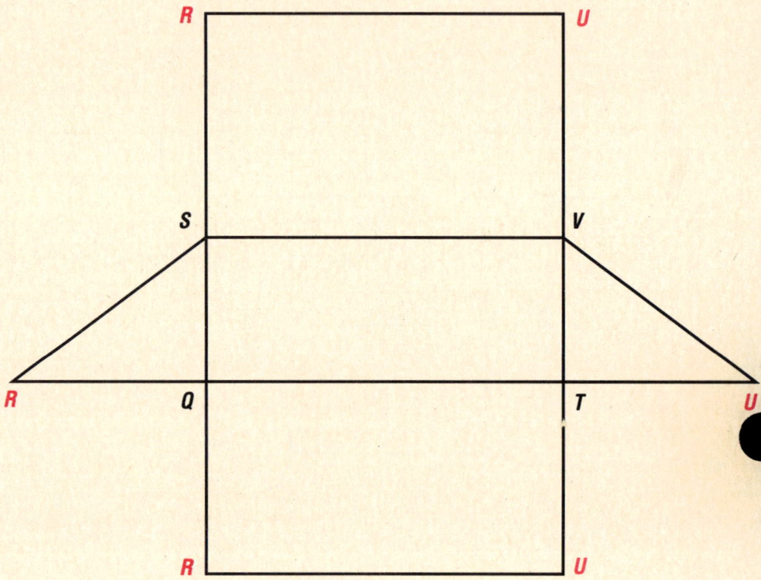

Identify the number and type of polygons that are faces in each polyhedron. Assume that edges that appear for all are parallel.

3.

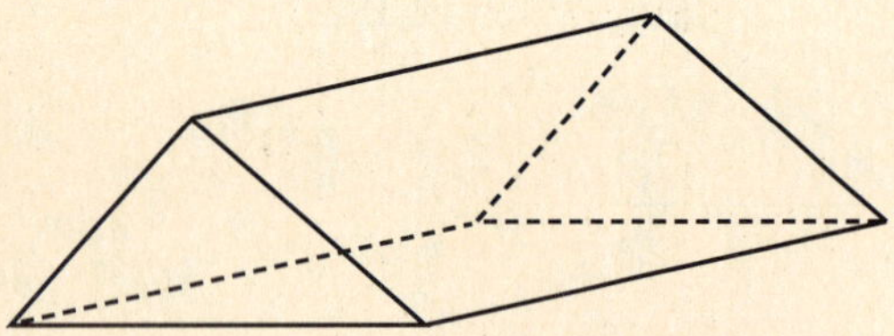

2 triangles
3 rectangles

4.

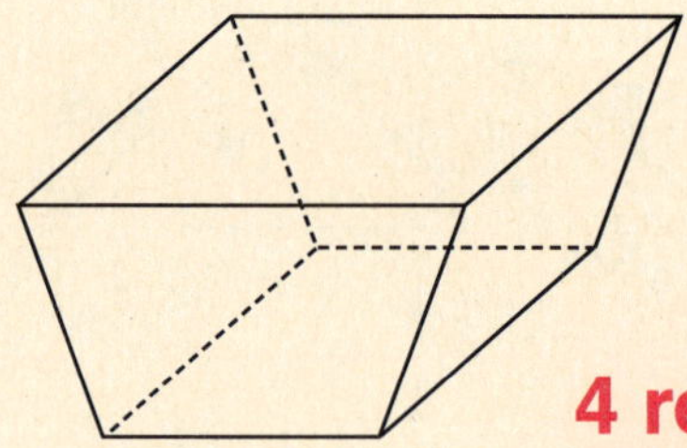

4 rectangles
2 trapezoids

5.

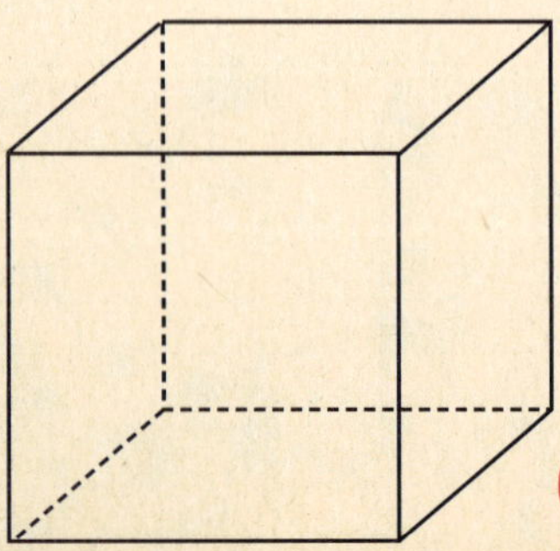

6 squares

6.

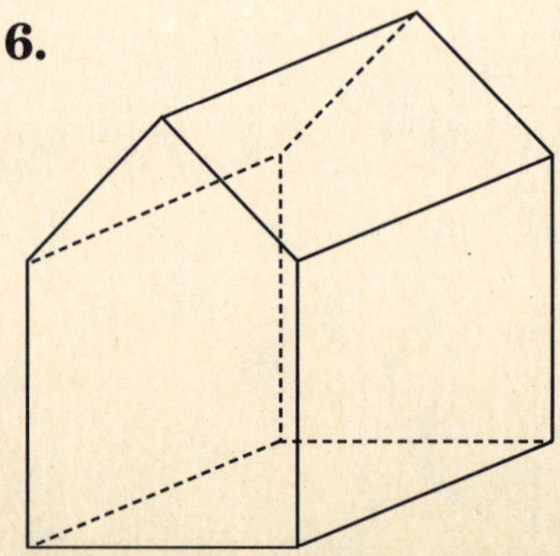

5 rectangles
2 pentagons

NAME ______________________ DATE __________

11-3 Practice Worksheet

Surface Area of Prisms and Cylinders

Use the right cylinders shown to answer each of the following. Express all answers in terms of π.

1. Find the circumference of the base.

2. Find the lateral area.

3. Find the area of a base.

4. Find the surface area.

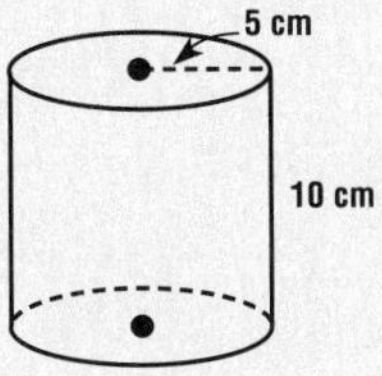

5. Find the circumference of the base.

6. Find the lateral area.

7. Find the area of a base.

8. Find the surface area.

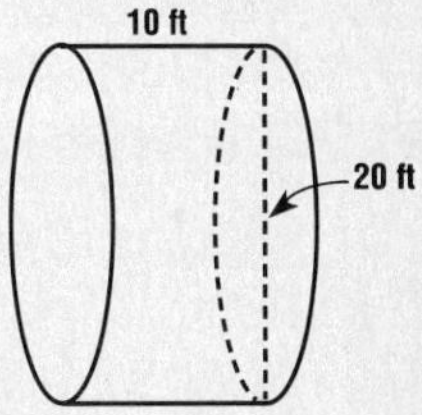

Find the lateral area and the surface area of each right prism. Round your answers to the nearest tenth.

9.

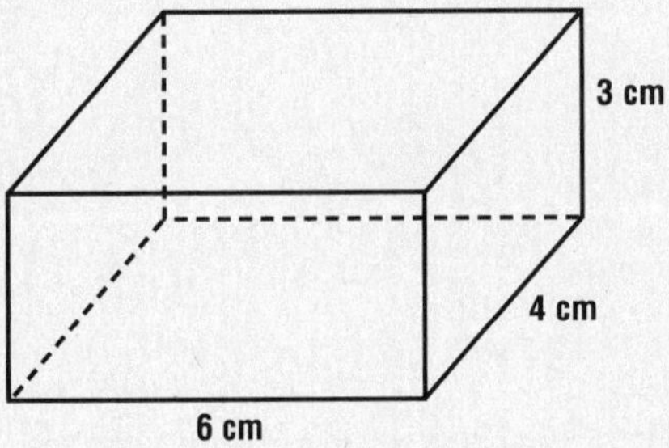

10.

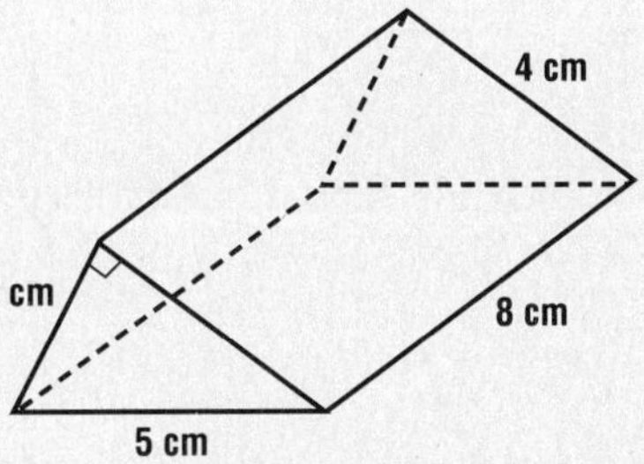

11.

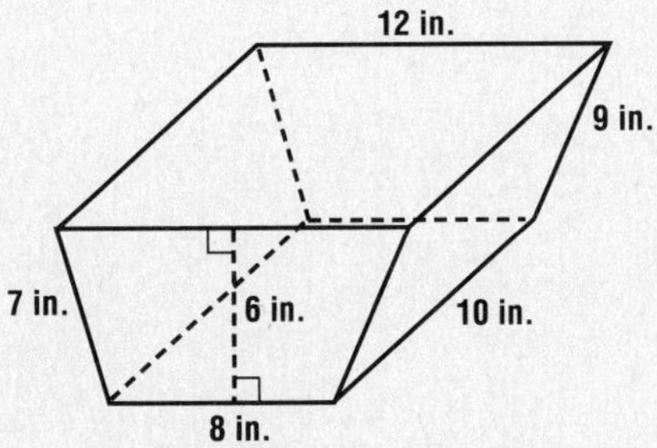

12.

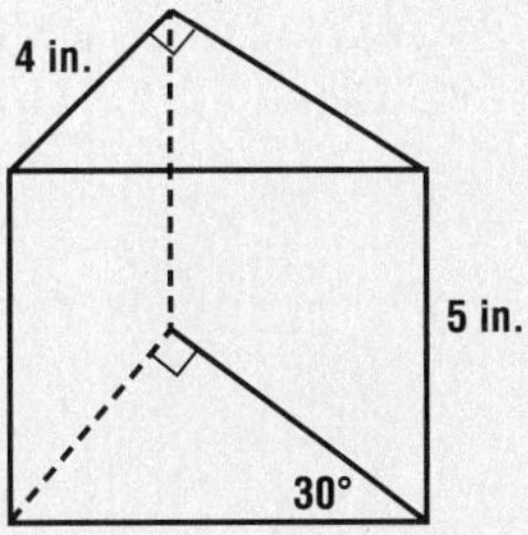

13.

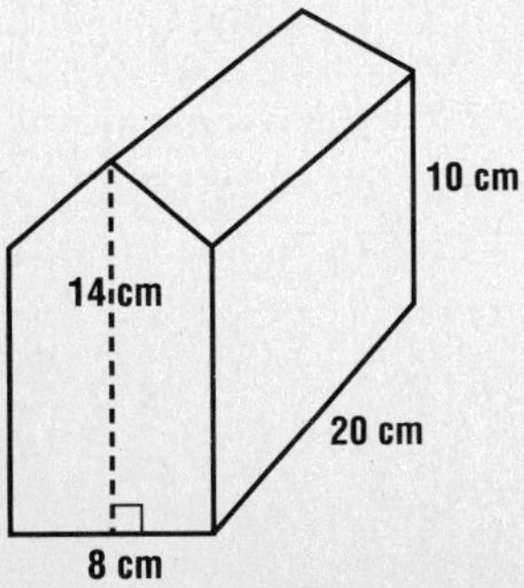

14.

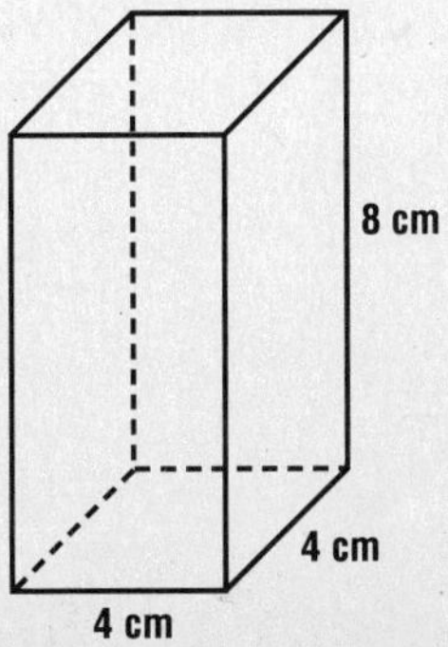

Glencoe Division, Macmillan/McGraw-Hill

NAME ____________________ DATE ____________

11-3 Practice Worksheet

Surface Area of Prisms and Cylinders

Use the right cylinders shown to answer each of the following. Express all answers in terms of π.

1. Find the circumference of the base.
10π cm

2. Find the lateral area.
100π cm^2

3. Find the area of a base.
25π cm^2

4. Find the surface area.
150π cm^2

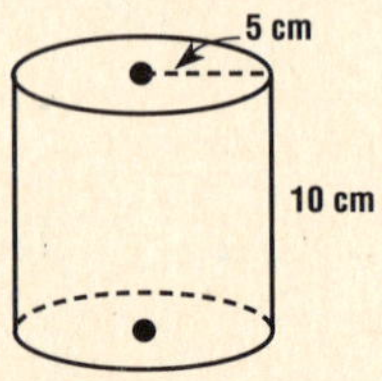

5. Find the circumference of the base.
20π ft

6. Find the lateral area.
200π ft^2

7. Find the area of a base.
100π ft^2

8. Find the surface area.
400π ft^2

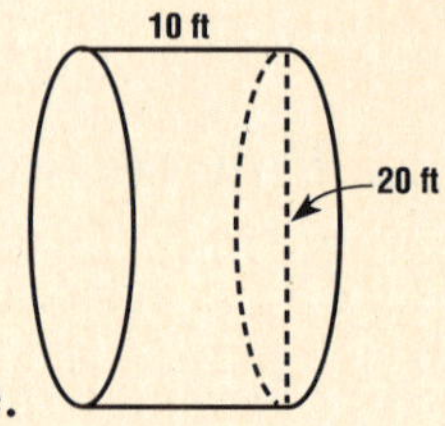

Find the lateral area and the surface area of each right prism. Round your answers to the nearest tenth.

9.

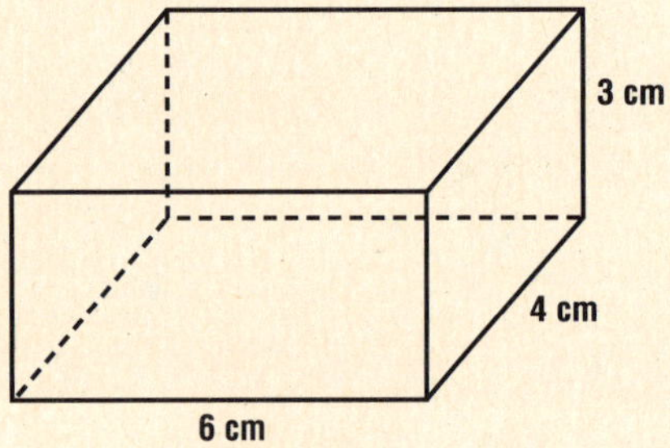

60 cm^2, 108 cm^2

10.

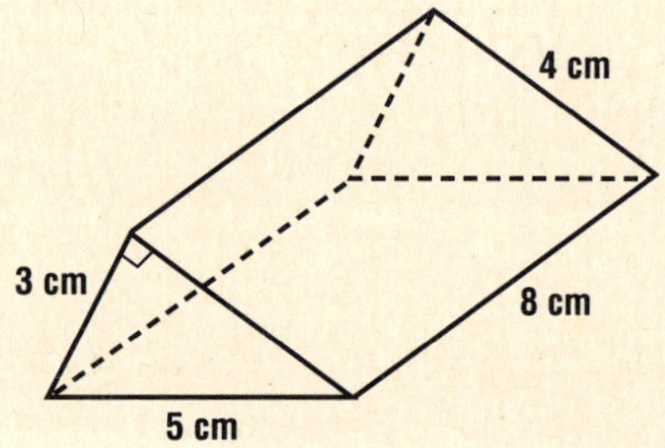

96 cm^2, 108 cm^2

11.

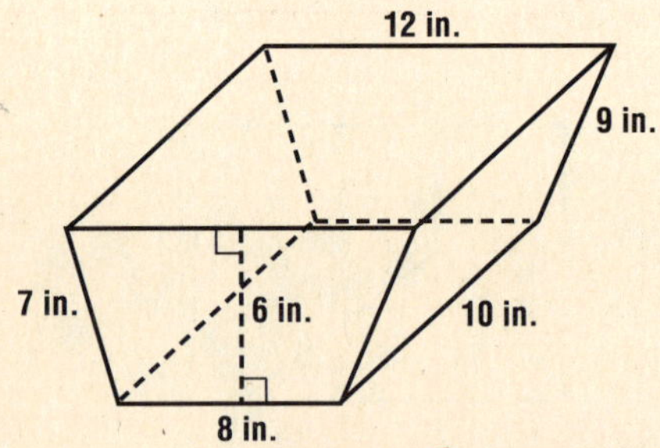

360 in^2, 480 in^2

12.

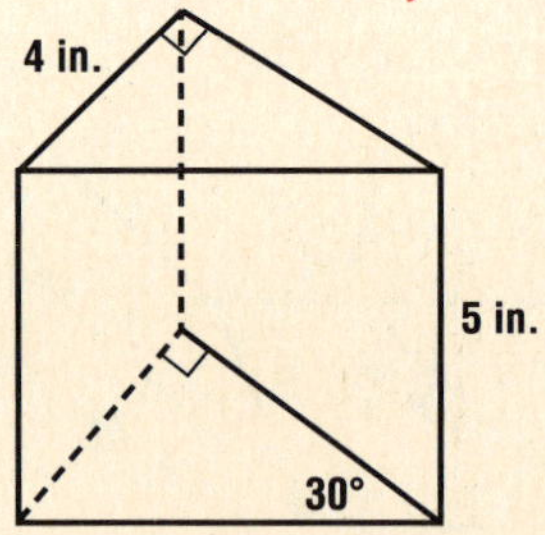

94.6 in^2, 122.4 in^2

13.

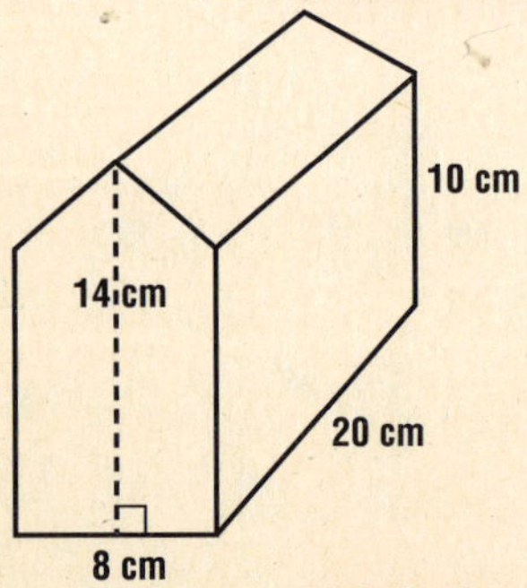

786.3 cm^2, 978.3 cm^2

14.

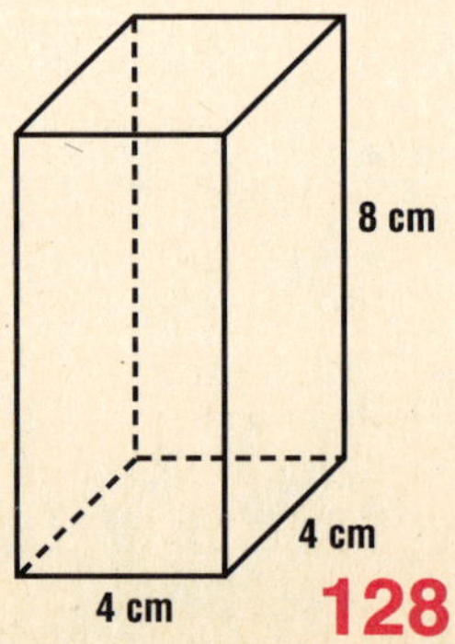

128 cm^2, 160 cm^2

NAME ______________________________ DATE ______________

11-4 Practice Worksheet

Surface Area of Pyramids and Cones

Find the lateral area of each regular pyramid or right cone. Round your answers to the nearest tenth.

1.

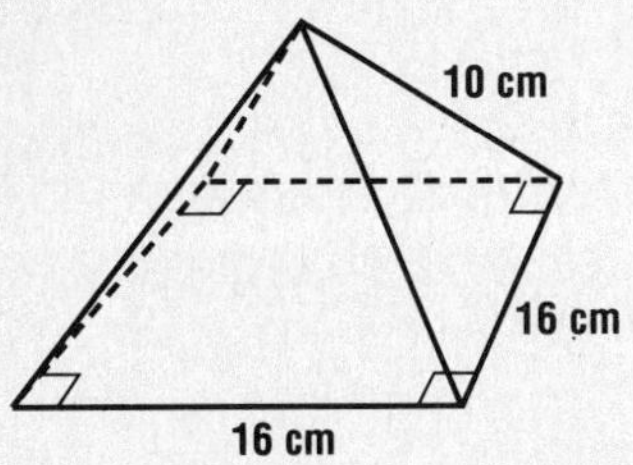

2.

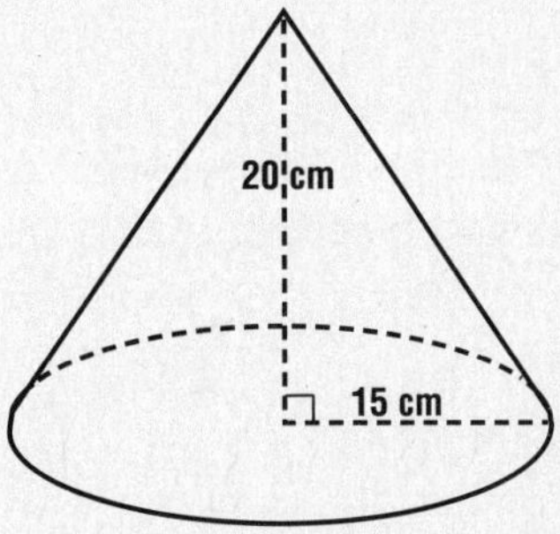

3.

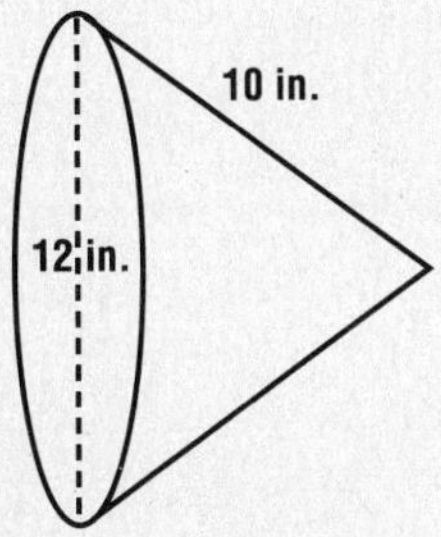

4.

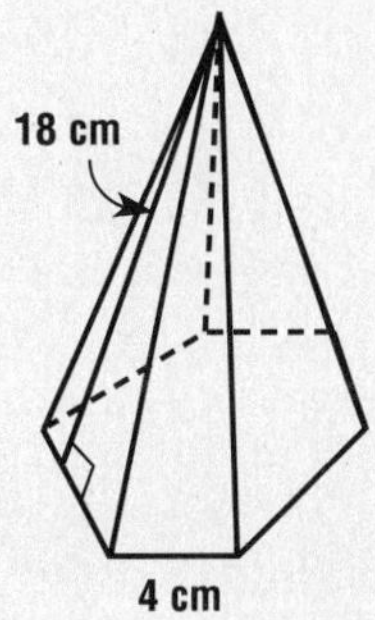

Find the surface area of each solid.

5.

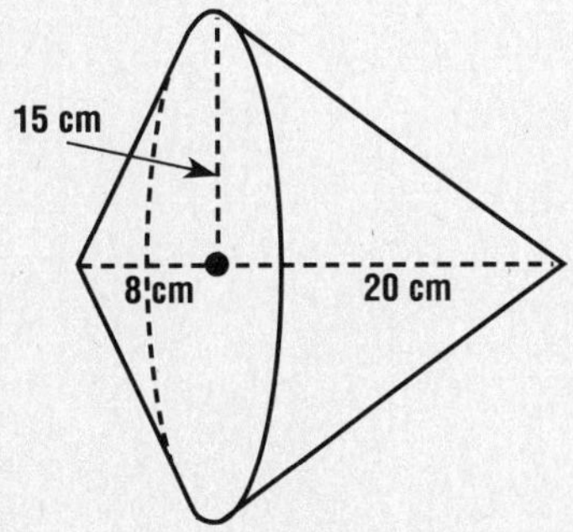

6.

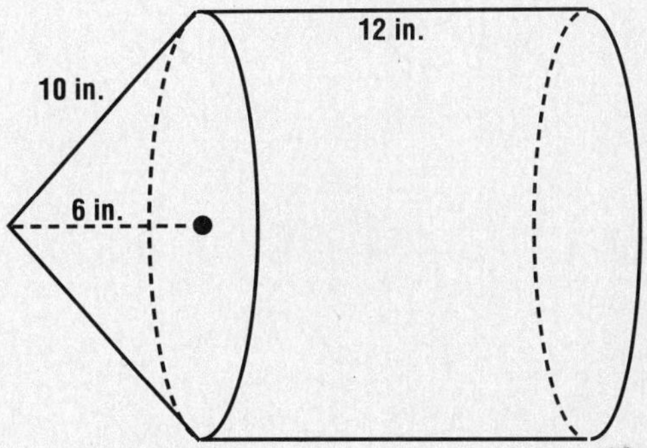

11-4

NAME ______________________ DATE ____________

Practice Worksheet

Surface Area of Pyramids and Cones

Find the lateral area of each regular pyramid or right cone. Round your answers to the nearest tenth.

1.

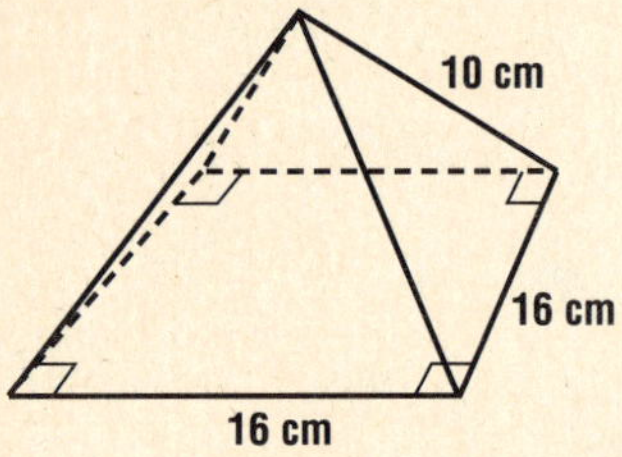

192 cm^2

2.

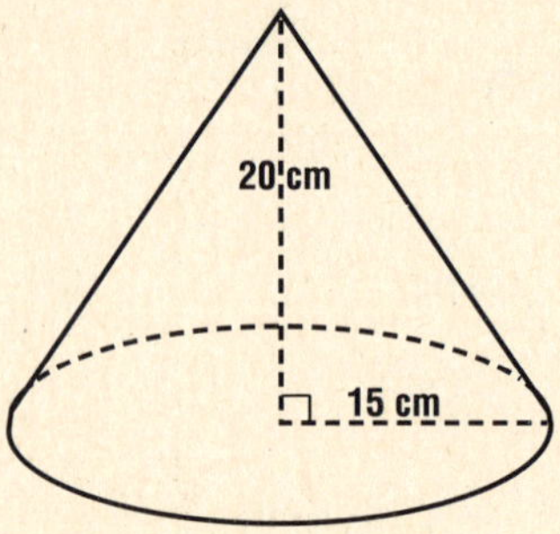

1178.1 cm^2

3.

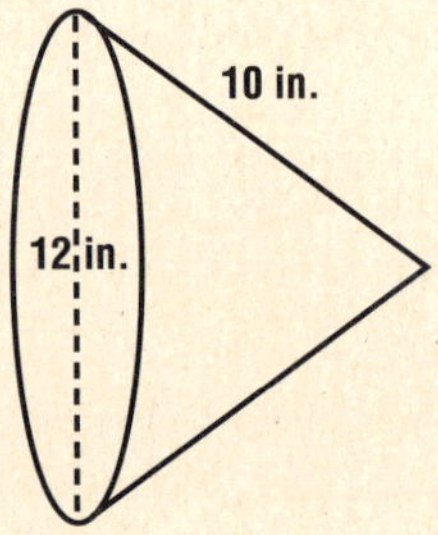

188.5 in^2

4.

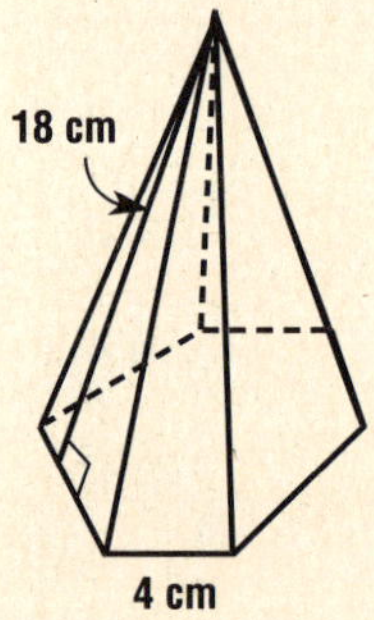

216 cm^2

Find the surface area of each solid.

5.

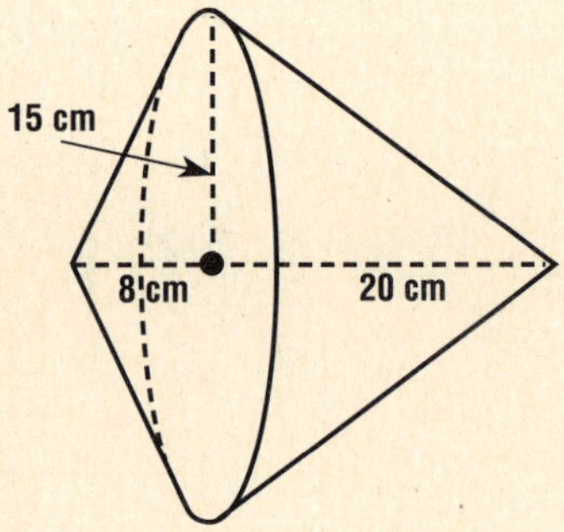

1979.2 cm^2

6.

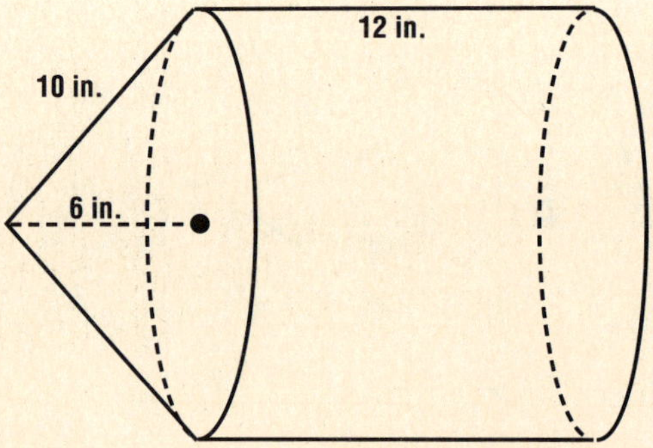

1055.6 in^2

NAME ______________________ DATE __________

11-5 Practice Worksheet

Volume of Prisms and Cylinders

Find each of the following. Round your answers to the nearest tenth.

1. the volume of a right prism whose square base has sides of 4 feet and whose height is 9 feet

2. the volume of a cylinder with a height of 2 meters and a radius of 0.5 meters

Find the volume of each solid. Round your answers to the nearest tenth.

3.

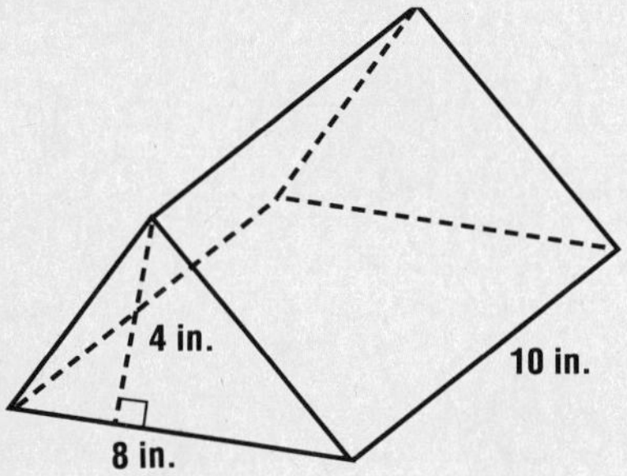

4.

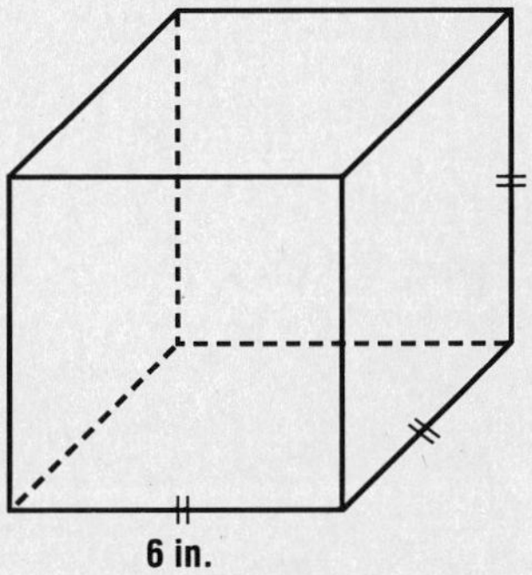

5.

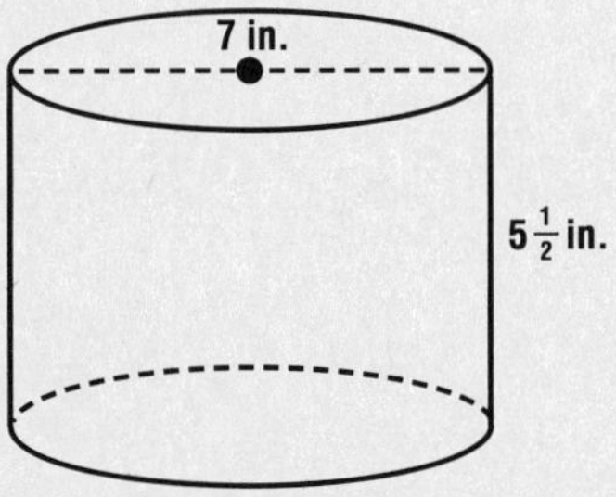

6.

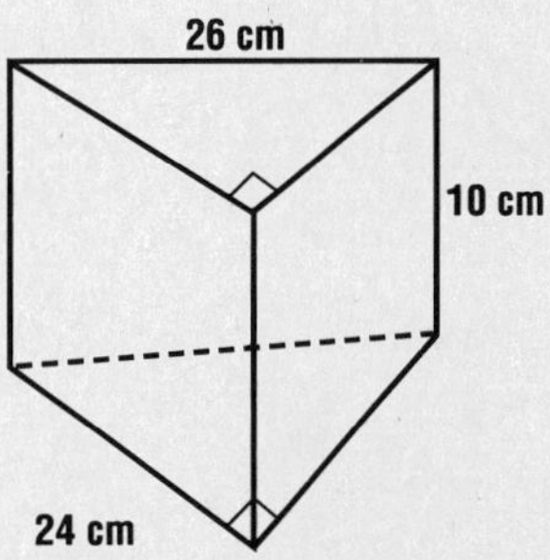

7.

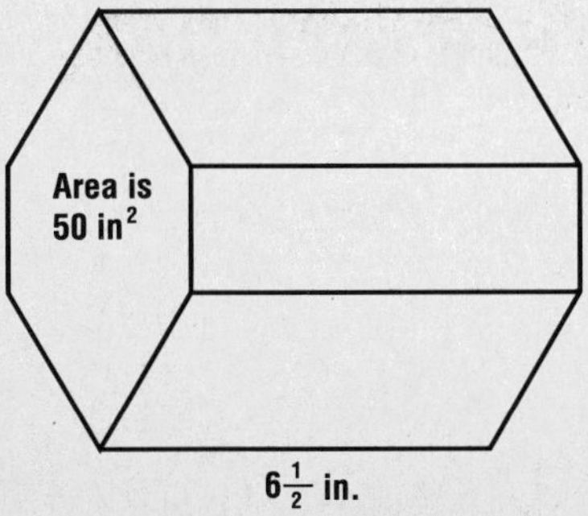

8.

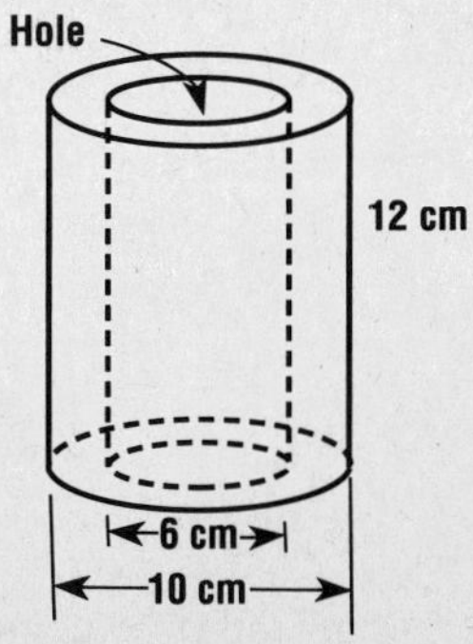

NAME ______________________ DATE ____________

11-5 Practice Worksheet

Volume of Prisms and Cylinders

Find each of the following. Round your answers to the nearest tenth.

1. the volume of a right prism whose square base has sides of 4 feet and whose height is 9 feet **144.0 ft^3**

2. the volume of a cylinder with a height of 2 meters and a radius of 0.5 meters **1.6 m^3**

Find the volume of each solid. Round your answers to the nearest tenth.

3.

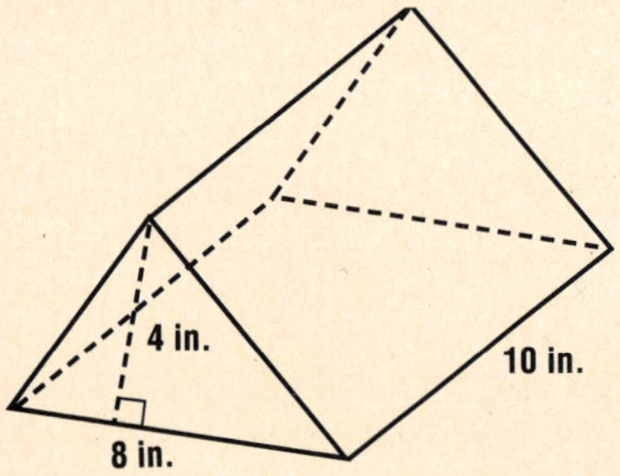

160 in^3

4.

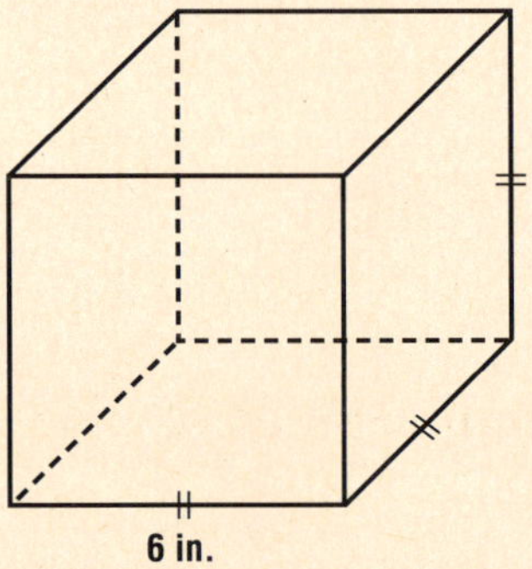

216 in^3

5.

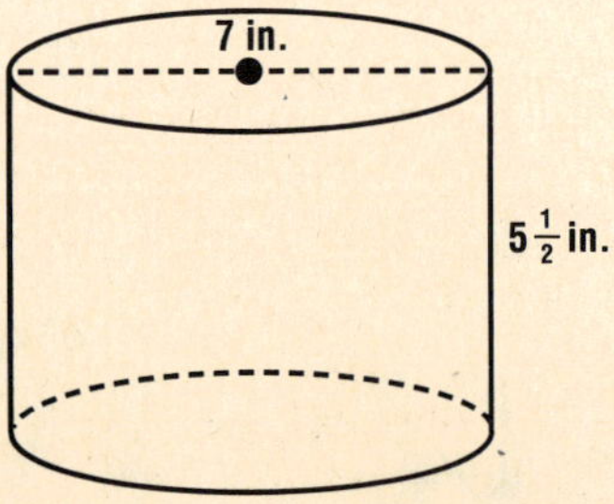

211.7 in^3

6.

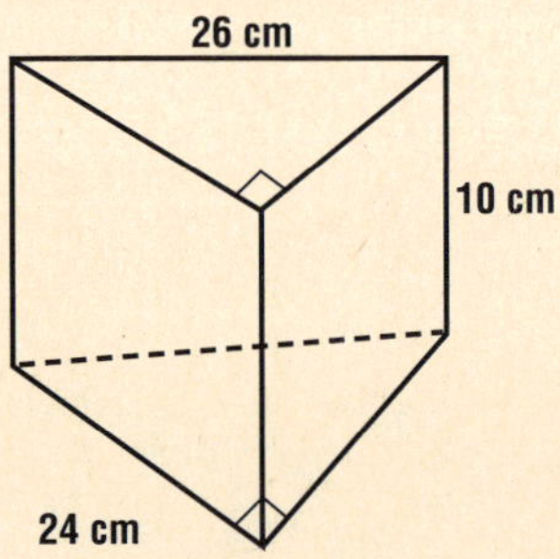

1200 cm^3

7.

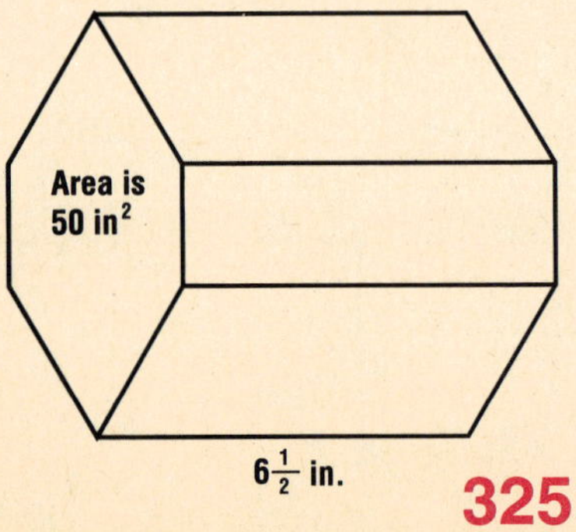

325 in^3

8.

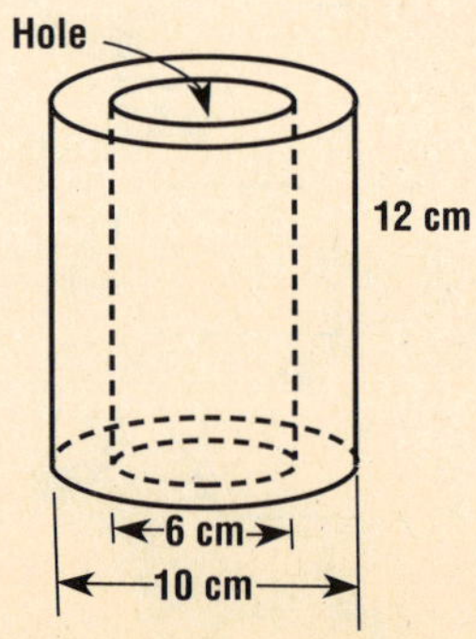

603.2 cm^3

NAME ____________________ DATE ____________

11-6 **Practice Worksheet**

Volume of Pyramids and Cones

Find the volume of each pyramid. Round your answers to the nearest tenth.

1. The base has an area of 84.3 square centimeters, and the height is 16.4 centimeters.

2. The base has an area of 17 square feet, and the height is 3 feet.

Find the volume of each cone. Round your answers to the nearest tenth.

3. The base has a radius of 16 centimeters, and the height is 12 centimeters.

4. The base has a diameter of 24 meters, and the height is 15.3 meters.

Find the volume of each solid. Round your answers to the nearest tenth.

5.

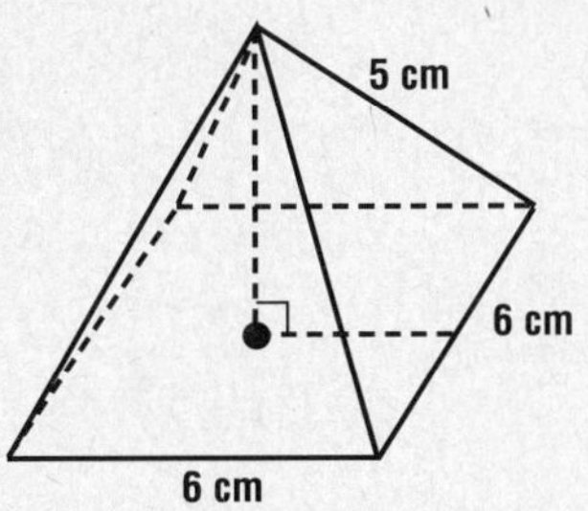

6.

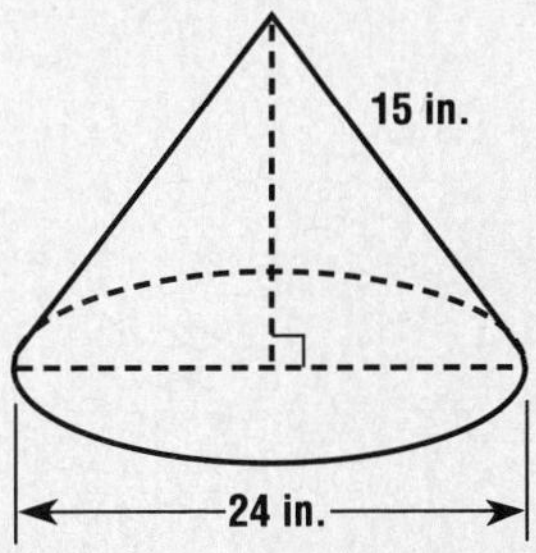

7.

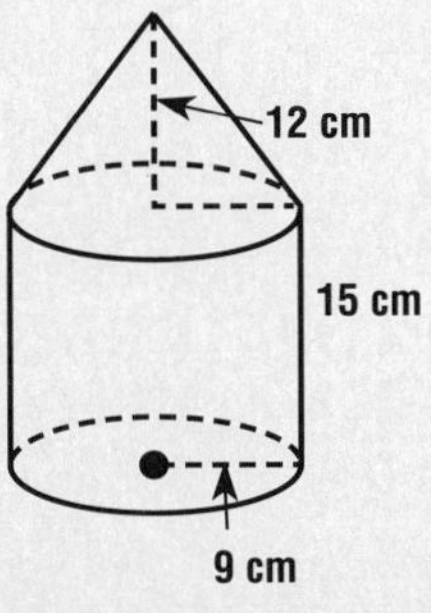

8.

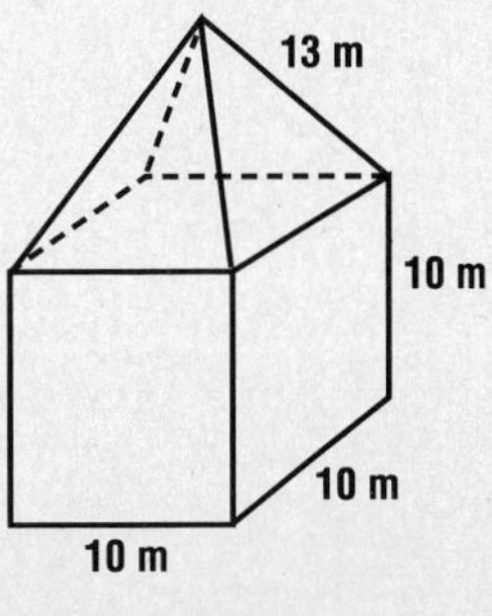

NAME ________________________________ DATE ______________

11-6 Practice Worksheet

Volume of Pyramids and Cones

Find the volume of each pyramid. Round your answers to the nearest tenth.

1. The base has an area of 84.3 square centimeters, and the height is 16.4 centimeters.
460.8 cm³

2. The base has an area of 17 square feet, and the height is 3 feet.
17.0 ft³

Find the volume of each cone. Round your answers to the nearest tenth.

3. The base has a radius of 16 centimeters, and the height is 12 centimeters.
3217.0 cm³

4. The base has a diameter of 24 meters, and the height is 15.3 meters.
2307.2 m³

Find the volume of each solid. Round your answers to the nearest tenth.

5.

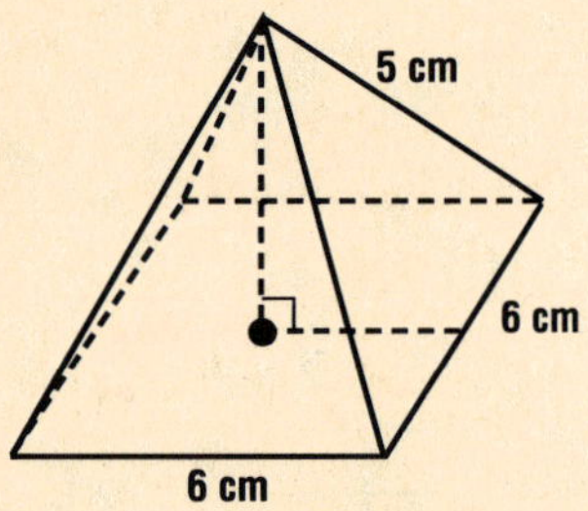

31.7 cm³

6.

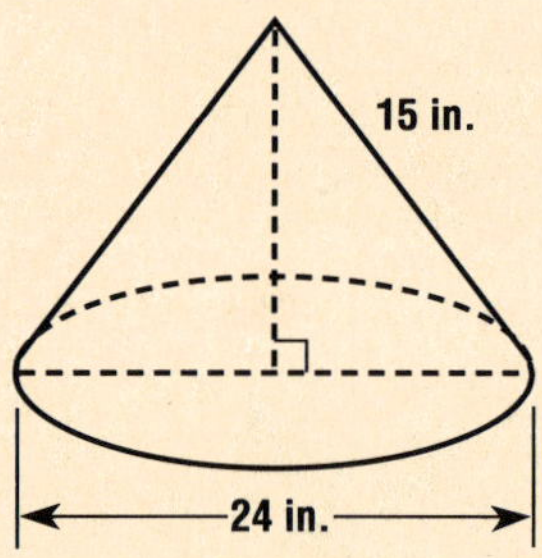

1357.2 in³

7.

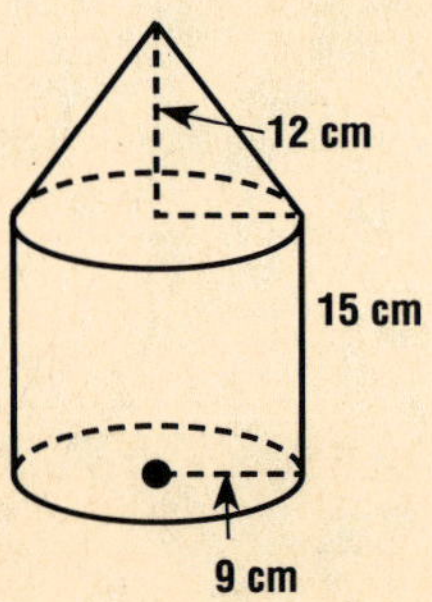

4834.9 cm³

8.

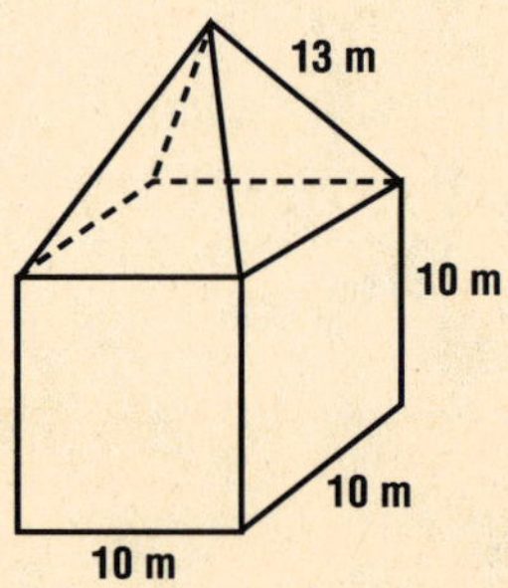

1363.6 m³

NAME ______________________ DATE ______________

12-1 Practice Worksheet

Graphing Lines On a Coordinate Plane

Graph each pair of linear equations on the same coordinate plane. Determine if the lines are parallel, perpendicular, or neither by finding the slope of each line.

1. $y = x + 2$
$y - x = 4$

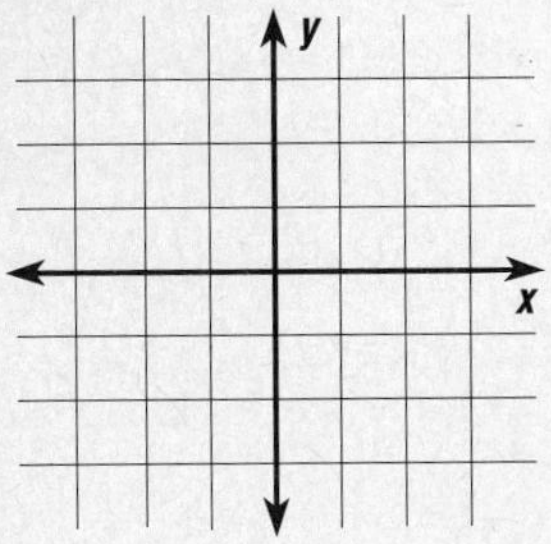

2. $2x - y = 3$
$y = 2x + 4$

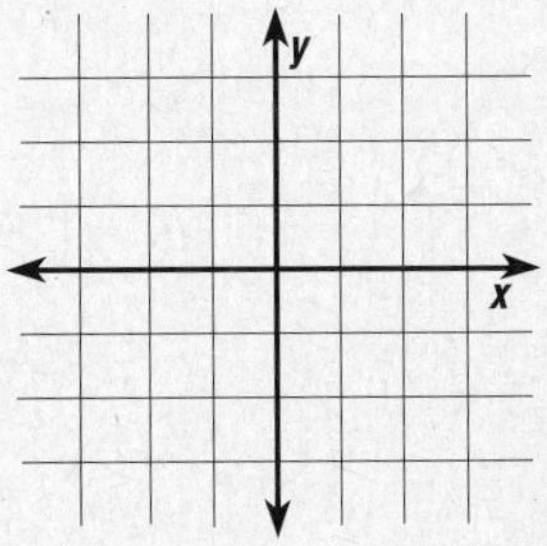

3. $3y = x + 1$
$y = -3x + 2$

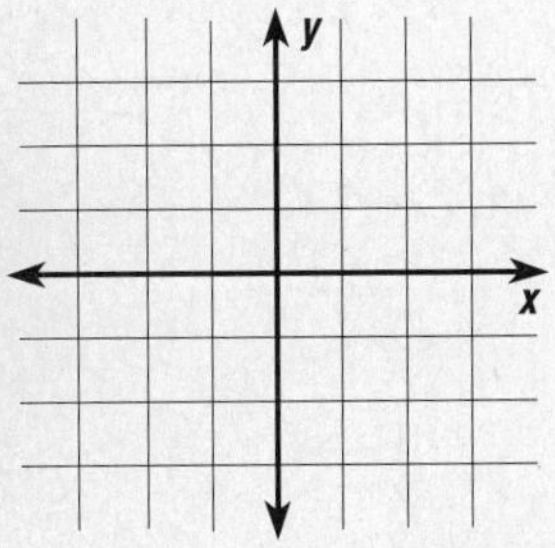

4. $y = 2x + 1$
$y = 3x - 1$

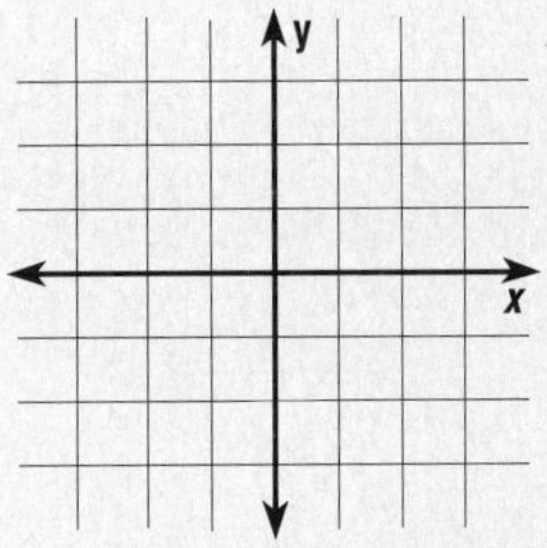

Find the slope and y-intercept of the graph of each equation.

5. $x + y = 8$

6. $2x - y = 4$

7. $2x - 3y = 10$

Determine the x- and y-intercept of each line. Then graph the equation.

8. $y = 2x - 3$

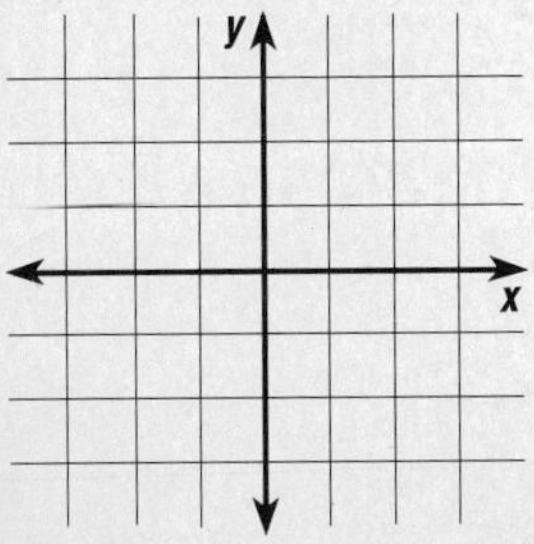

9. $y = 5$

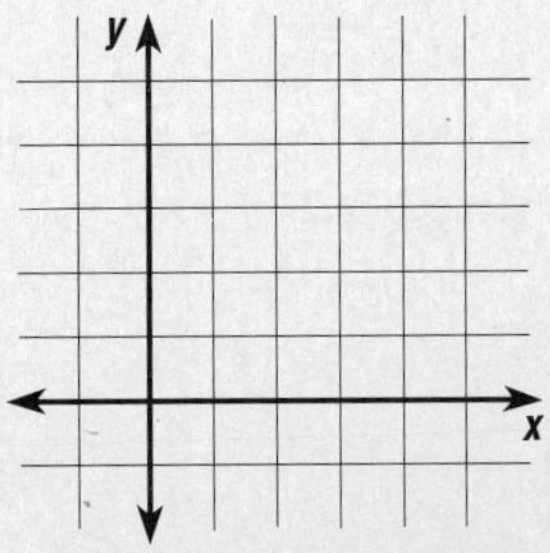

12-1 NAME ______________________ DATE __________

Practice Worksheet

Graphing Lines On a Coordinate Plane

Graph each pair of linear equations on the same coordinate plane. Determine if the lines are parallel, perpendicular, or neither by finding the slope of each line.

1. $y = x + 2$ **parallel; 1, 1**
$y - x = 4$

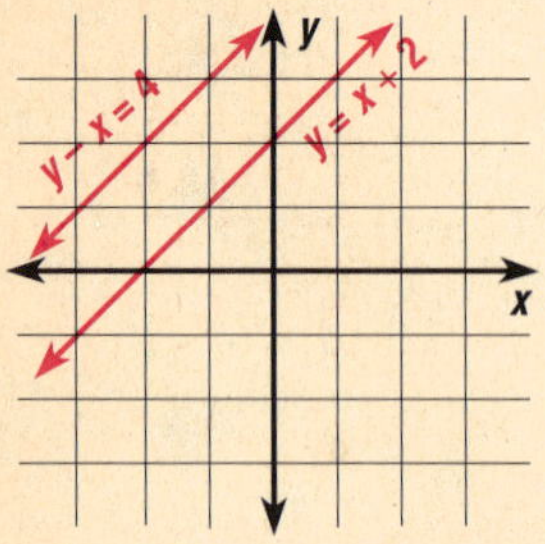

2. $2x - y = 3$ **parallel; 2, 2**
$y = 2x + 4$

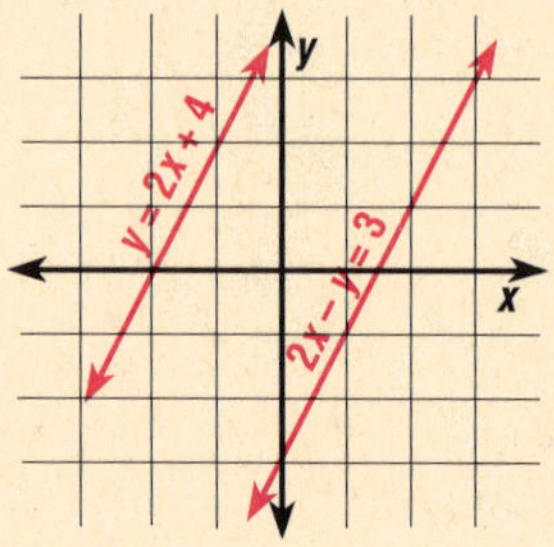

3. $3y = x + 1$ **perpendicular;** $\frac{1}{3}$**, −3**
$y = -3x + 2$

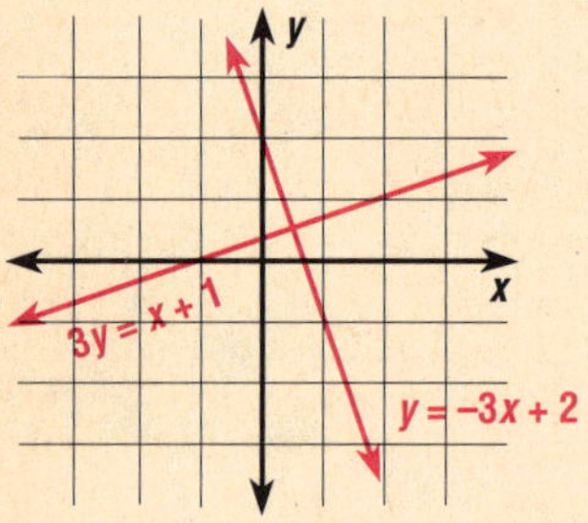

4. $y = 2x + 1$ **neither; 2, 3**
$y = 3x - 1$

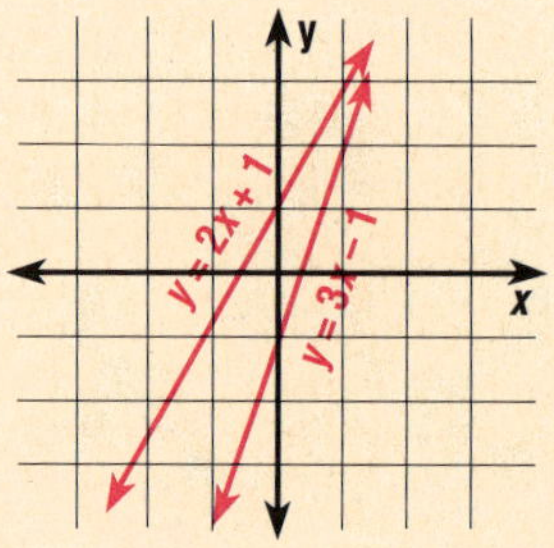

Find the slope and y-intercept of the graph of each equation.

5. $x + y = 8$
−1, 8

6. $2x - y = 4$
2, −4

7. $2x - 3y = 10$
$\frac{2}{3}, -\frac{10}{3}$

Determine the x- and y-intercept of each line. Then graph the equation.

8. $y = 2x - 3$

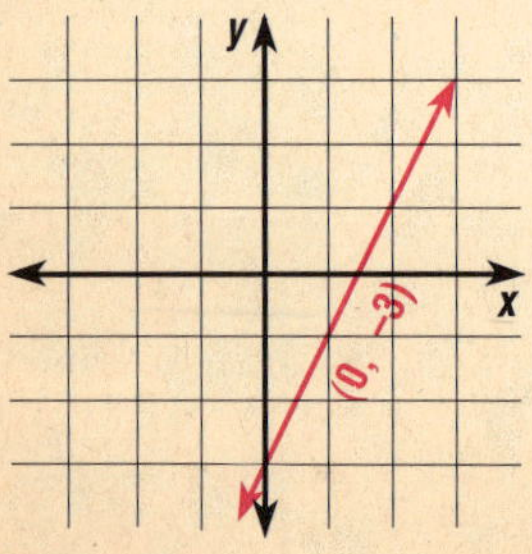

$\frac{3}{2}, -3$

9. $y = 5$

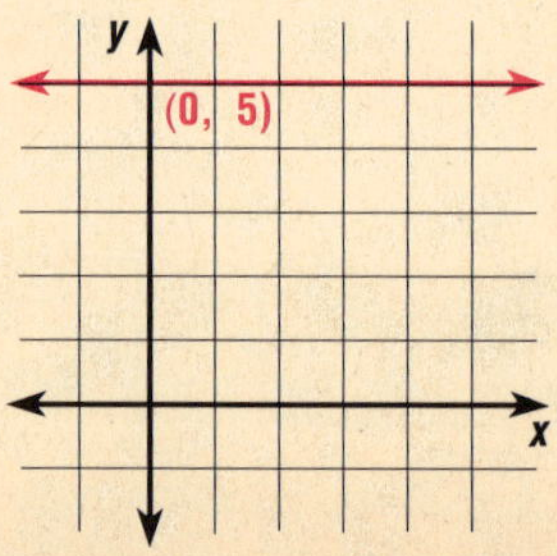

no *x*-intercept, 5

NAME ______________________________ DATE ______________

12-2 **Practice Worksheet**

Writing Equations of Lines

State the slope and y-intercept for each line. Then write the equation of the line in slope-intercept form.

1. $\overleftrightarrow{AB}$

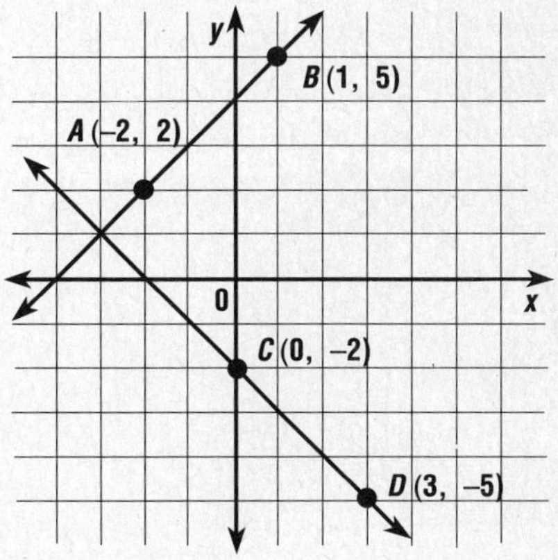

2. $\overleftrightarrow{CD}$

Write the equation in slope-intercept form of the line satisfying the given conditions.

3. parallel to $y = 3x - 2$, y-intercept = 1

4. perpendicular to $y = \frac{1}{2}x + 8$, passes through (0, 3)

5. perpendicular to the line containing (−2, 3) and (2, 1), y-intercept = −3

6. passes through (1, 5) and (−3, 5)

7. x-intercept = 5, y-intercept = −2

8. $m = -\frac{2}{3}$, passes through (−1, 2)

9. $m = -4$, x-intercept = 0

10. passes through (2, 7) and (−1, 9)

12-2

NAME ______________________ DATE ____________

Practice Worksheet

Writing Equations of Lines

State the slope and y-intercept for each line. Then write the equation of the line in slope-intercept form.

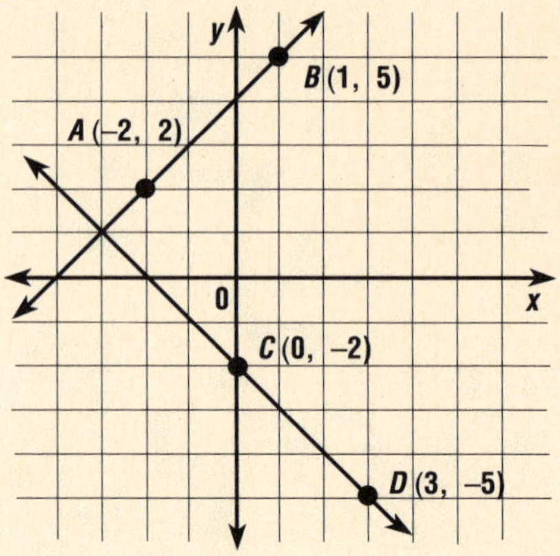

1. $\overleftrightarrow{AB}$

   1, 4, $y = x + 4$

2. $\overleftrightarrow{CD}$

   –1, –2, $y = -x - 2$

Write the equation in slope-intercept form of the line satisfying the given conditions.

3. parallel to $y = 3x - 2$, y-intercept = 1

   $y = 3x + 1$

4. perpendicular to $y = \frac{1}{2}x + 8$, passes through (0, 3)

   $y = -2x + 3$

5. perpendicular to the line containing (–2, 3) and (2, 1), y-intercept = –3

   $y = 2x - 3$

6. passes through (1, 5) and (–3, 5)

   $y = 5$

7. x-intercept = 5, y-intercept = –2

   $y = \frac{2}{5}x - 2$

8. $m = -\frac{2}{3}$, passes through (–1, 2)

   $y = -\frac{2}{3}x + \frac{4}{3}$

9. $m = -4$, x-intercept = 0

   $y = -4x$

10. passes through (2, 7) and (–1, 9)

    $y = -\frac{2}{3}x + \frac{25}{3}$

NAME ______________________ DATE ____________

12-3 Practice Worksheet

Connections to Algebra

Write an equation of the line that satisfies the given conditions.

1. $m = \frac{2}{3}$, passes through $(-1, 3)$

2. perpendicular to $y = -\frac{1}{3}x + 4$, passes through $(6, -1)$

3. Find the equation of the line that is the perpendicular bisector of the segment whose endpoints are $(3, 7)$ and $(-3, -1)$.

4. Find the equations of the lines that contain sides of an isosceles triangle if the vertex of the vertex angle is at the y-intercept of $y = -2x + 3$ and the vertex of a base angle is at $(6, 1)$.

The vertices of $\triangle DEF$ are $D(0, 12)$, $E(-2, 10)$, and $F(4, 6)$.

5. Write the equation of the line that contains the altitude to $\overline{EF}$.

6. Write the equation of the line that contains the perpendicular bisector of $\overline{ED}$.

NAME ______________________ DATE ____________

12-3 Practice Worksheet

Connections to Algebra

Write an equation of the line that satisfies the given conditions.

1. $m = \frac{2}{3}$, passes through (–1, 3)

$y = \frac{2}{3}x + \frac{11}{3}$

2. perpendicular to $y = -\frac{1}{3}x + 4$, passes through (6, –1)

$y = 3x - 19$

3. Find the equation of the line that is the perpendicular bisector of the segment whose endpoints are (3, 7) and (–3, –1).

$y = -\frac{3}{4}x + 3$

4. Find the equations of the lines that contain sides of an isosceles triangle if the vertex of the vertex angle is at the y-intercept of $y = -2x + 3$ and the vertex of a base angle is at (6, 1).

$y = 1$, $y = \frac{1}{3}x + 3$, $y = -\frac{1}{3}x + 3$

or $x = 6$, $y = \frac{1}{3}x + 3$, $y = -\frac{1}{3}x + 3$

The vertices of △DEF are D(0, 12), E(–2, 10), and F(4, 6).

5. Write the equation of the line that contains the altitude to $\overline{EF}$.

$y = \frac{3}{2}x + 12$

6. Write the equation of the line that contains the perpendicular bisector of $\overline{ED}$.

$y = -x + 10$

NAME ______________________________ DATE ______________

12-4 Practice Worksheet

Problem-Solving Strategy: Write an Equation

Solve. Use any strategy.

1. Karla must make a 530-mile trip in 10 hours. If she travels 50 miles per hour for 4 hours, how fast must she travel the rest of the way in order to make her destination on time?

2. The length of a rectangular garden is 30 feet more than 2 times its width. Its perimeter is 300 feet. Find its length and width.

3. Jeremy sold tickets for a football game. The tickets cost $1.50 for students and $2.50 for adults. If Jeremy sold 60 more student tickets than adult tickets and he has a total of $650.00 in sales, how many of each type of tickets did he sell?

4. Place the numbers 1, 2, 3, 4, 5, and 6 in the squares and circles so that the sum of the digits in the squares is 3 less than the sum of the digits in the circles.

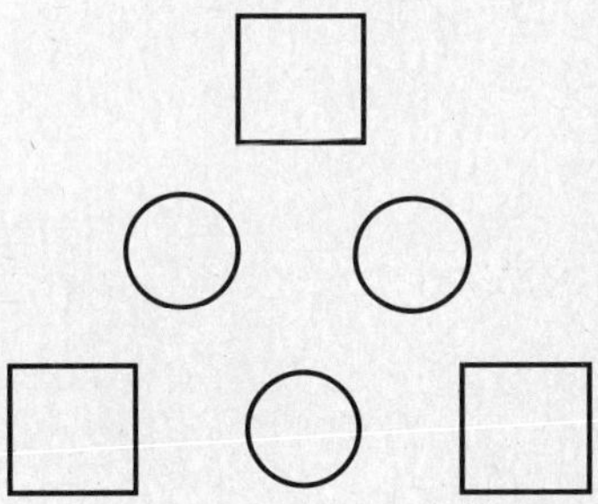

5. A certain bacteria doubles its population every 12 hours. After 3 days there are 4800 bacteria. How many bacteria were there at the beginning of the first day?

Glencoe Division, Macmillan/McGraw-Hill

12-4 NAME ____________________ DATE ____________

Practice Worksheet

Problem-Solving Strategy: Write an Equation

Solve. Use any strategy.

1. Karla must make a 530-mile trip in 10 hours. If she travels 50 miles per hour for 4 hours, how fast must she travel the rest of the way in order to make her destination on time?
55 mph

2. The length of a rectangular garden is 30 feet more than 2 times its width. Its perimeter is 300 feet. Find its length and width.
110 feet, 40 feet

3. Jeremy sold tickets for a football game. The tickets cost $1.50 for students and $2.50 for adults. If Jeremy sold 60 more student tickets than adult tickets and he has a total of $650.00 in sales, how many of each type of tickets did he sell?
140 adult tickets and 200 student tickets.

4. Place the numbers 1, 2, 3, 4, 5, and 6 in the squares and circles so that the sum of the digits in the squares is 3 less than the sum of the digits in the circles.

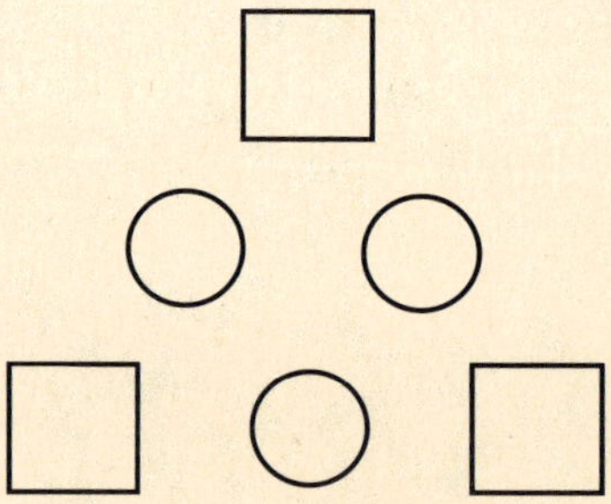

Answers may vary; possible solution:

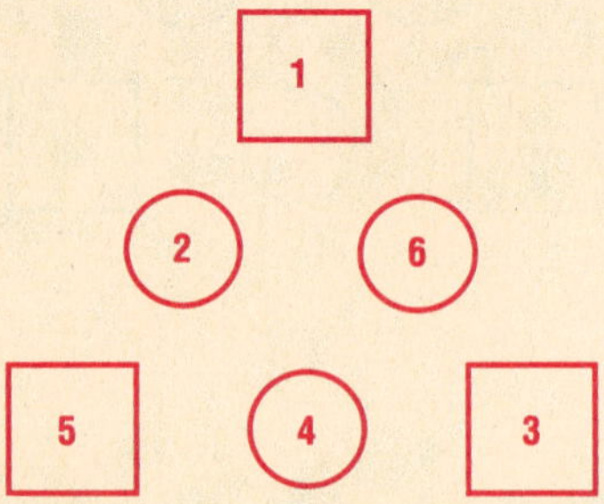

5. A certain bacteria doubles its population every 12 hours. After 3 days there are 4800 bacteria. How many bacteria were there at the beginning of the first day?
75 bacteria

Glencoe Division, Macmillan/McGraw-Hill

NAME ______ DATE ______

12-5 Practice Worksheet

Coordinate Proof

Name the missing coordinates in terms of the given variables.

1. $\triangle XYZ$ is isosceles and right.

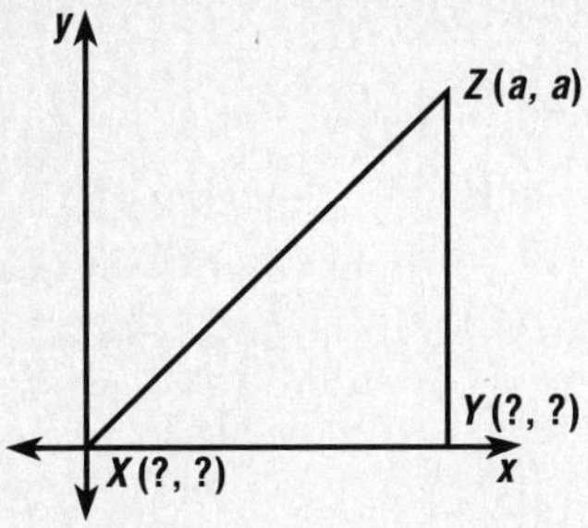

2. *MART* is a rhombus.

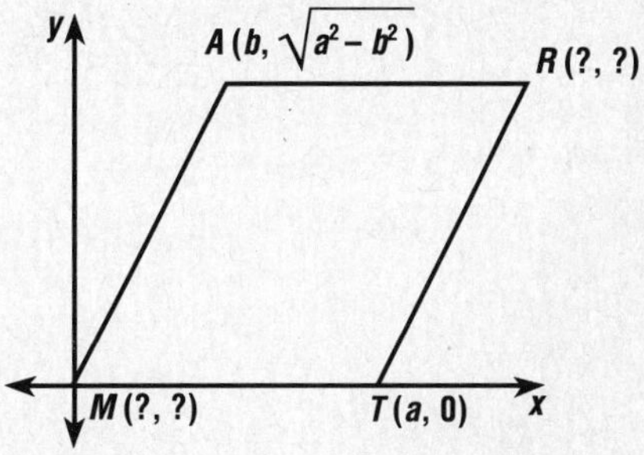

3. *RECT* is a rectangle.

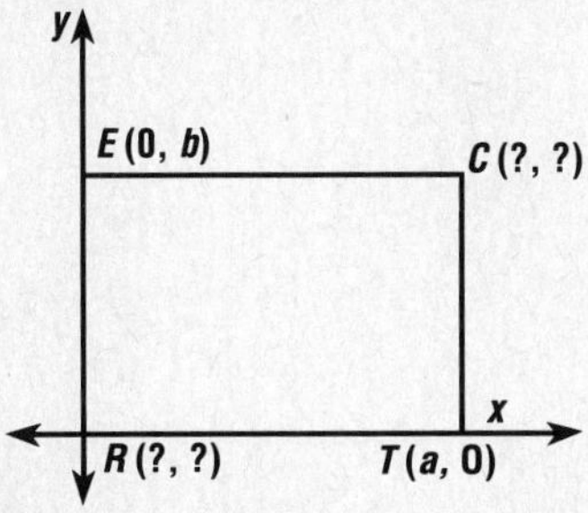

4. *DEFG* is a parallelogram.

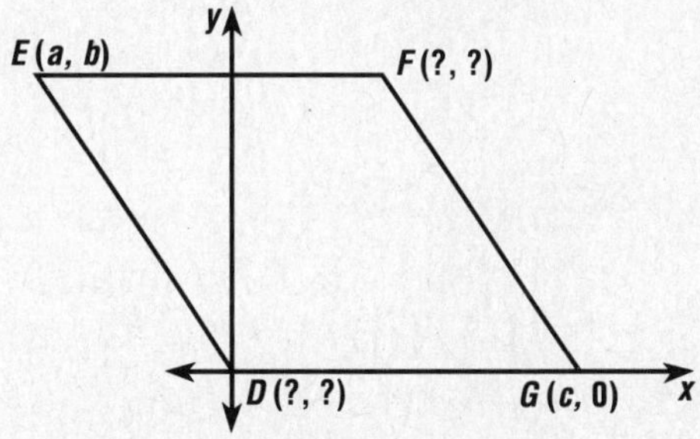

5. Use a coordinate proof to prove that the diagonals of a rhombus are perpendicular. Draw the diagram at the right.

NAME ________________ DATE ________

12-5

Practice Worksheet

Coordinate Proof

Name the missing coordinates in terms of the given variables.

1. $\triangle XYZ$ is isosceles and right.

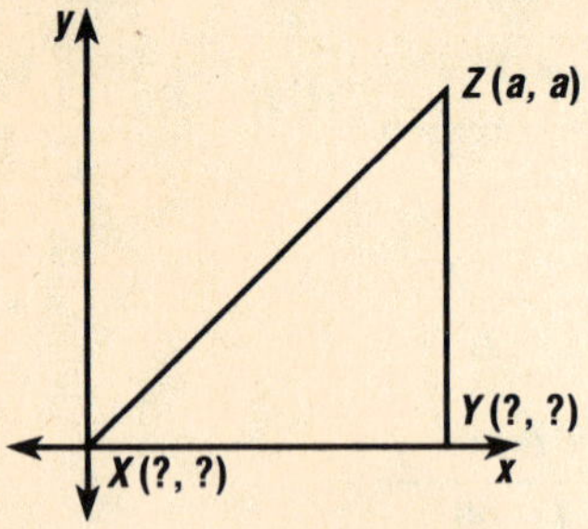

$X(0, 0)$ $Y(a, 0)$

2. *MART* is a rhombus.

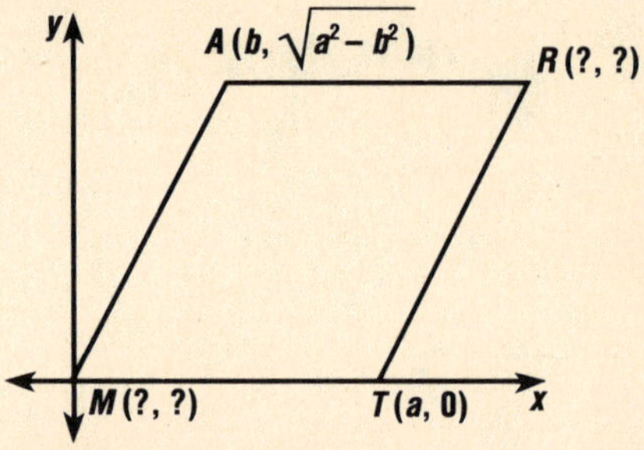

$M(0, 0)$, $R(a + b, \sqrt{a^2 - b^2})$

3. *RECT* is a rectangle.

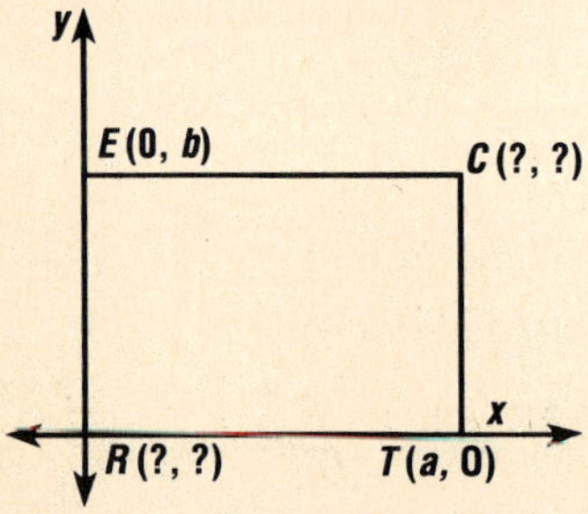

$R(0, 0)$, $C(a, b)$

4. *DEFG* is a parallelogram.

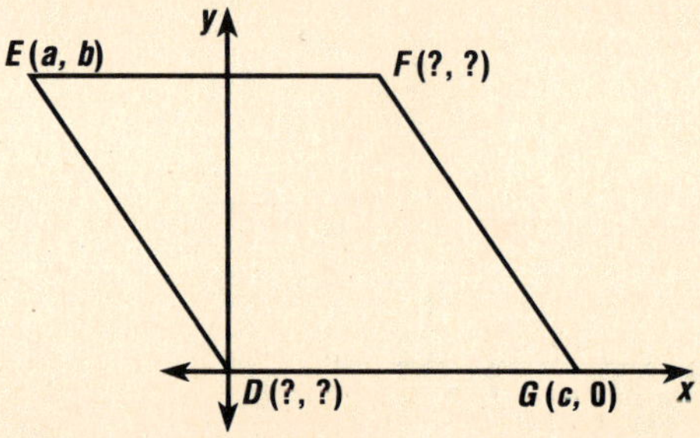

$D(0, 0)$, $F(a + c, b)$

5. Use a coordinate proof to prove that the diagonals of a rhombus are perpendicular. Draw the diagram at the right.

Typical proof:

$$\text{slope of } \overline{AC} = \frac{\sqrt{a^2 - b^2} - 0}{a + b - 0} = \frac{\sqrt{a^2 - b^2}}{a + b}$$

$$\text{slope of } \overline{BD} = \frac{\sqrt{a^2 - b^2} - 0}{b - a} = \frac{\sqrt{a^2 - b^2}}{b - a}$$

$$\frac{\sqrt{a^2 - b^2}}{a + b} \bullet \frac{\sqrt{a^2 - b^2}}{b - a} = \frac{a^2 - b^2}{b^2 - a^2} = -1$$

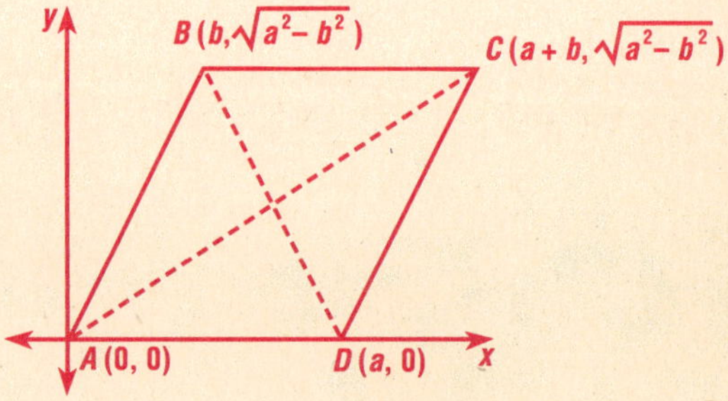

NAME ____________________ DATE ____________

12-6

Practice Worksheet

Vectors

Given $\vec{a} = (8, 6)$, $\vec{b} = (4, 3)$, $\vec{c} = (1, 2)$, and $\vec{d} = (-3, -6)$, answer each of the following.

1. Find the magnitude of $\vec{a}$.

2. Find the magnitude $\vec{c}$.

3. Determine if $\vec{b}$ and $\vec{d}$ are equal.

4. Determine if $\vec{c}$ and $\vec{d}$ are equal.

5. Find the coordinates of $\vec{a} + \vec{b}$.

6. Find the coordinates of $(\vec{b} + \vec{c}) + \vec{d}$.

7. Given $A(2, 5)$ and $B(7, 10)$, find the magnitude and direction of $\overrightarrow{AB}$.

8. Given $C(0, 1)$ and $D(8, 12)$, find the magnitude and direction of $\overrightarrow{CD}$.

Given path from A south 5 units to B, then east 12 units to C, answer each question.

9. What is the total length of the path?

10. What is the magnitude of $\overrightarrow{AC}$?

NAME ______________________ DATE ____________

12-6 Practice Worksheet

Vectors

Given $\vec{a} = (8, 6)$, $\vec{b} = (4, 3)$, $\vec{c} = (1, 2)$, and $\vec{d} = (-3, -6)$, answer each of the following.

1. Find the magnitude of $\vec{a}$.
10

2. Find the magnitude $\vec{c}$.
$\sqrt{5}$

3. Determine if $\vec{b}$ and $\vec{d}$ are equal.
no

4. Determine if $\vec{c}$ and $\vec{d}$ are equal.
no

5. Find the coordinates of $\vec{a} + \vec{b}$.
(12, 9)

6. Find the coordinates of $(\vec{b} + \vec{c}) + \vec{d}$.
(2, –1)

7. Given $A(2, 5)$ and $B(7, 10)$, find the magnitude and direction of $\overrightarrow{AB}$.
≈ 7.1, 45°

8. Given $C(0, 1)$ and $D(8, 12)$, find the magnitude and direction of $\overrightarrow{CD}$.
≈ 13.6, $\approx 54°$

Given path from A south 5 units to B, then east 12 units to C, answer each question.

9. What is the total length of the path?
17

10. What is the magnitude of $\overrightarrow{AC}$?
13

Glencoe Division, Macmillan/McGraw-Hill

NAME ______________________________ DATE ____________

12-7 Practice Worksheet

Determine the distance between each pair of points.

1. $A(0, 0, 0)$ and $B(1, 2, 3)$

2. $C(4, -2, 3)$ and $D(0, 2, 0)$

3. $E(1, -2, 5)$ and $F(1, 2, 5)$

4. $P(0, 1, 0)$ and $Q(-1, 0, 1)$

Determine the coordinates of the midpoint of each line segment whose endpoints are given.

5. $A(0, 0, 4)$, $B(4, -6, 6)$

6. $C(-1, 2, 4)$, $D(3, -6, 8)$

7. $E(-2, -3, 6)$, $F(0, -6, 8)$

8. $G(-1, 5, \frac{3}{2})$, $H(1, -5, \frac{1}{2})$

Write an equation of the sphere given the coordinates of the center and the measure of the radius.

9. $C(0, -3, 1)$, $r = 3$

10. $C(-2, 1, 3)$, $r = 1\frac{1}{2}$

11. $C(3, 0, -1)$, $r = \frac{\sqrt{2}}{2}$

12. $C(5, 5, 5)$, $r = 7$

13. Find the perimeter of a triangle whose vertices are $A(0, 2, 1)$, $B(-2, 2, 6)$, and $C(4, 2, -2)$.

The diameter of a sphere has endpoints A(–3, 2, 4) and B(1, –6, 5).

14. Determine the center of the sphere.

15. Determine the radius of the sphere.

16. Write the equation of the sphere.

17. Find the surface area of the sphere.

Glencoe Division, Macmillan/McGraw-Hill

NAME ______________________ DATE ______________

12-7 Practice Worksheet

Determine the distance between each pair of points.

1. $A(0, 0, 0)$ and $B(1, 2, 3)$
$\sqrt{14} \approx 3.7$

2. $C(4, -2, 3)$ and $D(0, 2, 0)$
$\sqrt{41} \approx 6.4$

3. $E(1, -2, 5)$ and $F(1, 2, 5)$
4

4. $P(0, 1, 0)$ and $Q(-1, 0, 1)$
$\sqrt{3} \approx 1.7$

Determine the coordinates of the midpoint of each line segment whose endpoints are given.

5. $A(0, 0, 4)$, $B(4, -6, 6)$
(2, –3, 5)

6. $C(-1, 2, 4)$, $D(3, -6, 8)$
(1, –2, 6)

7. $E(-2, -3, 6)$, $F(0, -6, 8)$
(–1, –4.5, 7)

8. $G(-1, 5, \frac{3}{2})$, $H(1, -5, \frac{1}{2})$
(0, 0, 1)

Write an equation of the sphere given the coordinates of the center and the measure of the radius.

9. $C(0, -3, 1)$, $r = 3$
$x^2 + (y+3)^2 + (z-1)^2 = 9$

10. $C(-2, 1, 3)$, $r = 1\frac{1}{2}$
$(x+2)^2 + (y-1)^2 + (z-3)^2 = \frac{9}{4}$

11. $C(3, 0, -1)$, $r = \frac{\sqrt{2}}{2}$
$(x-3)^2 + y^2 + (z+1)^2 = \frac{1}{2}$

12. $C(5, 5, 5)$, $r = 7$
$(x-5)^2 + (y-5)^2 + (z-5)^2 = 49$

13. Find the perimeter of a triangle whose vertices are $A(0, 2, 1)$, $B(-2, 2, 6)$, and $C(4, 2, -2)$.
$15 + \sqrt{29}$ units

The diameter of a sphere has endpoints A(–3, 2, 4) and B(1, –6, 5).

14. Determine the center of the sphere.
(–1, –2, 4.5)

15. Determine the radius of the sphere.
4.5 units

16. Write the equation of the sphere.
$(x+1)^2 + (y+2)^2 + (z-4.5)^2 = 20.25$

17. Find the surface area of the sphere.
81π square units or above 254.5 square units

NAME ______________________ DATE ______________

13-1 Practice Worksheet

Locus

For each exercise, draw a figure showing the locus of points. Then describe the locus.

1. all points in a plane that are midpoints of the radii of a given circle

2. all points in a plane that are equidistant from the endpoints of a given segment

3. all points in a plane that are equidistant from two parallel lines 12 centimeters apart

4. all points in a plane that are 4 centimeters away from $\overleftrightarrow{AB}$

5. all points in a plane that are equidistant from two concentric circles whose radii are 8 inches and 12 inches

6. all points in a plane that are centers of circles having a given line segment as a chord

7. all points in a plane that belong to a given angle or its interior and are equidistant from the sides of the given angle

NAME ______________________________ DATE ______________

13-1 Practice Worksheet

Locus

For each exercise, draw a figure showing the locus of points. Then describe the locus.

1. all points in a plane that are midpoints of the radii of a given circle

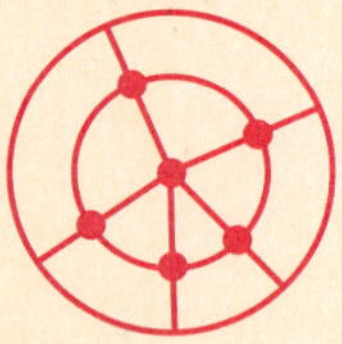

a circle whose radius is half the length of the radius of the given circle

2. all points in a plane that are equidistant from the endpoints of a given segment

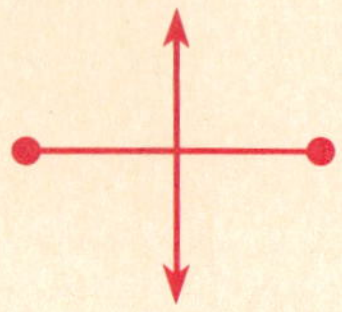

the perpendicular bisector of the segment

3. all points in a plane that are equidistant from two parallel lines 12 centimeters apart

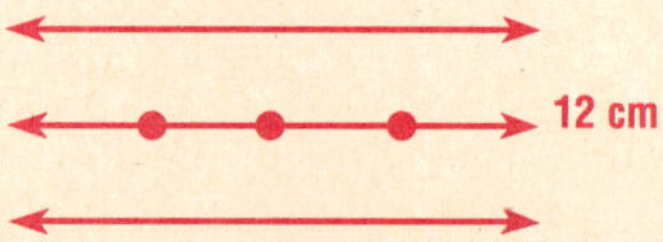

a line that is between the two given lines and 6 cm from each

4. all points in a plane that are 4 centimeters away from $\overleftrightarrow{AB}$

a pair of lines, one on each side of $\overleftrightarrow{AB}$ and 4 cm from $\overleftrightarrow{AB}$

5. all points in a plane that are equidistant from two concentric circles whose radii are 8 inches and 12 inches

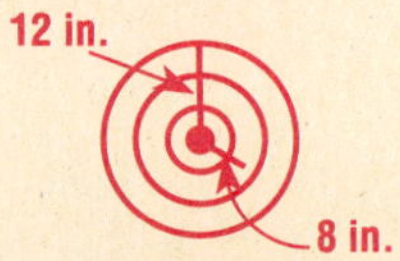

a circle of radius 10, concentric with the given circles

6. all points in a plane that are centers of circles having a given line segment as a chord

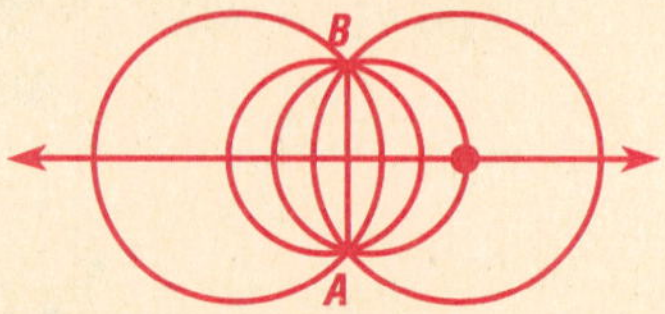

the perpendicular bisector of the line segment

7. all points in a plane that belong to a given angle or its interior and are equidistant from the sides of the given angle

the angle bisector

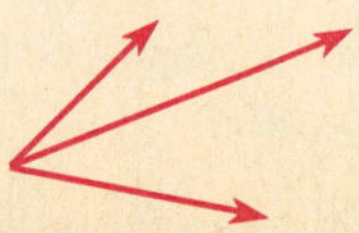

NAME ______________________ DATE __________

13-2 Practice Worksheet

Locus and Systems of Equations

Graph each pair of equations to find the locus of points that satisfy both equations.

1. $x + y = 3$
$x - y = 1$

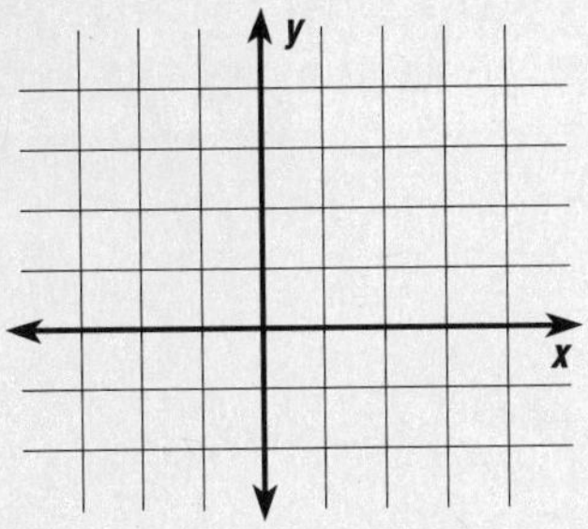

2. $y = 3 + x$
$x + y = 5$

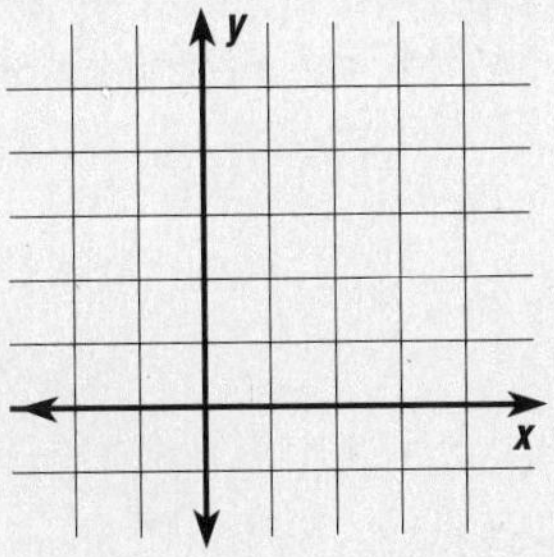

3. $y = 2x$
$x + y = 3$

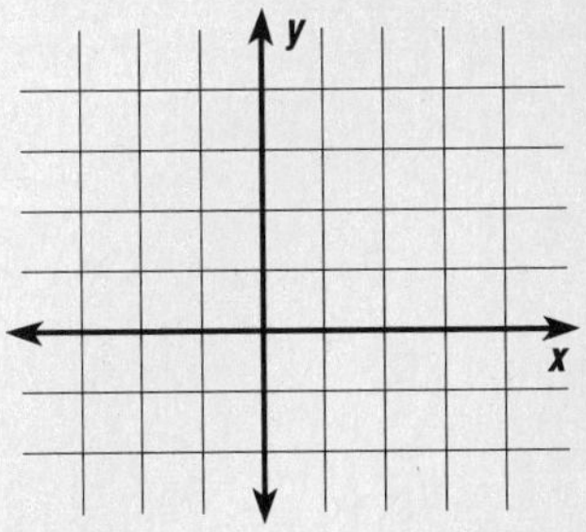

Use an algebraic method to find the locus of points that satisfy both equations.

4. $x + y = 7$
$x - y = 9$

5. $x + y = 3$
$3x - 5y = 17$

6. $y = 2x$
$3x + y = 5$

7. $4x - 3y = -1$
$x + 1 = y$

8. $2x + 3y = -1$
$3x + 5y = -2$

9. $3y = 2 - x$
$2x = 7 - 3y$

10. $y = 2x + 1$
$y = 4x + 7$

11. $x = 4$
$y = 3x - 5$

12. $3x + 2y = 10$
$6x - 3y = 6$

NAME ______________________ DATE ____________

13-2

Practice Worksheet

Locus and Systems of Equations

Graph each pair of equations to find the locus of points that satisfy both equations.

1. $x + y = 3$
$x - y = 1$

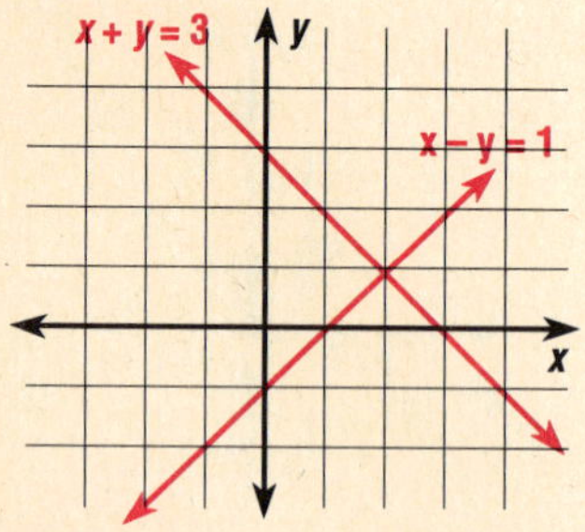

(2, 1)

2. $y = 3 + x$
$x + y = 5$

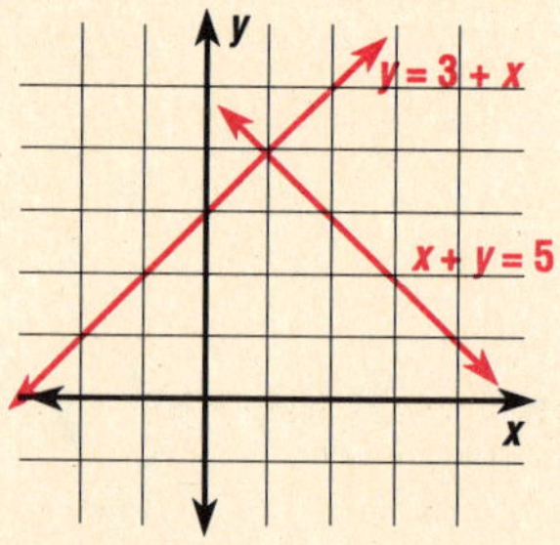

(1, 4)

3. $y = 2x$
$x + y = 3$

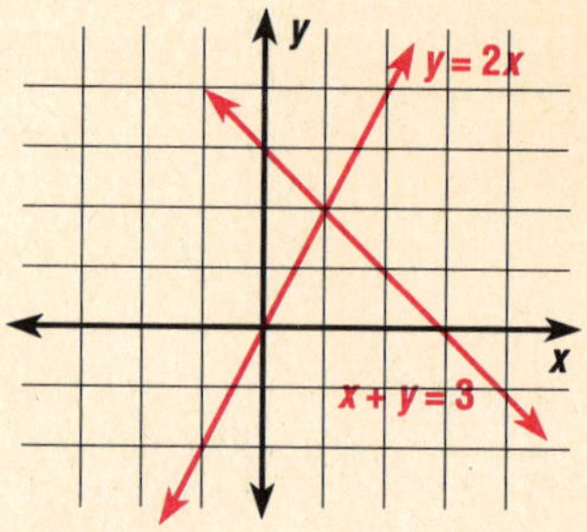

(1, 2)

Use an algebraic method to find the locus of points that satisfy both equations.

4. $x + y = 7$
$x - y = 9$ (8, −1)

5. $x + y = 3$
$3x - 5y = 17$ (4, −1)

6. $y = 2x$
$3x + y = 5$ (1, 2)

7. $4x - 3y = -1$
$x + 1 = y$ (2, 3)

8. $2x + 3y = -1$
$3x + 5y = -2$ (1, −1)

9. $3y = 2 - x$
$2x = 7 - 3y$ (5, −1)

10. $y = 2x + 1$
$y = 4x + 7$ (−3, −5)

11. $x = 4$
$y = 3x - 5$ (4, 7)

12. $3x + 2y = 10$
$6x - 3y = 6$ (2, 2)

13-3

NAME ________________________ DATE ____________

Practice Worksheet

Intersection of Loci

Describe the locus of points in a plane that satisfy each condition.

1. $x + y = 4$

2. $x = y$

3. $y = 5$

4. $x^2 + (y - 2)^2 = 25$

Describe the locus of points in space that satisfy each condition.

5. $(x + 2)^2 + (y - 3)^2 + z^2 = 64$

6. $(x - 2)^2 + (y + 3)^2 + (z - 1)^2 = 81$

Draw a diagram to find the locus of points that satisfy the conditions. Then describe the locus.

7. all points in a plane that are 3 inches from a given segment and equidistant from the two endpoints

8. all points in a plane that are 3 inches from a given line and 3 inches from a given point on the line

9. all points in a plane that are equidistant from the vertices of a given square

10. all points in a plane that are 5 centimeters from a given point A and equidistant from points A and B that are 8 centimeters apart

13-3

NAME ______________________________ DATE ______________

Practice Worksheet

Intersection of Loci

Describe the locus of points in a plane that satisfy each condition.

1. $x + y = 4$
a line with slope –1 that passes through (0, 4)

2. $x = y$
a line with slope 1 that passes through (0, 0)

3. $y = 5$
a line with slope 0 that passes through (0, 5)

4. $x^2 + (y - 2)^2 = 25$
a circle with center (0, 2) and radius 5

Describe the locus of points in space that satisfy each condition.

5. $(x + 2)^2 + (y - 3)^2 + z^2 = 64$
a sphere with center (–2, 3, 0) and radius 8

6. $(x - 2)^2 + (y + 3)^2 + (z - 1)^2 = 81$
a sphere with center (2, –3, 1) and radius 9

Draw a diagram to find the locus of points that satisfy the conditions. Then describe the locus.

7. all points in a plane that are 3 inches from a given segment and equidistant from the two endpoints
two points on the perpendicular bisector of the segment, one on each side, whose perpendicular distance to the segment is 3 in.

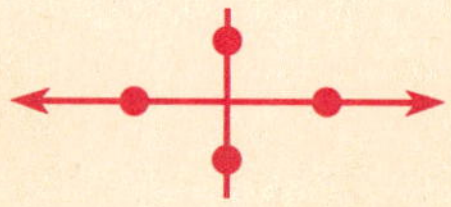

8. all points in a plane that are 3 inches from a given line and 3 inches from a given point on the line
two points, one on each side of the line, whose perpendicular distance from the point on the line is 3 in.

9. all points in a plane that are equidistant from the vertices of a given square
the center of the square; the point of intersection of the diagonals of the square

10. all points in a plane that are 5 centimeters from a given point A and equidistant from points A and B that are 8 centimeters apart
two points, the intersection of the perpendicular bisector of $\overline{AB}$ and the circle with center A and radius 5

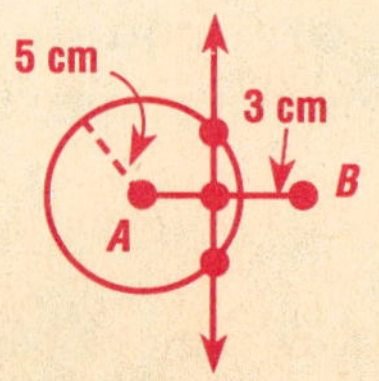

Glencoe Division, Macmillan/McGraw-Hill

NAME ______________________ DATE ____________

13-4

Practice Worksheet

Mappings

Answer the following, if hexagon ABCDEF ↔ hexagon MNOPQR.

1. Name the image of $\overline{BC}$.

2. Name the image of $\angle F$.

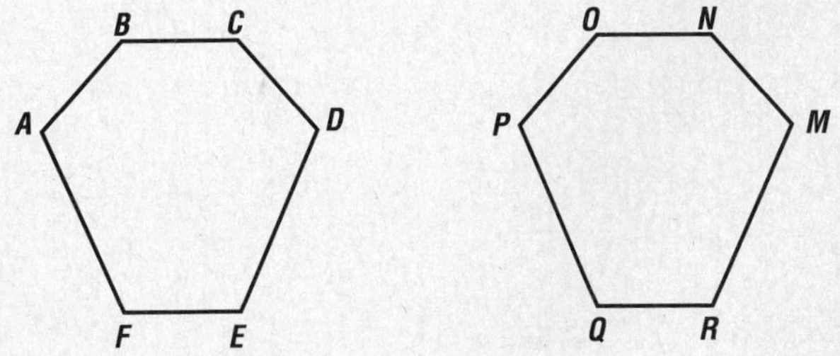

3. Name the image of $\angle D$.

4. Name the preimage of $\overline{PQ}$.

5. Name the preimage of $\overline{RM}$.

6. Name the preimage of $\angle M$.

Each figure below is the preimage or the image of an isometry. Write the image of each given preimage.

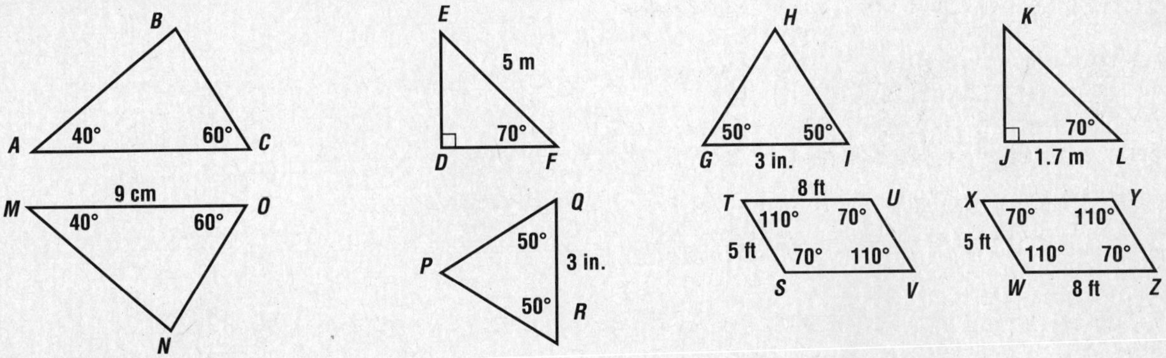

7. $\triangle ABC$

8. $\triangle DEF$

9. $\triangle GHI$

10. $\triangle ONM$

11. quadrilateral $STUV$

12. quadrilateral $XYZW$

13-4

NAME ______________________ DATE ____________

Practice Worksheet

Mappings

Answer the following, if hexagon ABCDEF ↔ hexagon MNOPQR.

1. Name the image of $\overline{BC}$.
   $\overline{NO}$

2. Name the image of $\angle F$.
   $\angle R$

3. Name the image of $\angle D$.
   $\angle P$

4. Name the preimage of $\overline{PQ}$.
   $\overline{DE}$

5. Name the preimage of $\overline{RM}$.
   $\overline{FA}$

6. Name the preimage of $\angle M$.
   $\angle A$

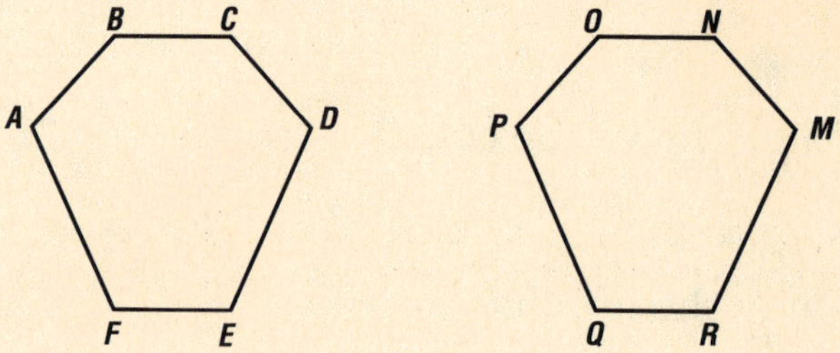

Each figure below is the preimage or the image of an isometry. Write the image of each given preimage.

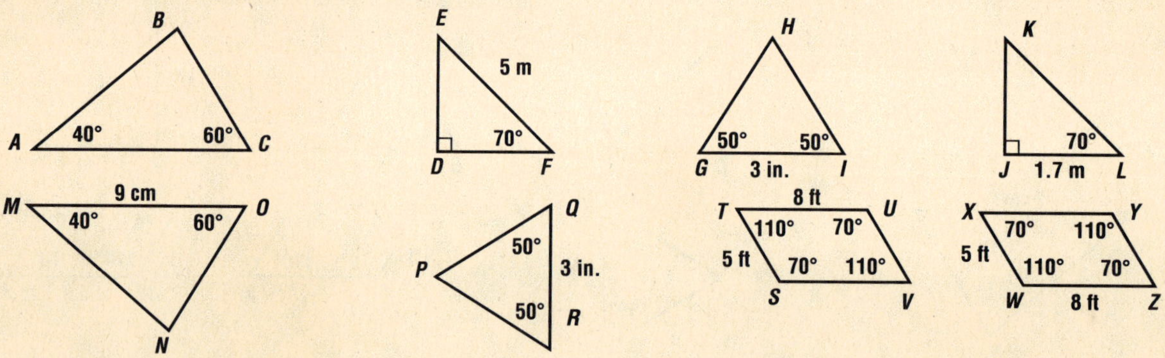

7. $\triangle ABC$
   $\triangle MNO$

8. $\triangle DEF$
   $\triangle LKJ$

9. $\triangle GHI$
   $\triangle RPQ$ or $\triangle QPR$

10. $\triangle ONM$
    $\triangle CBA$

11. quadrilateral *STUV*
    quadrilateral *ZWXY*, quadrilateral *XYZW*

12. quadrilateral *XYZW*
    quadrilateral *UVST*, quadrilateral *STUV*

NAME ______________________ DATE ______________

13-5 Practice Worksheet

Reflections

For each figure, name the reflection image with respect to line ℓ. Use the diagram.

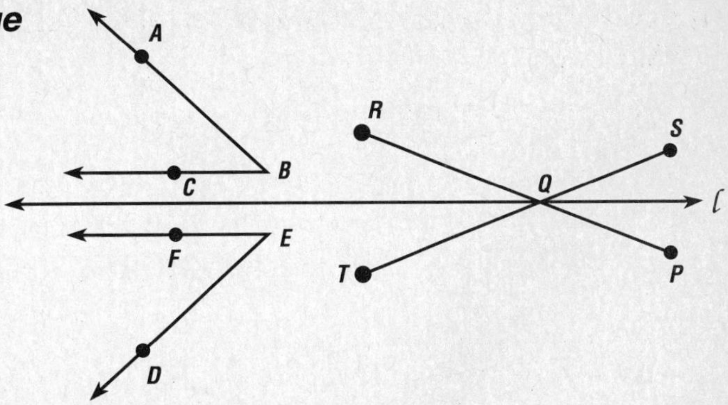

1. A **2.** C

3. T **4.** Q

5. $\overline{BC}$ **6.** $\angle CBA$

7. $\overline{PR}$ **8.** $\overline{QT}$

For each of the following figures, indicate if the figure has line symmetry, point symmetry, or both.

9.

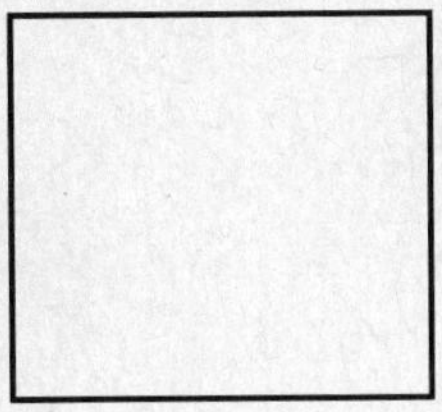

10.

11.

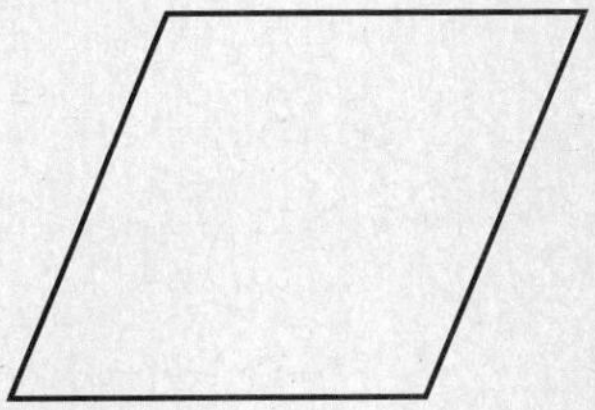

Draw the reflection image of each figure with respect to line m.

12.

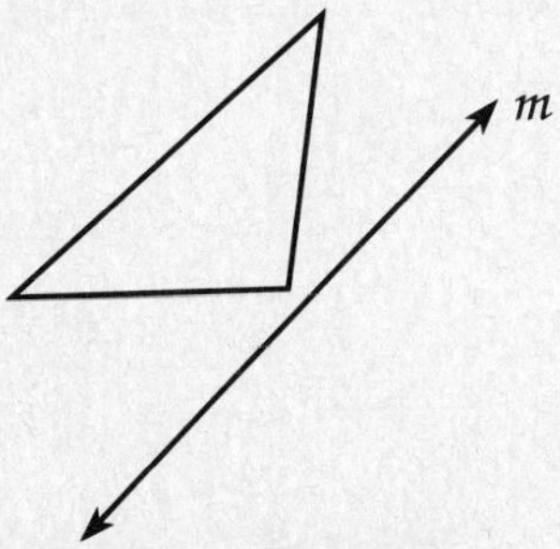

13.

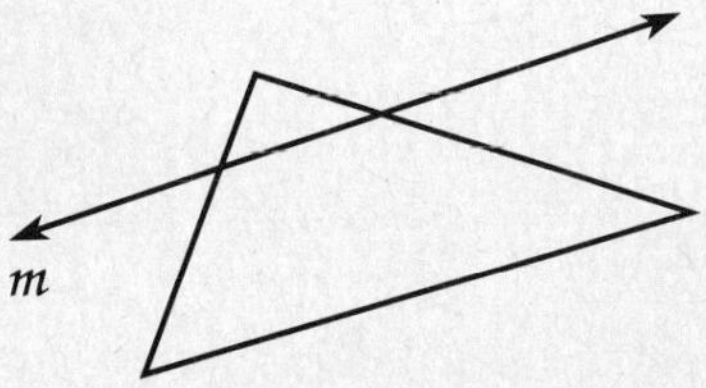

14.

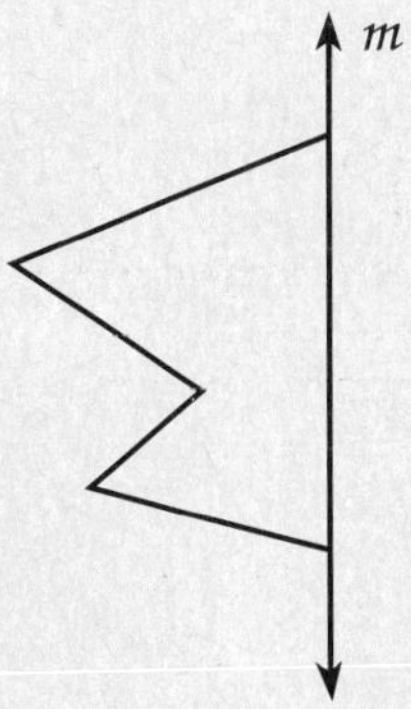

Draw all possible lines of symmetry. If none exist, write none.

15.

16.

17.

13-5

NAME ______________________ DATE ____________

Practice Worksheet

Reflections

For each figure, name the reflection image with respect to line ℓ. Use the diagram.

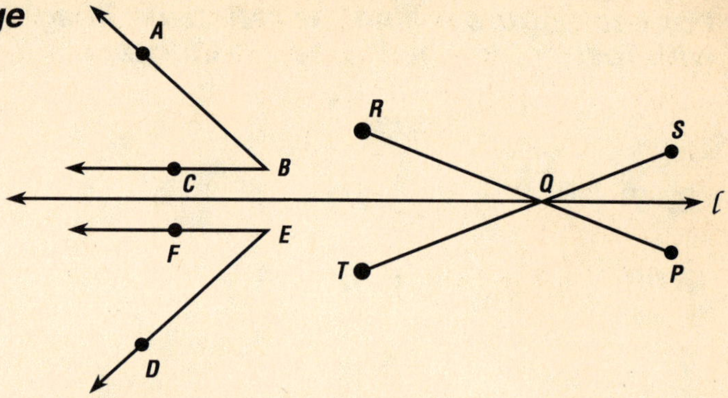

1. *A* **D**
2. *C* **F**
3. *T* **R**
4. *Q* **Q**
5. $\overline{BC}$ $\overline{EF}$
6. $\angle CBA$ $\angle FED$
7. $\overline{PR}$ $\overline{ST}$
8. $\overline{QT}$ $\overline{QR}$

For each of the following figures, indicate if the figure has line symmetry, point symmetry, or both.

9.

both

10.

line

11.

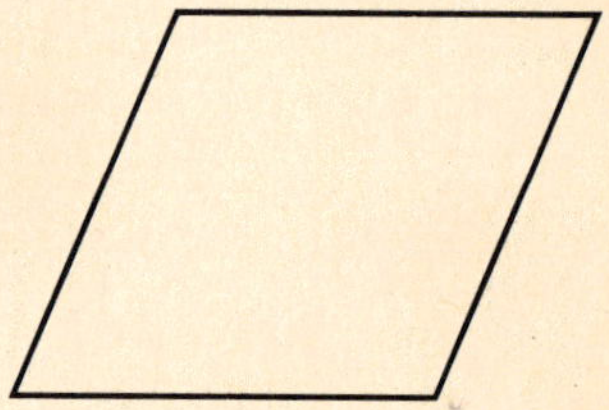

point

Draw the reflection image of each figure with respect to line m.

12.

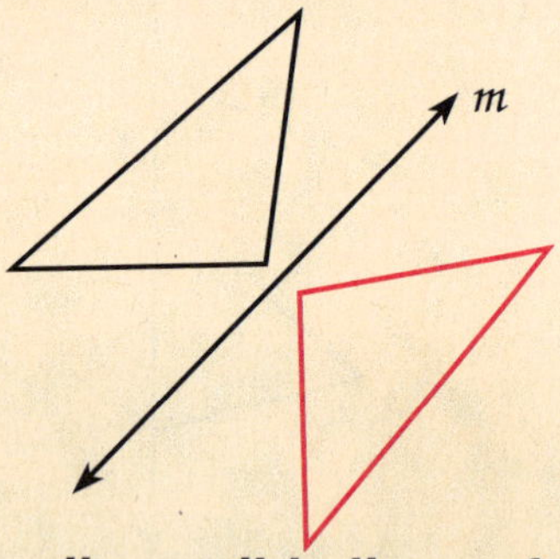

13.

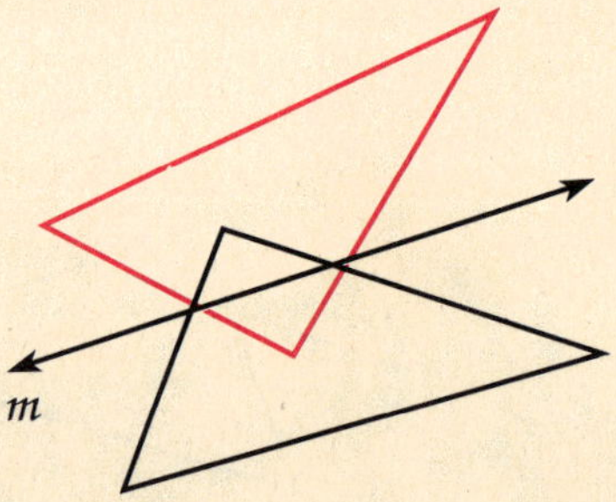

14.

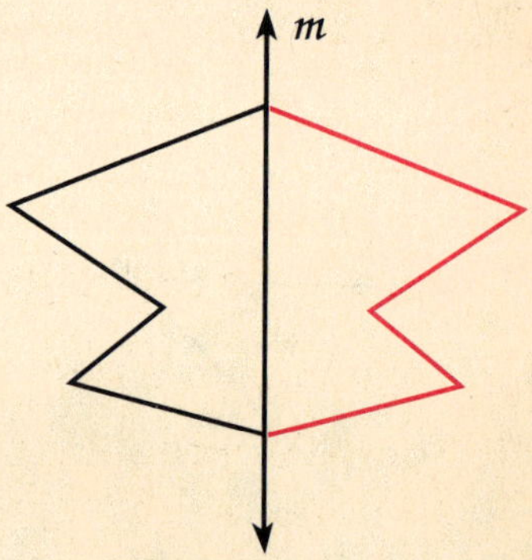

Draw all possible lines of symmetry. If none exist, write none.

15.

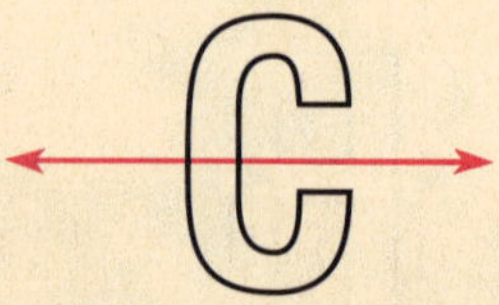

16.

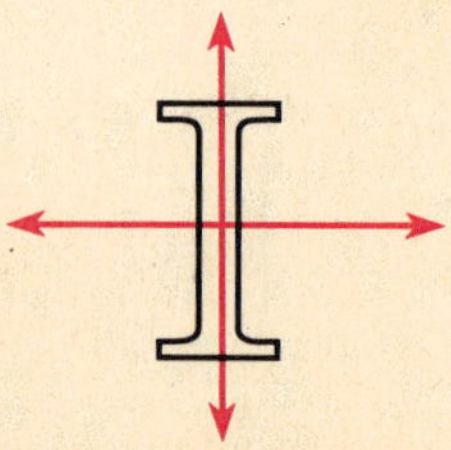

17.

none

NAME ______________________ DATE ____________

13-6 Practice Worksheet

Translations

For each of the following, lines ℓ and m are parallel. Determine whether figure 3 is a translation image of figure 1 for lines ℓ and m. Write yes or no.

1.

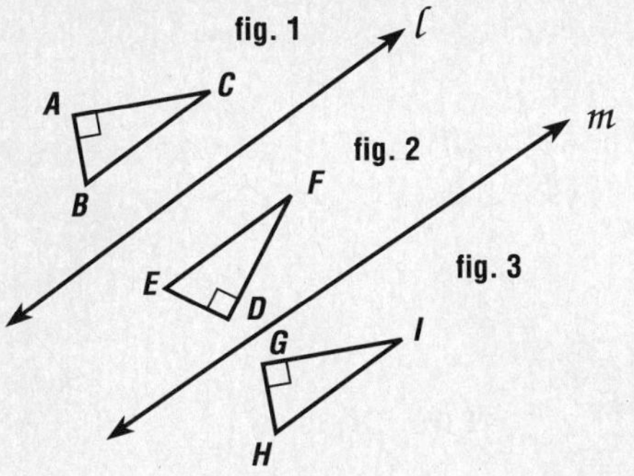

2.

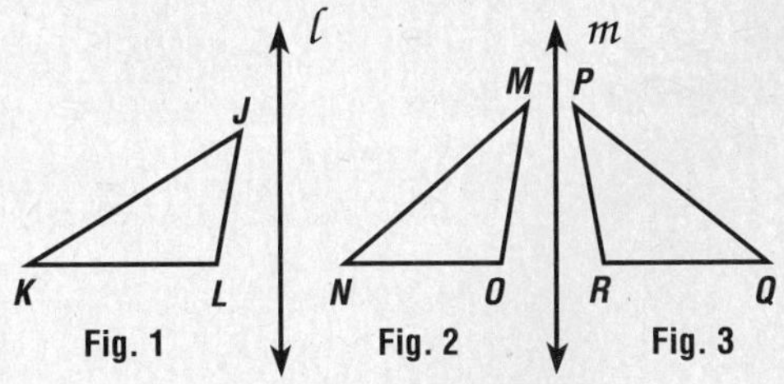

3.

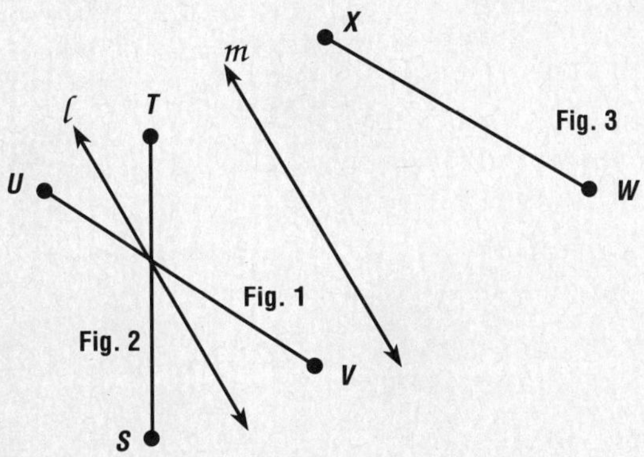

4.

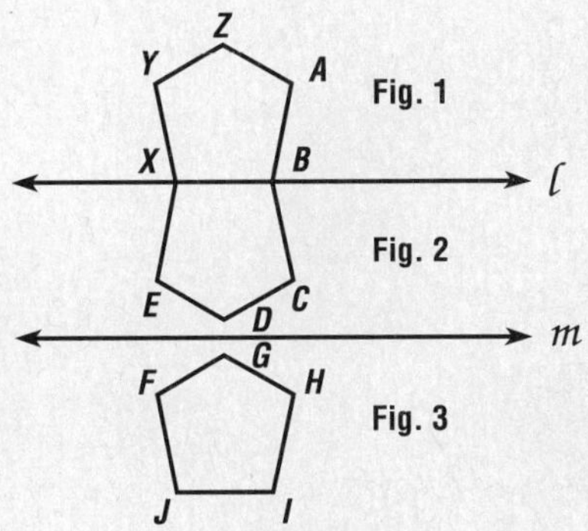

For exercises 5-8, refer to the figures above.

5. Name the reflection of $\overline{ST}$ with respect to line ℓ. If none is drawn, write *none.*

6. Name the reflection of $\triangle JKL$ with respect to line ℓ. In none is drawn, write *none.*

7. Name the reflection of pentagon $XYZAB$ with respect to line ℓ. In none is drawn, write *none.*

8. Name the reflection of $\overline{UV}$ with respect to line m. If none is drawn, write *none.*

Find the translation image of each geometric figure with respect to the parallel lines ℓ and m.

9.

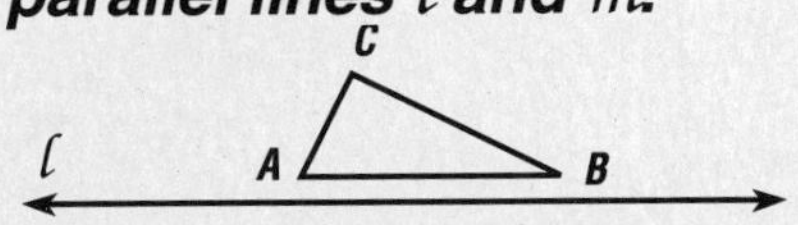

10.

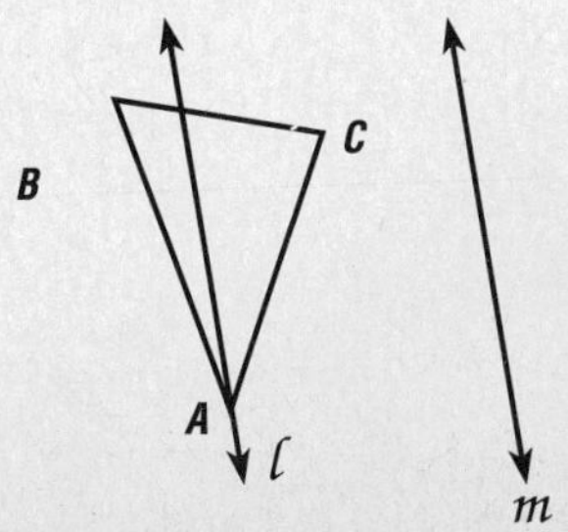

13-6

NAME ________________________________ DATE ______________

Practice Worksheet

Translations

For each of the following, lines ℓ and m are parallel. Determine whether figure 3 is a translation image of figure 1 for lines ℓ and m. Write yes or no.

1.

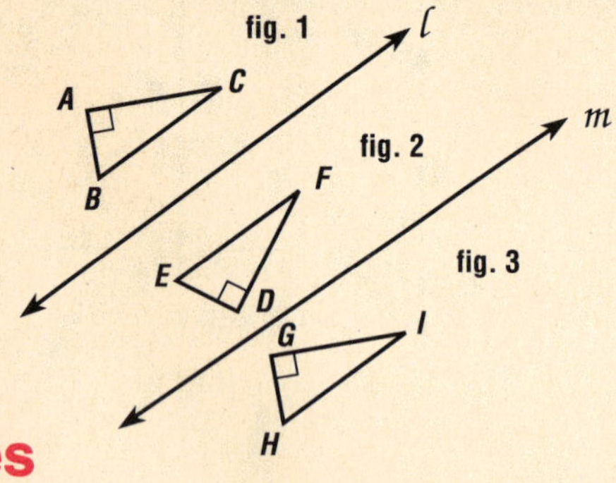

yes

2.

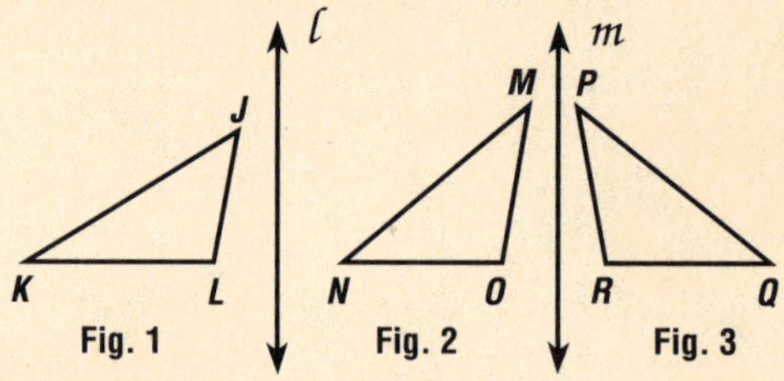

no; Fig. 2 is not the reflection of Fig. 1 with respect to ℓ.

3.

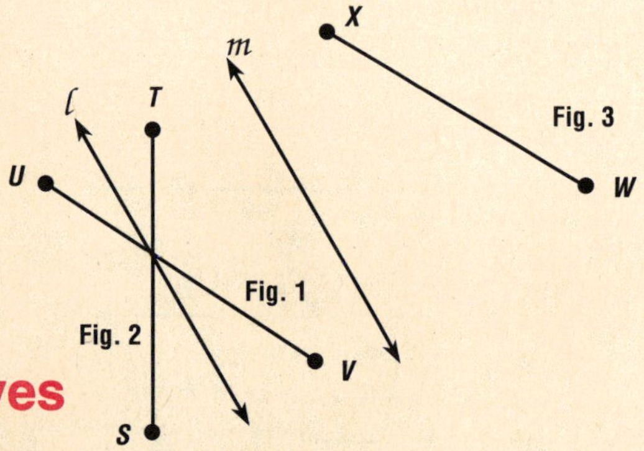

yes

4.

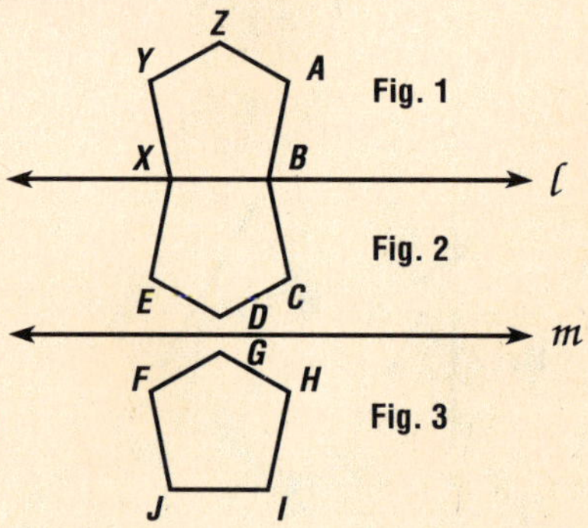

yes

For exercises 5-8, refer to the figures above.

5. Name the reflection of $\overline{ST}$ with respect to line ℓ. If none is drawn, write *none*. $\overline{VU}$

6. Name the reflection of △*JKL* with respect to line ℓ. In none is drawn, write *none*. none

7. Name the reflection of pentagon *XYZAB* with respect to line ℓ. In none is drawn, write *none*. pentagon *XEDCB*

8. Name the reflection of $\overline{UV}$ with respect to line m. If none is drawn, write *none*. none

Find the translation image of each geometric figure with respect to the parallel lines ℓ and m.

9.

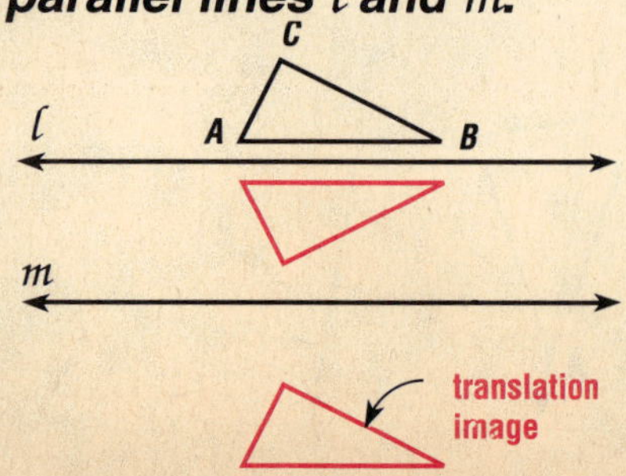

10.

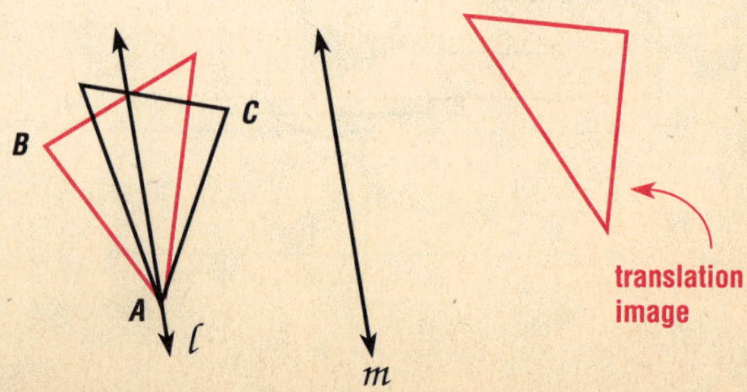

NAME ______________________ DATE ____________

13-7 Practice Worksheet

Rotations

Two lines intersect to form an angle with the following measure. Find the measure of each angle of rotation.

1. 40°   **2.** 72°   **3.** 23°

4. 35°   **5.** 55°   **6.** 63°

Use the angles of rotation to find the rotation image of each geometric figure with respect to lines s and t.

7.

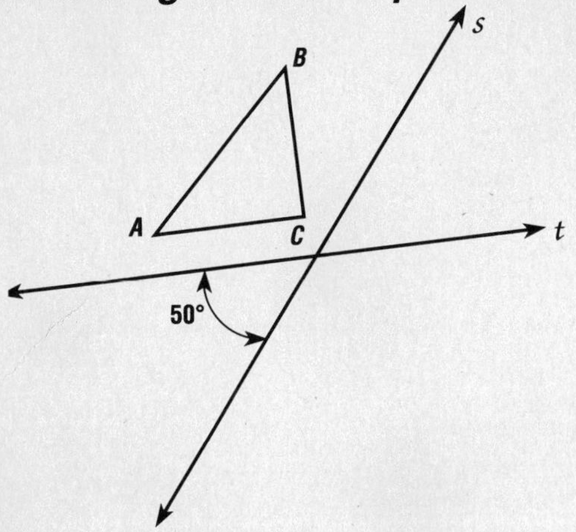

8.

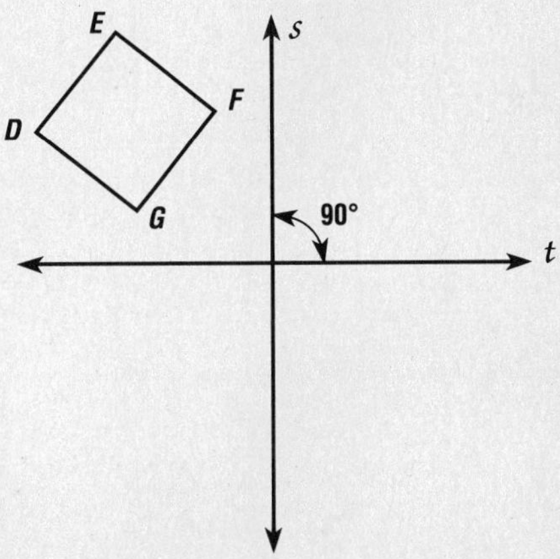

9.

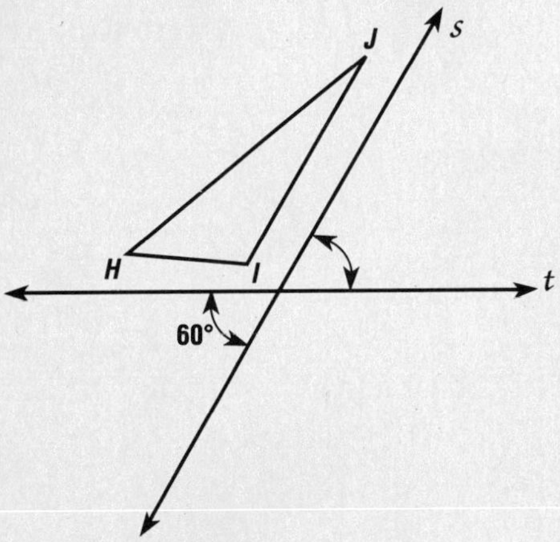

10.

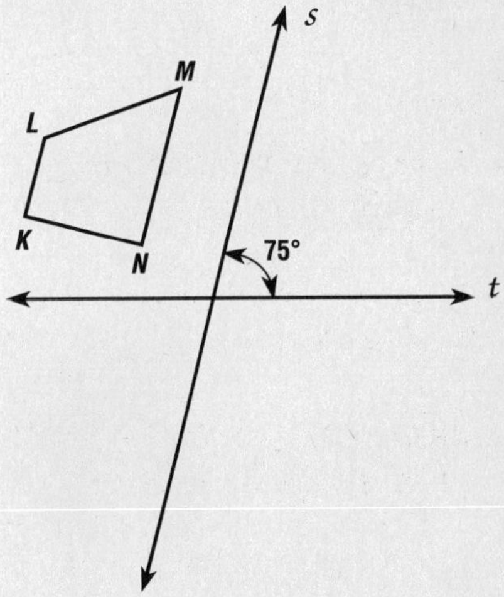

11.

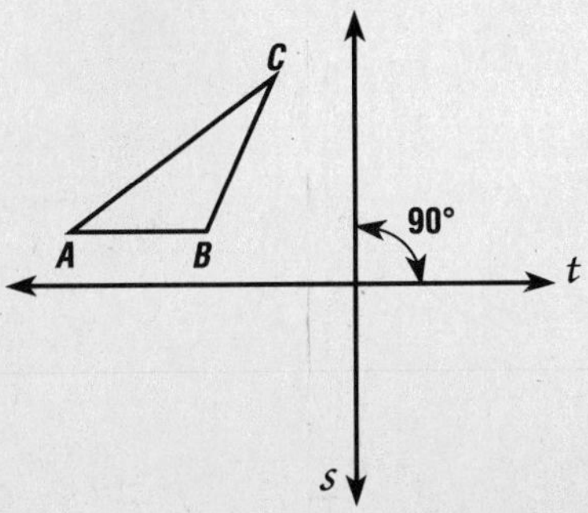

12.

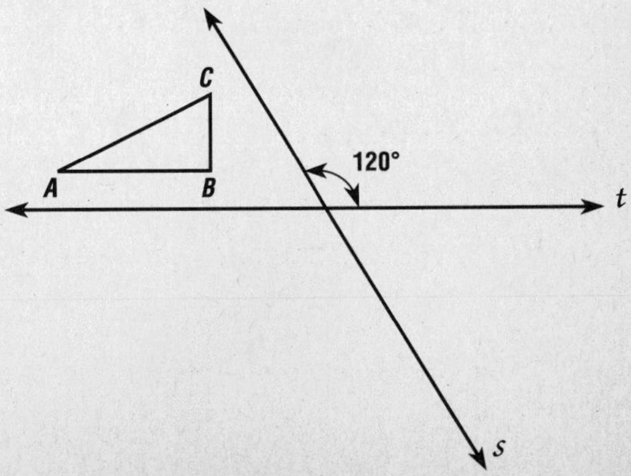

Glencoe Division, Macmillan/McGraw-Hill

NAME ______________________ DATE ____________

13-7 Practice Worksheet

Rotations

Two lines intersect to form an angle with the following measure. Find the measure of each angle of rotation.

1. 40° **80°**

2. 72° **144°**

3. 23° **46°**

4. 35° **70°**

5. 55° **110°**

6. 63° **126°**

Use the angles of rotation to find the rotation image of each geometric figure with respect to lines s and t.

7.

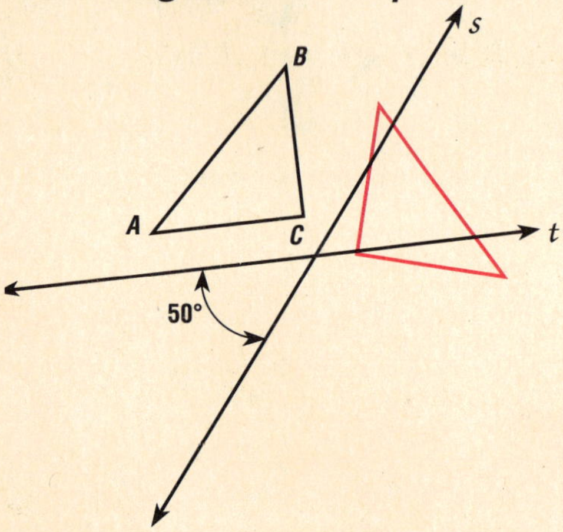

8.

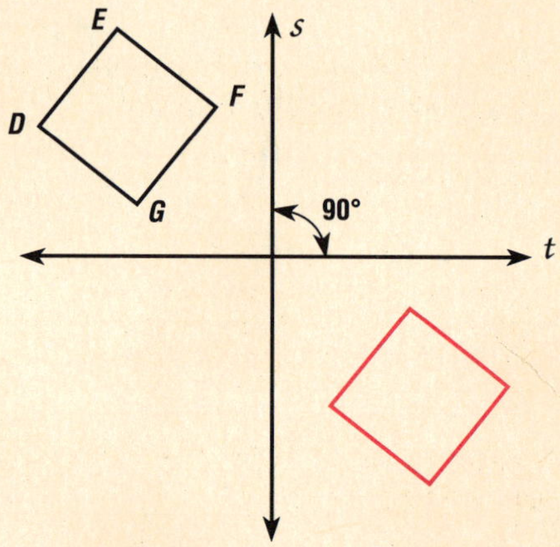

9.

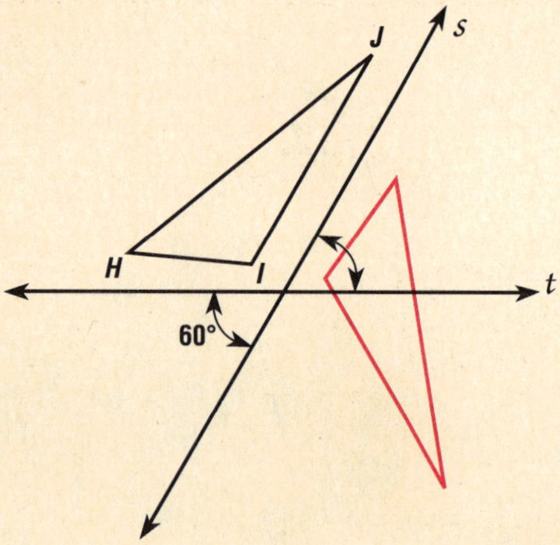

10.

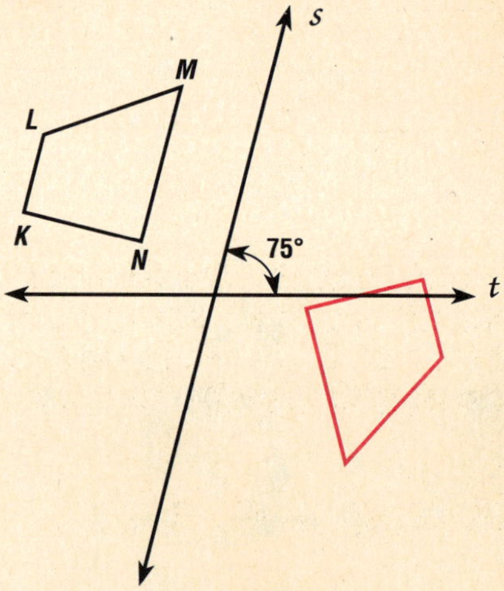

11.

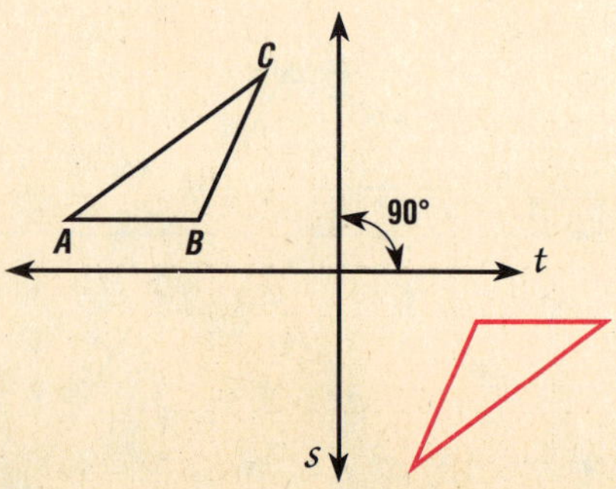

12.

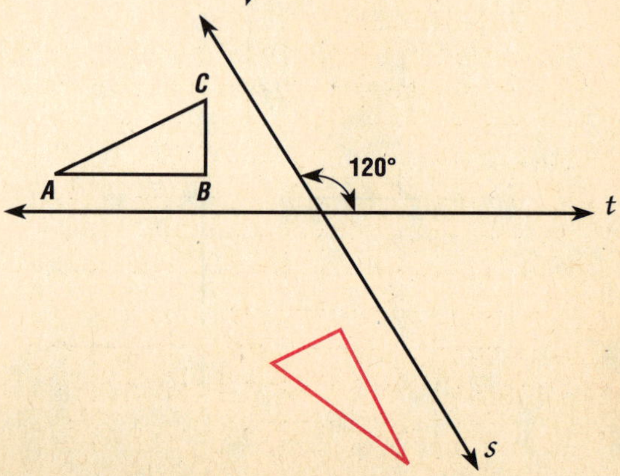

NAME ________________________________ DATE ______________

13-8 Practice Worksheet

Dilations

A dilation with center C and a scale factor k maps X onto Y. Find k for each dilation. Then determine whether each dilation is an enlargement, a reduction, or a congruence transformation.

1. $CY = 15, CX = 10$

2. $CY = 2, CX = 2$

3. $CX = 5, CY = 2$

4. $CY = 20, CX = \frac{1}{2}$

Find the measure of the image of $\overline{AB}$ with respect to a dilation with the given scale factor.

5. $AB = 6$ in., $k = -\frac{2}{3}$

6. $AB = 4$ in., $k = 1$

7. $AB = 1\frac{1}{2}$ in., $k = \frac{1}{2}$

8. $AB = 20$ in., $k = -2\frac{1}{2}$

For each scale factor, find the image of A with respect to a dilation with center C.

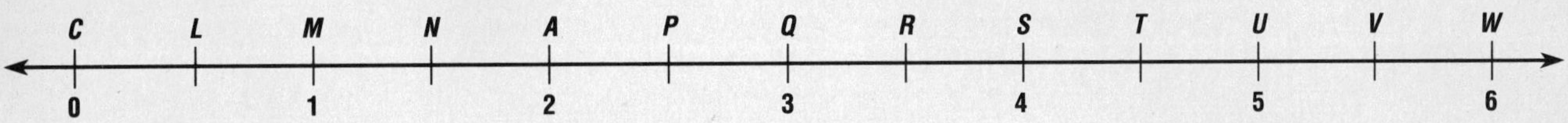

9. 3

10. $\frac{1}{4}$

11. $2\frac{1}{4}$

12. $\frac{3}{4}$

Graph each of the following ordered pairs. Then connect the points in order. Using (0, 0) as the center of dilation and a scale factor of 2, draw the dilation image of the triangle. Then draw the dilation image using a scale factor of $\frac{1}{2}$.

13. (2, 2), (4, 6), (6, –2)

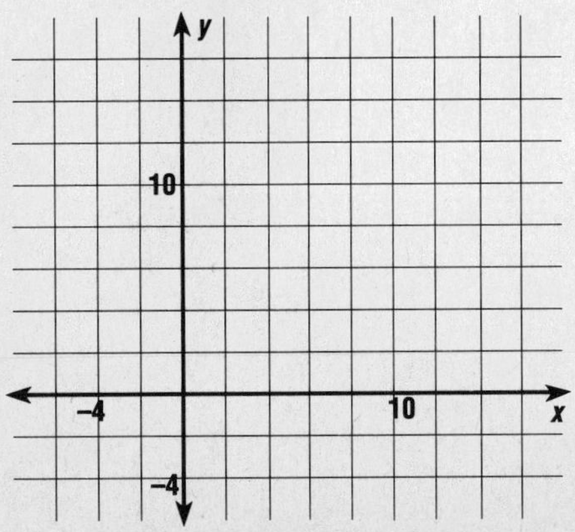

14. (0, 2), (–4, 2), (–4, –2)

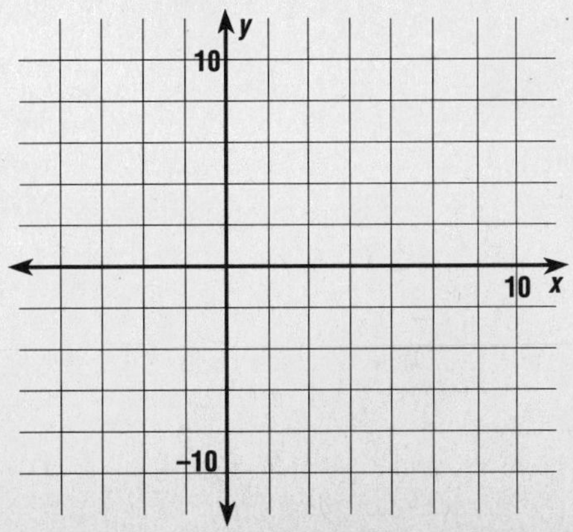

NAME ______________________ DATE __________

13-8 Practice Worksheet

Dilations

A dilation with center C and a scale factor k maps X onto Y. Find k for each dilation. Then determine whether each dilation is an enlargement, a reduction, or a congruence transformation.

1. $CY = 15$, $CX = 10$
$\frac{3}{2}$; enlargement

2. $CY = 2$, $CX = 2$
1; congruence

3. $CX = 5$, $CY = 2$
$\frac{2}{5}$; reduction

4. $CY = 20$, $CX = \frac{1}{2}$
40; enlargement

Find the measure of the image of $\overline{AB}$ with respect to a dilation with the given scale factor.

5. $AB = 6$ in., $k = -\frac{2}{3}$ **4 in.**

6. $AB = 4$ in., $k = 1$ **4 in.**

7. $AB = 1\frac{1}{2}$ in., $k = \frac{1}{2}$ **$\frac{3}{4}$ in.**

8. $AB = 20$ in., $k = -2\frac{1}{2}$ **50 in.**

For each scale factor, find the image of A with respect to a dilation with center C.

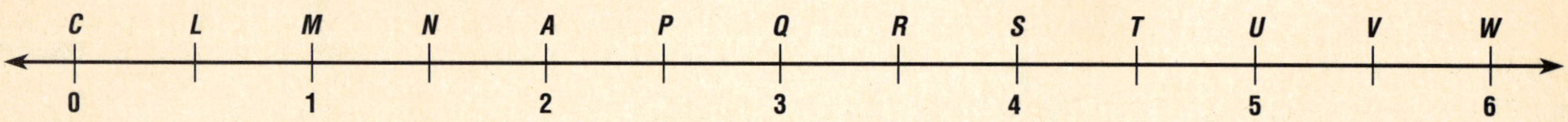

9. 3 ***W***

10. $\frac{1}{4}$ ***L***

11. $2\frac{1}{4}$ ***T***

12. $\frac{3}{4}$ ***N***

Graph each of the following ordered pairs. Then connect the points in order. Using (0, 0) as the center of dilation and a scale factor of 2, draw the dilation image of the triangle. Then draw the dilation image using a scale factor of $\frac{1}{2}$.

13. (2, 2), (4, 6), (6, –2)

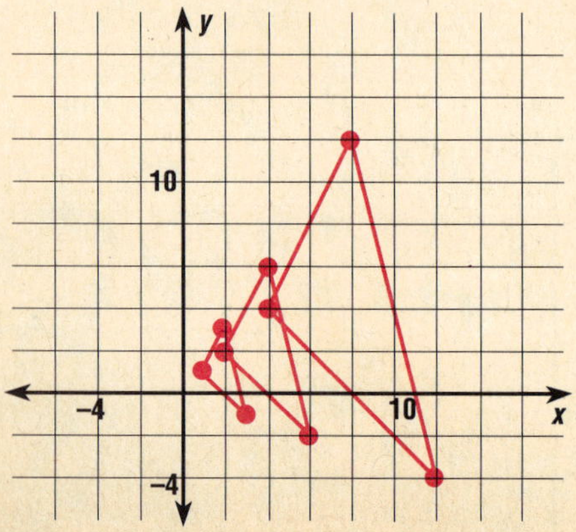

14. (0, 2), (–4, 2), (–4, –2)

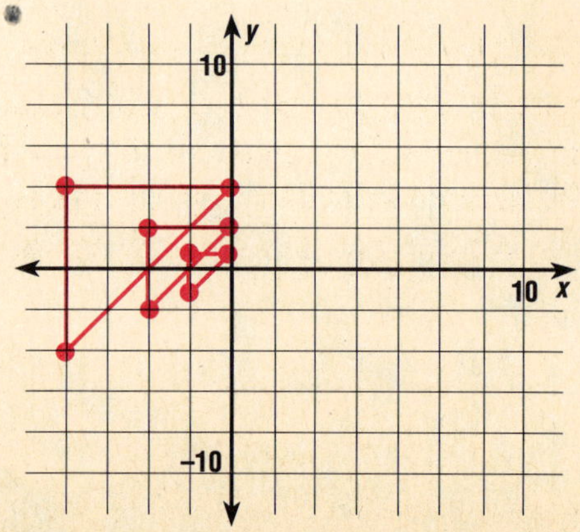

NAME ______________________ DATE ____________

13-9

Practice Worksheet

Problem-Solving Strategy: Make a Table

Solve. Use any strategy.

1. Benjamin has seven employees, Janell, Pedro, David, Elyse, Faye, Gordon, and Irene, who work in his store. Janell and Gordon work full time, 5 days in a row per week. Pedro and Faye each work 4 days per week. Pedro cannot work on Thursdays. David and Elyse also work 4 days per week each. David cannot work on weekends, and Elyse cannot work on Wednesdays. Irene works any three days assigned each week. Benjamin needs three people to work on Monday–Thursday, five to work on Friday, and six to work on weekends. Make a possible schedule for Benjamin to use.

2. Find four positive integers w, x, y, and z such that $\frac{1}{w} + \frac{1}{x} + \frac{1}{y} + \frac{1}{z} = \frac{5}{4}$.

3. At a recent open house, 10 cousins met for the first time in years. If each cousin shook hands with each of the other cousins exactly once, how many handshakes occurred?

4. Calvin, Lisa, Marita, and Ramon all ordered different meals at a restaurant. One person ordered chicken. Calvin and the person who ordered fish are neighbors. Lisa is allergic to fish. Marita only orders salads. Lisa is dating the person who chose steak. Which meal did each person order?

Glencoe Division, Macmillan/McGraw-Hill

13-9

NAME ______________________ DATE ____________

Practice Worksheet

Problem-Solving Strategy: Make a Table

Solve. Use any strategy.

1. Benjamin has seven employees, Janell, Pedro, David, Elyse, Faye, Gordon, and Irene, who work in his store. Janell and Gordon work full time, 5 days in a row per week. Pedro and Faye each work 4 days per week. Pedro cannot work on Thursdays. David and Elyse also work 4 days per week each. David cannot work on weekends, and Elyse cannot work on Wednesdays. Irene works any three days assigned each week. Benjamin needs three people to work on Monday–Thursday, five to work on Friday, and six to work on weekends. Make a possible schedule for Benjamin to use.

Possible answer:

	M	T	W	T	F	S	S
Janell	X	X	X			X	X
Pedro	X				X	X	X
David		X	X	X	X		
Elyse				X	X	X	X
Faye				X	X	X	X
Gordon	X	X	X			X	X
Irene					X	X	X

2. Find four positive integers w, x, y, and z such that $\frac{1}{w} + \frac{1}{x} + \frac{1}{y} + \frac{1}{z} = \frac{5}{4}$. **2, 3, 4, 6**

3. At a recent open house, 10 cousins met for the first time in years. If each cousin shook hands with each of the other cousins exactly once, how many handshakes occurred?
45 handshakes

4. Calvin, Lisa, Marita, and Ramon all ordered different meals at a restaurant. One person ordered chicken. Calvin and the person who ordered fish are neighbors. Lisa is allergic to fish. Marita only orders salads. Lisa is dating the person who chose steak. Which meal did each person order?
Calvin–steak; Lisa–chicken; Marita–salad; Ramon–fish